High-Performance Programming in C# and .NET

Understand the nuts and bolts of developing robust, faster, and resilient applications in C# 10.0 and .NET 6

Jason Alls

BIRMINGHAM—MUMBAI

High-Performance Programming in C# and .NET

Associate Group Product Manager: Gebin George
Publishing Product Manager: Alok Dhuri
Senior Editor: Kinnari Chohan
Technical Editor: Maran Fernandes
Copy Editor: Safis Editing
Project Coordinator: Manisha Singh
Proofreader: Safis Editing
Indexer: Subalakshmi Govindhan
Production Designer: Sinhayna Bais
Marketing Coordinator: Sonakshi Bubbar

First published: July 2022

Production reference: 1290722

Published by Packt Publishing Ltd.
Livery Place
35 Livery Street
Birmingham
B3 2PB, UK.

ISBN 978-1-80056-471-8

www.packt.com

About the reviewers

Gianni Rosa Gallina is an Italian senior software engineer and architect who has been focused on emerging technologies, AI, and virtual/augmented reality since 2013. Currently, he works at Deltatre's Innovation Lab, prototyping solutions for next-generation sports experiences and business services. Besides that, he has more than 10 years of certified experience as a consultant on Microsoft and .NET technologies (including technologies such as the Internet of Things, the cloud, and desktop/mobile apps). Since 2011, he has been awarded Microsoft MVP in the Windows Development category. He has been a Pluralsight Author since 2013 and is a speaker at national and international conferences.

Joy Rathnayake is a solutions architect with over 19 years of industry experience and is part of the **Digital & Emerging Technology** (**DET**) team at EY GDS, based in Colombo, Sri Lanka. He is primarily responsible for understanding customer requirements, identifying required products/ technologies, and defining the overall solution design/ architecture.

Before he joined EY GDS Sri Lanka, Joy worked as a solutions architect at WSO2 Inc., Totalamber Ltd, Virtusa Ltd, Solid Quality Mentors, IronOne Technologies, and Sri Lankan Airlines. He was responsible for architecting, designing, and developing software solutions primarily using Microsoft and related technologies.

Joy has been recognized as both a Microsoft **Most Valuable Professional** (**MVP**) and a **Microsoft Certified Trainer** (**MCT**). He has contributed to developing content for Microsoft Certifications and has worked as a **Subject Matter Expert** (**SME**) for many Microsoft exam development projects.

He has contributed a lot to the community by presenting at various events, such as Microsoft Tech-Ed, Southeast Asia SharePoint Conference, and SharePoint Saturday.

Joy enjoys traveling, speaking at public events/conferences, and reading.

Contributors

About the author

Jason Alls is the author of *Clean Code in C#* and has been programming for over 21 years. Working with an Australasian company, he started his career developing call center management reporting software used by global clients, including telecom providers, banks, airlines, and the police. He then moved on to develop GIS marketing applications and worked in the banking sector, performing data migrations between Oracle and SQL Server. Certified as an MCAD in C# since 2005, he has been involved in the development of various desktop, web, and mobile applications.

Currently employed by a leading software house, he develops and supports order processing and warehouse management software written in C#.

I would like to send a warm thank you to Joy and Gianni, who reviewed the chapters for me. Their attention to detail was excellent, and they would often let me know when things needed improving or when I had left something out. Because of them, the content of this book, I feel, will be of great benefit to programmers and developers at all levels. I happily recommend them as reviewers to any budding authors.

I would like to dedicate this book to the team at Packt, who helped me through each step of the way and were very accommodating and understanding of the difficulties I was going through in my personal life while working on this book. They have been a truly great team that has encouraged me each step of the way. I could not have written this book without their expertise and guidance.

I would also like to dedicate this book to my wonderful parents, who had to suffer long periods without me as I spent many a long night and many weekends on this book, but who encouraged me to keep going through difficult times.

– Jason Alls

Table of Contents

2
Implementing C# Interoperability

3
Predefined Data Types and Memory Allocations

4

Memory Management

5

Application Profiling and Tracing

Part 2: Writing High-Performance Code

6
The .NET Collections

7
LINQ Performance

8

File and Stream I/O

9

Enhancing the Performance of Networked Applications

10

Setting Up Our Database Project

11

Benchmarking Relational Data Access Frameworks

12

Responsive User Interfaces

13

Distributed Systems

Part 3: Threading and Concurrency

14

Multi-Threaded Programming

15
Parallel Programming

16
Asynchronous Programming

Assessments

Preface

Writing high-performance code while building an application is crucial, and over the years, Microsoft has focused on delivering various performance-related improvements within the .NET ecosystem. This book will help you understand the aspects involved in designing responsive, resilient, and high-performance applications with the new versions of C# and .NET.

You will start by understanding the foundation of high-performance code and the latest performance-related improvements in C# 10.0 and .NET 6. Next, you'll learn how to use tracing and diagnostics to track down performance issues and the cause of memory leaks. The chapters that follow then show you how to enhance the performance of your networked applications and various ways to improve directory tasks, file tasks, and more. You'll go on to improve data querying performance and write responsive user interfaces. You'll also discover how you can use cloud providers such as Microsoft Azure to build scalable distributed solutions. Finally, you'll explore various ways to process code synchronously, asynchronously, and in parallel to reduce the time it takes to process a series of tasks.

By the end of this C# programming book, you'll have the confidence you need to build highly resilient, high-performance applications that meet your customer's demands.

Who this book is for

This book is for software engineers, professional software developers, performance engineers, and application profilers looking to improve the speed of their code or take their skills to the next level to gain a competitive advantage. You should be a proficient C# programmer who can already put the language to good use and is also comfortable using Microsoft Visual Studio 2022.

What this book covers

Chapter 1, Introducing C# 10.0 and .NET 6, talks about the **Common Language Runtime** (**CLR**). You will start by learning about what's new in C# 10.0 and .NET 6. Then you will learn about the .NET native runtime and CoreCLR. Next, you will learn about the unified BCL followed by Windows Store performance. Finally, you will learn about ASP.NET 5 performance.

Chapter 2, Implementing C# Interoperability, introduces Microsoft .NET interoperability. You will learn how to call and dispose of unsafe code. You will also learn how to migrate legacy COM programs to .NET using COM interoperability. In this chapter, you will learn how to create .NET libraries and components and use them in legacy COM applications. By the end of the chapter, you will have learned how to consume COM components in .NET and how to consume .NET applications in COM components. This will help you to migrate COM applications over to the .NET platform.

Chapter 3, Predefined Data Types and Memory Allocations, explores C# primitive types and C# object types. You will learn about the stack and the heap and about passing data by reference and by value. Then you will learn about boxing and unboxing and their implications on application performance. You will also be refreshed on the C# primitive type, and how to build objects that perform well.

Chapter 4, Memory Management, talks about the garbage collector. You will learn how to use tracing and diagnostics to track down performance issues and the cause of memory leaks. Then you will learn about object generations and how the garbage collector decides what to dispose of. You will also learn about weak references and how to correctly dispose of objects in order to prevent memory leaks.

Chapter 5, Application Profiling and Tracing, teaches you how to profile your applications to identify poor areas of performance. You will learn about code metrics and how to perform static code analysis. In your drive to write code that is more performant, you will learn to make use of memory dumps, the loaded modules viewer, debugging, tracing, and dotnet-counters. By the time you have completed this chapter, you will have the skills and experience you need to profile your own applications.

Chapter 6, The .NET Collections, explores the collections framework. You will learn about the different collections and how to best use them to get maximum performance from them. You will access the various collections in the `System.Collection`, `System.Collection.Concurrent`, and `System.Collections.Generic` namespaces. You will also create your own custom exceptions and learn how to query collections using LINQ.

Chapter 7, *LINQ Performance*, explains how to perform LINQ queries with performance in mind. Depending on how you use LINQ, different methods that return the same result can behave and perform differently. And so, in this chapter, you will learn how best to perform queries on LINQ to improve the performance of your applications.

Chapter 8, *File and Stream I/O*, explains how to improve file and directory performance. You will learn ways to improve directory tasks, file tasks, memory tasks, and isolated storage tasks. In this book, you will learn how to write to files asynchronously and read from files asynchronously.

Chapter 9, *Enhancing the Performance of Networked Applications*, breaks down how to speed up the performance of network applications. You will learn how to communicate over a network using the TCP and UDP network protocols. Then you will learn how to perform network tracing processes with the OSI Network Layer Reference Model and a selection of TCP and UDP networking protocols. Cache management will also be covered so that you can improve the efficiency of resource retrieval.

Chapter 10, *Setting Up Our Database Project*, sets up the Northwind database project on SQL Server as we will be using this database in the next section to benchmark data access methods.

Chapter 11, *Benchmarking Relational Data Access Frameworks*, benchmarks three different ways to manipulate SQL Server database data. We will be performing a side-by-side comparison of Entity Framework, ADO.NET, and Dapper.NET. After running the benchmarks for each of these data access and object mappers, you will be able to make an educated judgment call on the best form of data access and object mapping for your projects.

Chapter 12, *Responsive User Interfaces*, explains how to write responsive user interfaces. You will write responsive **Windows Forms** (**WinForms**), **Windows Presentation Foundation** (**WPF**), ASP.NET, .NET MAUI, and WinUI applications. Using background worker threads, you will see how you can update and work with the user interface in real time by running long-running tasks in the background.

Chapter 13, *Distributed Systems*, describes distributed applications and explains how to improve their performance. You will learn how to build performant distributed applications using the **Command Query Responsibility Separation** (**CQRS**) software design pattern, event sourcing, and microservices. You will see how to use cloud providers such as Microsoft Azure to build scalable distributed solutions using Cosmos DB, Azure Functions, and the open source Pulumi infrastructure tool.

Chapter 14, Multi-Threaded Programming, explores what threads and threading are and discusses background and foreground threads. Then you will learn how to pass data into threads before you run them. You will also learn how to pause, interrupt, destroy, schedule, and cancel threads.

Chapter 15, Parallel Programming, explains how to take advantage of the multiple CPU cores that are available in today's modern computers. You will learn how to process your code by distributing the work between processes concurrently.

Chapter 16, Asynchronous Programming, demystifies the **Task Asynchronous Programming** (**TAP**) model. You will learn how to program tasks asynchronously and access web resources using `async`, `await`, and `WhenAll`. You will also look at different return types, how to extract the required results, and how to correctly cancel asynchronous operations and perform asynchronous file reading and writing.

To get the most out of this book

You will need to be proficient in C# and know how to use Visual Studio 2022 to create, run, and debug C# programs and install NuGet packages. You will get the most from this book if you follow along, write the code, and use the tools specified. But if you are too busy, follow Microsoft's guidance for obtaining and installing the following software.

Software/hardware covered in the book	Operating system requirements
.NET 6 SDK	Windows, macOS, or Linux
Visual Studio 2022/Visual Studio 2022 Preview	
SQL Server	
SQL Server Management Studio	

If you are using the digital version of this book, we advise you to type the code yourself or access the code from the book's GitHub repository (a link is available in the next section). Doing so will help you avoid any potential errors related to the copying and pasting of code.

Please try and answer the questions, read the external resources provided at the end of each chapter, and put what you have learned into action in your own programming and performance training exercises. This will help to reinforce what you have learned throughout this book.

Download the example code files

You can download the example code files for this book from GitHub at `https://github.com/PacktPublishing/High-Performance-Programming-in-CSharp-and-.NET`. If there's an update to the code, it will be updated in the GitHub repository.

We also have other code bundles from our rich catalog of books and videos available at `https://github.com/PacktPublishing/`. Check them out!

Download the color images

We also provide a PDF file that has color images of the screenshots and diagrams used in this book. You can download it here: `https://packt.link/hQmsb`.

Conventions used

There are a number of text conventions used throughout this book.

`Code in text`: Indicates code words in text, database table names, folder names, filenames, file extensions, pathnames, dummy URLs, user input, and Twitter handles. Here is an example: "The `enum` data type is 4 bytes (32 bits) in size, nullable, and has a minimum value of 0. You can measure the size of a value type using `sizeof(Type type)`."

A block of code is set as follows:

```
static void Main(string[] _)
{
Console.WriteLine("Chapter 3: Strings are immutable");
var greeting1 = "Hello, world!";
var greeting2 = greeting1;
Console.WriteLine($"greeting1={greeting1}");
Console.WriteLine($"greeting2={greeting2}");
greeting1 += " Isn't life grand!";
Console.WriteLine($"greeting1={greeting1}");
Console.WriteLine($"greeting1={greeting2}");
}
```

When we wish to draw your attention to a particular part of a code block, the relevant lines or items are set in bold:

```
git clone https://github.com/dotnet/roslyn.git
```

Any command-line input or output is written as follows:

```
csc /help
csc -langversion:10.0 /out:HelloWorld.exe Program.cs
csc HelloWorld
cd css
```

Bold: Indicates a new term, an important word, or words that you see onscreen. For instance, words in menus or dialog boxes appear in **bold**. Here is an example: "Make sure the project is set to **Debug** mode, and then step through the code."

> **Tips or important notes**
> Appear like this.

Get in touch

Feedback from our readers is always welcome.

General feedback: If you have questions about any aspect of this book, email us at `customercare@packtpub.com` and mention the book title in the subject of your message.

Errata: Although we have taken every care to ensure the accuracy of our content, mistakes do happen. If you have found a mistake in this book, we would be grateful if you would report this to us. Please visit `www.packtpub.com/support/errata` and fill in the form.

Piracy: If you come across any illegal copies of our works in any form on the internet, we would be grateful if you would provide us with the location address or website name. Please contact us at `copyright@packt.com` with a link to the material.

If you are interested in becoming an author: If there is a topic that you have expertise in and you are interested in either writing or contributing to a book, please visit `authors.packtpub.com`.

Share Your Thoughts

Once you've read *High-Performance Programming in C# and .NET*, we'd love to hear your thoughts! Scan the QR code below to go straight to the Amazon review page for this book and share your feedback.

`https://packt.link/r/1-800-56471-6`

Your review is important to us and the tech community and will help us make sure we're delivering excellent quality content.

Part 1: High-Performance Code Foundation

Part 1 covers the foundation of high-performance code. We cover what's new in C# 10.0 and .NET 6, including performance improvements. Next, we look at the interoperability that is available that allows the gradual porting of Python systems to C#, followed by the garbage collector. You will learn how types can negatively impact performance, as can manually calling the garbage collector. And finally, we look at how we can use profiling tools to identify and address performance issues.

This part contains the following chapters:

- *Chapter 1, Implementing C# 10.0 and .NET 6*
- *Chapter 2, Introducing C# Interoperability*
- *Chapter 3, Predefined Data Types and Memory Allocations*
- *Chapter 4, Memory Management*
- *Chapter 5, Application Profiling and Tracing*

1
Introducing C# 10.0 and .NET 6

Microsoft .NET 6 and C# 10.0 are the latest incarnations of the .NET platform and C# programming language. They bring many performance enhancements to the C# and .NET programmer community. We will start this book with an overview of the new versions of C# and .NET.

In this chapter, you will start by downloading, restoring, building, and testing the latest version of the .NET compiler called **Roslyn**. Then, you will review what's new in .NET 6, including the areas where performance has been greatly enhanced. Then, you will review what's new in C# 10.0 by looking at some code examples that demonstrate these features.

In the *Native compilation* section, you will build a project and run it as an MSIL project with multiple binaries, then compile and run it as a single native binary. Finally, you will learn how to improve the performance of Windows Store applications and ASP.NET websites.

In this chapter, we will cover the following topics:

- **Overview of .NET 6**: In this section, we will cover, at a high level, what's new in .NET 6. You will learn about the various performance improvements that will be part of .NET 6.
- **Overview of C# 10.0**: Having learned how to obtain the latest Roslyn code in the *Technical requirements* section, in this section, you will learn about the various features that will be part of C# 10.0. This will include code examples.
- **Native compilation**: In this section, you will learn how to compile a .NET Core application into a single native executable. You will write a simple console application that recursively converts audio files from one format into another.
- **Improving Windows Store performance**: This is a brief section that provides standard guidelines for improving the performance of applications that target the Windows Store.
- **Improving ASP.NET performance**: This is a brief section that provides some standard guidelines for improving ASP.NET applications.

By the end of this chapter, you will have the following skills:

- You will understand what's new in Microsoft .NET 6.
- You will be able to apply the new C# 10.0 code features within your source code.
- You will be able to compile your source code to native assemblies (also known as binaries).
- You will know what, how, and where to look for information on improving the performance of applications that target the Windows Store.
- You will know what, how, and where to look for information on improving the performance of ASP.NET applications.

Let's begin this chapter by looking at Microsoft .NET 6.

Technical requirements

You will need the following prerequisites to complete this chapter:

- The latest preview version of Visual Studio Community Edition or higher.
- Microsoft .NET 6 SDK.

- This book's source code: `https://github.com/PacktPublishing/High-Performance-Programming-in-CSharp-and-.NET/tree/master/CH01`.
- Optional: The latest Roslyn compiler built from source. The source code is available on GitHub at `https://github.com/dotnet/roslyn`. This should be automatically installed when you install the latest preview versions of Visual Studio.

> **Note**
>
> You can find the latest complete and up-to-date C# 10.0 feature set at `https://github.com/dotnet/roslyn/blob/master/docs/Language%20Feature%20Status.md`. At the time of writing, C# 10.0 is still undergoing much development and change. So, the contents of this book may not work as expected. If this turns out to be the case, then please refer to the preceding URL for the most relevant information to help you start working.

Obtaining and building the latest Roslyn compiler from the source code

> **Note**
>
> The build system of all .NET-related repositories has been in flux for several years now. We will provide the instructions for compiling Roslyn here; these were correct at the time of writing. For the latest instructions, please read the `README.md` file located at `https://github.com/dotnet/roslyn`.

The following instructions are for downloading and building the latest version of the Roslyn compiler source on Windows 10:

1. In the root of the `C:\` drive, clone the Roslyn source code by using the following command in the Windows Command Prompt:

```
git clone https://github.com/dotnet/roslyn.git
```

2. Then, run the following command:

```
cd Roslyn
```

3. Restore the Roslyn dependencies by running the following command:

```
restore.cmd
```

4. Build the Roslyn source code by running the following command:

```
build.cmd
```

5. Test the Roslyn build by running the following command:

```
test.cmd
```

6. Once all the tests have finished running, check the versions of C# that are accessible to the new computer. Do this by opening a Command Prompt window and navigating to `C:\roslyn\artifacts\bin\csc\Debug\net472`.
7. Then, run the following command:

```
csc /langversion:?
```

> **Note**
>
> I always run my Command Prompt as an administrator. Hence, the screenshots will show Command Prompt in administrative mode. But running Command Prompt as an administrator is not necessary for this exercise. Where Command Prompt must be executed as an administrator, this will be made clear as needed.

You should see something equivalent to the following:

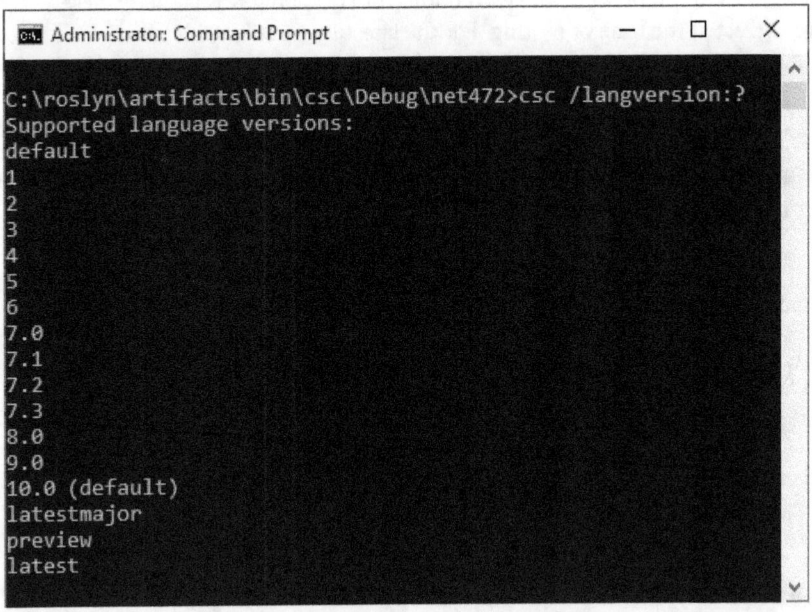

```
Administrator: Command Prompt

C:\roslyn\artifacts\bin\csc\Debug\net472>csc /langversion:?
Supported language versions:
default
1
2
3
4
5
6
7.0
7.1
7.2
7.3
8.0
9.0
10.0 (default)
latestmajor
preview
latest
```

Figure 1.1 – The versions of the C# programming language supported by the compiler

As you can see, at the time of writing, version 10.0 of the C# language is available via the C# compiler. C# 10.0 is set as the default. The preview is still under development. The default version may be different on your computer.

> **Note**
>
> The latest version of Visual Studio 2022 should allow you to use the latest available C# 10.0 code features. If it doesn't, then compile the latest source and overwrite the files located at `C:\Program Files (x86)\Microsoft Visual Studio\2022\Preview\MSBuild\Current\Bin\Roslyn`.

The following three sets of instructions provide compiler help for compiling a program that targets a specific C# version and then runs the program. These commands are for demonstrative purposes only, and you do not have to run them now:

```
csc /help
csc -langversion:10.0 /out:HelloWorld.exe Program.cs
csc HelloWorld
```

Now that you can build C# 10.0 from the command line and from within Visual Studio 2022, let's learn what kind of new development is taking place with Microsoft .NET 6.

Overview of Microsoft .NET 6

Microsoft .NET 6 is the latest incarnation of .NET. You can access the downloads at `https://dotnet.microsoft.com/download/dotnet/6.0`. The downloads are available for Windows, macOS, and Linux users.

> **Note**
>
> To get the most out of .NET 6 and C# 10.0, it is best that you have Visual Studio 2022 or later installed.

The .NET 6 API documentation is available at `https://docs.microsoft.com/dotnet/api/?view=net-6.0`.

Microsoft .NET 5 and later will no longer carry the *Core or Framework suffix*, as per the following article: `https://redmondmag.com/articles/2019/12/31/coming-in-2020-net-5.aspx`. Microsoft's goal with version 5 and later of the .NET platform is to create a single platform for the .NET development of WinForms, WPF, Xamarin. Forms, ASP.NET Core, and all other forms of .NET development. Xamarin.Forms becomes Microsoft MAUI, with the main difference between versions being that the new Microsoft MAUI will only use a single project to target all operating systems and devices.

Moving to one unified platform

The infrastructure for .NET 6 consists of runtime components, compilers, and languages. Microsoft .NET SDK will sit on top of this infrastructure. The tools that will be available include the command-line interface, Visual Studio Code, Visual Studio for Mac, and, of course, Visual Studio.

With the unified platform, you can write desktop applications using WinForms, WPF, and UWP. Web applications can be written using ASP.NET. Cloud applications will target Microsoft Azure. Mobile applications will be written using Microsoft MAUI. Games, **virtual reality** (**VR**), and **augmented reality** (**AR**) applications will be developed in Unity, using Visual Studio 2022 or higher as the C# code editor. IoT will target ARM32 and ARM64 architectures. Finally, you will be able to develop **artificial intelligence** (**AI**) applications using ML.NET and .NET for Apache Spark.

Microsoft is planning on producing a single .NET runtime and framework that is uniform in its developer experience and runtime behavior across applications and devices. This will be accomplished by building a single code base that combines the best elements of.NET Framework, .NET Core, Mono, and Xamarin.Forms.

The main features of.NET 6 are as follows:

- Unified developer experiences, regardless of the applications being developed and the devices being targeted.
- Unified runtime experiences across all devices and platforms.
- Java interoperability will be available on all platforms. This is stated in the Redmond Magazine article called *Coming in 2020: .NET 5, The Next Phase of Microsoft's .NET Framework*: `https://redmondmag.com/articles/2019/12/31/coming-in-2020-net-5.aspx`.
- Multiple operating systems will be supported for Objective-C and Swift.
- AOT will be supported by CoreFX to provide static .NET compilation, support multiple operating systems, and produce assemblies that are smaller in size.

Now, let's look at some of the new features of .NET 6 from a high-level viewpoint.

Garbage collection

The garbage collector's performance regarding marking and stealing has been improved. When a thread has finished its marking allotment, it can steal outstanding marking work from other threads. This speeds up the process of collecting items to be garbage collected. Reduced lock contentions on computers with higher core counts, improved de-committing, avoidance of costly memory resets, and vectorized sorting are just some of the new garbage collection performance improvements in .NET 6.

Just-In-Time compiler

In .NET 6, the **Just-In-Time** (**JIT**) compiler has also been improved. You can apply various optimizations to the JIT, and it has an unlimited amount of time to implement those optimizations. **Ahead-Of-Time** (**AOT**) is just one of the various techniques provided to the JIT so that it can compile as much code as it can before executing the application. The JIT now sees the length of an array as unsigned, which improves the performance of mathematical operations carried out on an array's length. There are still many changes being made.

Suffice to say that between the JIT and the GC, the performance improvements that have been made to JIT and GC concerning memory and compilation optimizations are just two reasons alone to migrate to .NET 6.

The JIT also recognizes more than a thousand new hardware intrinsic methods. These methods allow you to target various hardware instruction sets from C#. You are no longer tied to just x86_x64 hardware instruction sets.

Several runtime helper functions are available in the JIT. These helper functions enable the JIT compiler to manipulate the source code so that the code runs must faster. Generic lookups are much faster now, as they no longer need to employ slower lookup tables.

Text-based processing

Performance enhancements have also been made within the text-based processing elements of .NET 6. These include (but are not limited to) processing whitespace in the `System.Char` class, which requires less branching and fewer arguments. Because this class is used in various text-processing objects and methods within .NET 6, the speed of processing text in .NET 6 will be generally improved. `DateTime` processing is also at least 30% faster due to optimizations in extracting the date and time components from the raw tick count. Performance improvements have also been made to string operations due to culture-aware modifications of `StartsWith` and `EndsWith`. By utilizing stack allocation and JIT devirtualization, the performance of data encoding, such as *UTF8* and *Latin1* encoding, has also been enhanced.

Regular expression (**RegEx**) performance has also been improved in .NET 6. The RegEx engine has had performance improvements that increase textual processing by up to *three to six times* and even more. The `CharInClass` method is more intelligent in determining if characters appear within the specified character class. Character and digit comparisons use lookup tables and various method calls are inlined, providing improved RegEx processing. Generated code for various expressions has been improved. Searching for RegExes is carried out using span-based searching with vectorized methods. The need for backtracking has been eliminated as it analyzes RegExes during the node tree optimization phase and adds atomic groups that do not change the semantics but do prevent backtracking. These are only some of the improvements to RegEx performance. But there are many more.

> **Note**
> For more in-depth knowledge on .NET 5 performance improvements to RegExes, please read the following very detailed post by Stephen Toub: `https://devblogs.microsoft.com/dotnet/regex-performance-improvements-in-net-5/`.

Threading and asynchronous operations

Threading and asynchronous operations have also received a performance boost in .NET 5 with the experimental addition of async `ValueTask` pooling. You can turn on pooling by setting `DOTNET_SYSTEM_THREADING_POOLASYNCVALUETASK` to `true` or `1`. Pooling creates state machine box objects that implement the interfaces, `IvalueTaskSource`, and `IValueTaskSource<TResult>`. The runtime adds these objects to the pool. Volatility has also received performance improvements in `ConcurrentDictionary`, with performance improving as much as 30% on some ARM architectures.

Collections and LINQ

The collections have also seen several performance enhancements, mainly to `Dictionary<TKey, TValue>`, `HashSet<T>`, `ConcurrentDictionary<TKey, TValue>`, and `System.Collections.Immutable`. The `HashSet<T>` collection's implementation has been rewritten and re-synchronized with `Dictionary<TKey,` the `TValue>` implementation, and moved further down the stack. The performance of `foreach` when iterating through an `ImmutableArray<T>` has been improved, and the generated code has been reduced in size by the addition of the `[MethodImpl(MethodImplOptions.AggressiveInlining)]` annotation to the `GetEnumerator` method of `ImmutableArray<T>`. Other elements of the .NET collections, such as `BitArray`, have also seen performance improvements.

In .NET 5, LINQ has also seen further performance improvements, including `OrderBy`, `Comparison<T>`, `Enumerable.SkipLast`, and by making implementing `Enumerable.Any` more consistent with `Enumerable.Count`. These are only a few performance improvements that have been to the collections.

Networking and Blazor

Networking has received a lot of work on performance improvement, especially the `System.Uri` class (especially in its construction). The `System.Net.Sockets` and `System.Net.Http` namespaces have also seen performance improvements. Many improvements have been made to how JSON is processed with `JsonSerializer` in the `System.Text.Json` library for .NET.

As Blazor uses the .NET mono runtime and .NET 5 libraries, a linker has been added that trims code from the assembly that is not used down to the member level. The code to be trimmed is identified by *static code analysis*. User interface response times are also improved in Blazor Web Assembly applications, as the client-side code is downloaded before being executed, and behaves just like a desktop application – but from within the browser.

Furthermore, general improvements that have gone into .NET 5 include faster assembly loading, faster mathematical operations, faster encryption and decryption, faster interoperability, faster reflection emitting, faster I/O, and various allocations in various libraries.

New performance-based APIs and analyzers

A few new performance-focused APIs have been added to .NET 5. Internally, some of these APIs are already being used to reduce code size and improve the performance of .NET 5 itself. They focus on helping the programmer to concentrate on writing performant code and removing the complexity of tasks that have been previously hard to accomplish. These new APIs and improvements to existing APIs include `Decimal`, `GC`, `MemoryExtensions`, `StringSplitOptions`, `BinaryPrimitives`, `MailAddress`, `MemoryMarshall`, `SslStream`, `HttpClient`, and more.

The .NET 5 SDK has also seen the addition of some new performance-based analyzers. These analyzers can detect accidental allocations as a part of range indexing and offer ways to eliminate the allocation. Analyzers will detect the old overloads for the `Stream.Read/WriteAsync` methods and will offer fixes to enable automatic switching to the newer overload methods that prefer `Memory` overloads. In `StringBuilder`, it is more performant to use *typed overloads* to append non-string values such as `int` and `long` values. When situations are encountered by the analyzer where the programmer has called `ToString()` on a type that's being appended for which a typed overload exists, the fixer will detect these situations and automatically switch to using the correct typed overload. With LINQ, it is now more efficient to check if `(collection.Count != 0)` using the `(!collection.IsEmpty)` syntax. The old way will be detected by the analyzer and fixed to use the more performant new way. Finally, when you have worked to make your code faster, your code is made correct, as the analyzer flags cases that use loops to allocate memory from the stack using `stackalloc`. This helps prevent stack overflow exceptions from being raised.

To see the road ahead in terms of .NET's new development, you can view the .NET Core roadmap located at `https://github.com/dotnet/core/blob/master/roadmap.md`.

Now, let's look at C# 10.0.

Overview of C# 10.0

You can find the features that will become part of C# 10.0 on the Roslyn GitHub page at `https://github.com/dotnet/roslyn/blob/master/docs/Language%20Feature%20Status.md`.

Not all these features are available at the time of writing. However, we will look at some of the available features. With that, let's start with top-level programs.

Writing top-level programs

Before C# 9.0, the **Hello, World!** console application was always the starting point for learning C#. The file that students would update was called `Program.cs`. In this file, you would have something akin to the following:

```
using System;
namespace HelloWorld
{
class Program
{
        static void Main(string[] args)
        {
            Console.WriteLine("Hello, World!");
        }
}
}
```

As you can see, first, we import our `System` library. Then, we have a namespace definition followed by our class definition. Then, in the class definition, we have our `Main` method, in which we output the phrase `"Hello, World!"` to the console window.

In version 10.0 of the C# programming language, this can be simplified down to a single line:

```
System.Console.WriteLine("Hello, World");
```

Here, we have eradicated 10 lines of code. Running the program will output the following:

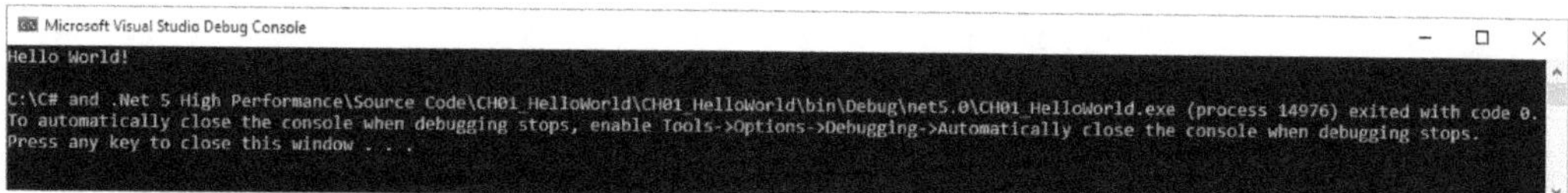

Figure 1.2 – The console window showing the output "Hello World!"

If we open the generated DLL in IL DASM, we will see the following:

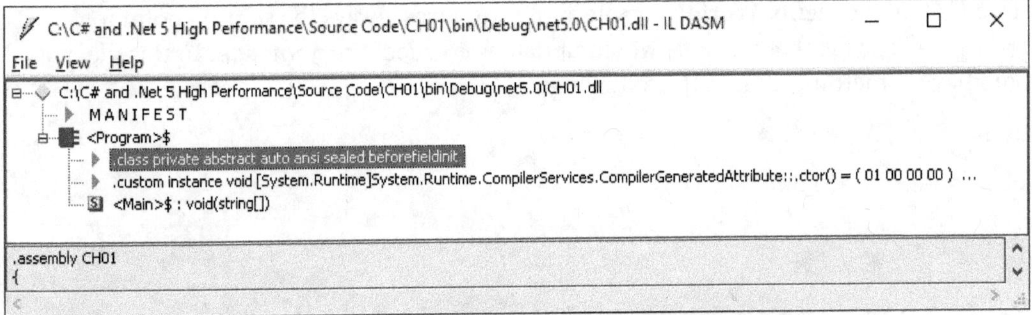

Figure 1.3 – ILDASM showing the internals of the hello world program

You will see from the decompilation that the compiler adds the `Main` method at compile time. The next addition to C# 10.0 that we will look at is init-only properties.

Using init-only properties

Init-only properties allow you to use object initializers with immutable fields. For our little demonstration, we will use a `Book` class that holds the name of a book and its author:

```
namespace CH01_Books
{
    internal class Book
    {
        public string Title { get; init; }
        public string Author { get; init; }
    }
}
```

The properties can be initialized when the book is created. But once created, they can only be read, not updated, making the `Book` type immutable. Now, let's look at init-only properties. In the `Program` class, replace its contents with the following:

```
using System;
using CH01_Books;
var bookName = new Book { Title = "Made up book name",
    Author = "Made Up Author" };

Console.WriteLine($"{bookName.Title} is written by
    {bookName.Author}. Well worth reading!");
```

Here, we imported the `System` and `CH01_Books` namespaces. Then, we declared a new immutable variable of the `Book` type. After that, we output the contents of that `Book` type using an interpolated string. Run the program; you should see the following output:

```
Microsoft Visual Studio Debug Console
Made up book name is written by Made Up Author. Well worth reading!

C:\_book\C-9-and-.NET-5-High-Performance\CH01\CH01_Books\CH01_Books\bin\Debug
\net5.0\CH01_Books.exe (process 25852) exited with code 0.
To automatically close the console when debugging stops, enable Tools->Option
s->Debugging->Automatically close the console when debugging stops.
Press any key to close this window . . .
```

Figure 1.4 – The output of our init-only properties example

Now that we have been introduced to init-only properties, let's look at records.

Using records

When updating data, you do not want that data to be changed by another thread. So, in multi-threaded applications, you will want to use thread-safe objects when making updates. Records allow complete objects to be immutable and behave as values. The advantage of using records over structs is that they require less memory to be allocated to them. This reduction in memory allocation is accomplished by compiling records to reference types. They are then accessed via references and not as copies. Due to this, other than the original record allocation, no further memory allocation is required.

Let's learn how to use records. Start a new console application.

To demonstrate the use of records, we will use the following `Book` example:

```
internal record Book
{
public string Title { get; init; }
     public string Author { get; init; }
}
```

The only change to the `Book` class is that `class` has been replaced with `record`. Everything else remains the same. Now, let's put the record to work:

1. Replace the contents of the `Program` class with the following code:

```
using System;
using CH01_Records;

var bookOne = new Book {
    Title = "Made Up Book",
    Author = "Made Up Author
};

var bookTwo = bookOne with {
    Title = "And Another Made Up Book"
};

var bookThree = bookTwo with {
    Title = "Yet Another Made Up Book"
};

var bookFour = bookThree with {
    Title = "And Yet Another Made Up Book: Part 1",
};

var bookFive = bookFour with {
    Title = "And Yet Another Made Up Book: Part 2"
};

var bookSix = bookFive with {
    Title = "And Yet Another Made Up Book: Part 3"
};

Console.WriteLine($"Some of {bookThree.Author}'s
    books include:\n");
Console.WriteLine($"- {bookOne.Title}");
Console.WriteLine($"- {bookTwo.Title}");
```

```
Console.WriteLine($"- {bookThree.Title}");
Console.WriteLine($"- {bookFour.Title}");
Console.WriteLine($"- {bookFive.Title}");
Console.WriteLine($"- {bookSix.Title}");
Console.WriteLine($"\nMy favourite book by {bookOne.
    Author} is {bookOne.Title}.");
```

2. As you can see, we are creating immutable record types. We can create new immutable types from them and change any fields we like using the `with` expression. The original record is not mutated in any way. Run the code; you will see the following output:

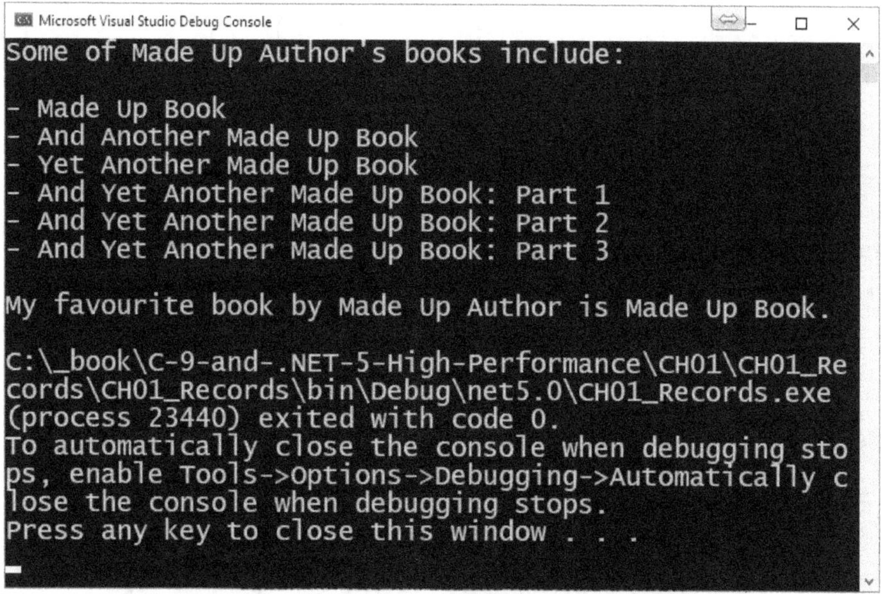

Figure 1.5 – Init-only properties showing their immutability

Despite changing the title during the assignment, the original record has not been mutated at all.

3. Records can also use *inheritance*. Let's add a new record that contains the publisher's name:

```
internal record Publisher
{
    public string PublisherName { get; init; }
}
```

4. Now, let's have our `Book` inherit this `Publisher` record:

```
internal record Book : Publisher
{
    public string Title { get; init; }
    public string Author { get; init; }
}
```

5. `Book` will now include `PublisherName`. When we initialize a new book, we can now set its `PublisherName`:

```
var bookOne = new Book {
    Title = "Made Up Book",
    Author = "Made Up Author",
    PublisherName = "Made Up Publisher Ltd."
};
```

6. Here, we have created a new `Book` that contains `Publisher.PublisherName`. Let's print the publisher's name. Add the following line to the end of the `Program` class:

```
Console.WriteLine($"These books were originally published
    by {bookSix.PublisherName}.");
```

7. Run the code; you should see the following output:

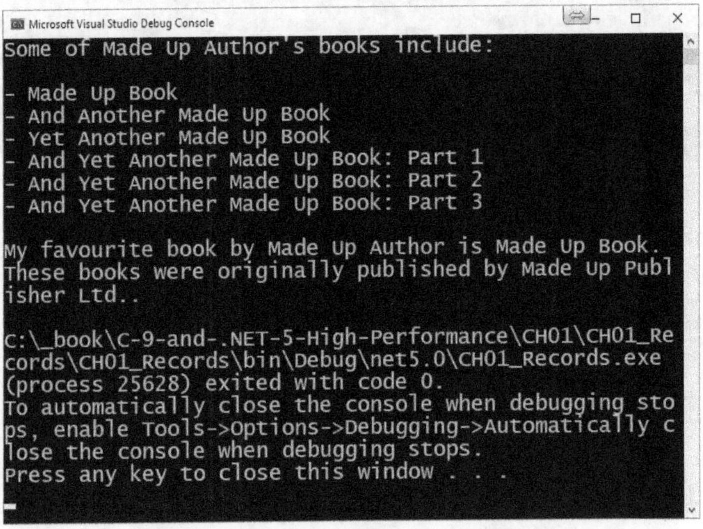

```
Microsoft Visual Studio Debug Console
Some of Made Up Author's books include:

- Made Up Book
- And Another Made Up Book
- Yet Another Made Up Book
- And Yet Another Made Up Book: Part 1
- And Yet Another Made Up Book: Part 2
- And Yet Another Made Up Book: Part 3

My favourite book by Made Up Author is Made Up Book.
These books were originally published by Made Up Publ
isher Ltd..

C:\_book\C-9-and-.NET-5-High-Performance\CH01\CH01_Re
cords\CH01_Records\bin\Debug\net5.0\CH01_Records.exe
(process 25628) exited with code 0.
To automatically close the console when debugging sto
ps, enable Tools->Options->Debugging->Automatically c
lose the console when debugging stops.
Press any key to close this window . . .
```

Figure 1.6 – Init-only properties using inheritance

8. As you can see, we never set the publisher's name for `bookTwo` to `bookSix`. However, the inheritance has followed through from when we set it for `bookOne`.
9. Now, let's perform object equality checking. Add the following code to the end of the `Program` class:

```
var book = bookThree with { Title = "Made Up Book" };
var booksEqual = Object.Equals(book, bookOne) ?
    "Yes" : "No";
Console.WriteLine($"Are {book.Title} and
    {bookOne.Title} equal? {booksEqual}");
```

10. Here, we created a new `Book` from `bookThree` and set the title to `Made Up Book`. Then, we performed an equality check and output the result to the console window. Run the code; you will see the following output:

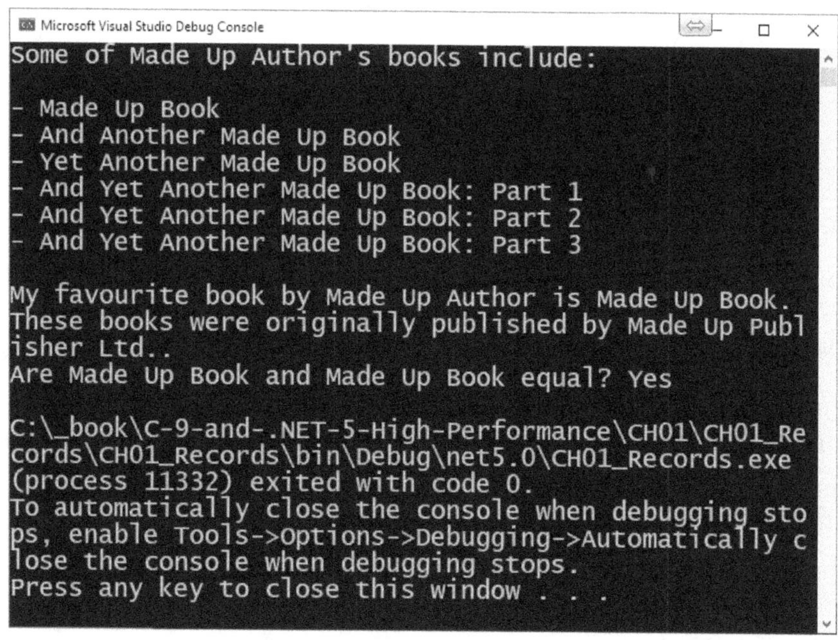

Figure 1.7 – Init-only properties showing the result of an equality check

It is clear to see that the equality check works with both book instances being equal.

11. Our final look at records considers positional records. Positional records set data via the constructor and extract data via the deconstructor. The best way to understand this is with code. Add a class called `Product` and replace the class with the following:

```
public record Product
{
    readonly string Name;
    readonly string Description;

    public Product(string name, string
        description)
        => (Name, Description) = (name,
        description);

    public void Deconstruct(out string name, out
        string description)
        => (name, description) = (Name,
            Description);
}
```

12. Here, we have an immutable record. The record has two private and `readonly` fields. They are set in the constructor. The `Deconstruct` method is used to return the data. Add the following code to the `Program` class:

```
var ide = new Product("Awesome-X", "Advanced Multi-
    Language IDE");
var (product, description) = ide;

Console.WriteLine($"The product called {product} is an
    {description}.");
```

In this code, we created a new product with parameters for the name and description. Then, we declared two fields called `product` and `description`. The fields are set by assigning the product. Then, we output the product and description to the console window, as shown here:

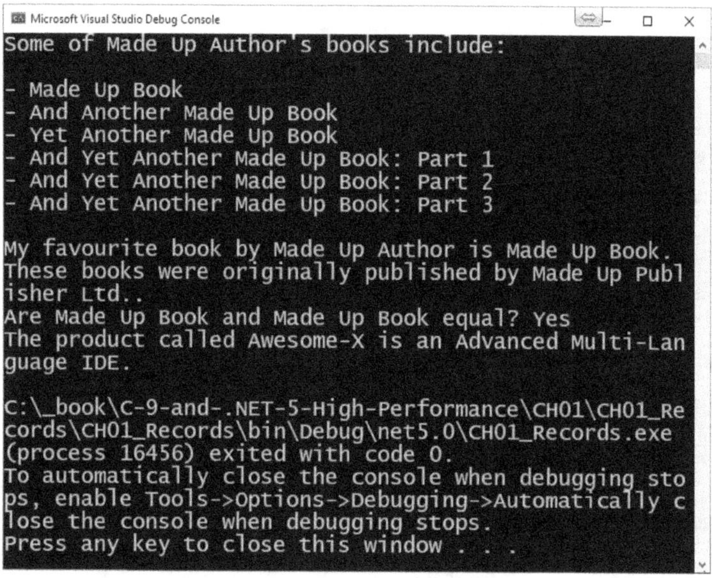

```
Microsoft Visual Studio Debug Console
Some of Made Up Author's books include:

- Made Up Book
- And Another Made Up Book
- Yet Another Made Up Book
- And Yet Another Made Up Book: Part 1
- And Yet Another Made Up Book: Part 2
- And Yet Another Made Up Book: Part 3

My favourite book by Made Up Author is Made Up Book.
These books were originally published by Made Up Publ
isher Ltd..
Are Made Up Book and Made Up Book equal? Yes
The product called Awesome-X is an Advanced Multi-Lan
guage IDE.

C:\_book\C-9-and-.NET-5-High-Performance\CH01\CH01_Re
cords\CH01_Records\bin\Debug\net5.0\CH01_Records.exe
(process 16456) exited with code 0.
To automatically close the console when debugging sto
ps, enable Tools->Options->Debugging->Automatically c
lose the console when debugging stops.
Press any key to close this window . . .
```

Figure 1.8 – Init-only positional records

Now that we have finished looking at records, let's look at the improved pattern matching capabilities of C# 10.0.

Using the new pattern matching features

Now, let's look at what's new for pattern matching in C# 10.0, starting with simple patterns. With simple pattern matching, you no longer need the discard (_) operator to just declare the type. In our example, we will apply discounts to orders:

1. Add a new record called `Product` to a new file called `Product.cs` in a new console application and add the following code:

```
internal record Product
{
    public string Name { get; init; }
    public string Description { get; init; }
    public decimal UnitPrice { get; init; }
}
```

2. Our `Product` record has three init-only properties for `Name`, `Description`, and `UnitPrice`. Now, add the `OrderItem` record that inherits from `Product`:

```
internal record OrderItem : Product
{
    public int QuantityOrdered { get; init; }
}
```

3. Our `OrderItem` record inherits the `Product` record and adds the `QuantityOrdered` init-only property. In the `Program` class, we will add three variables of the `OrderItem` type and initialize them. Here is the first `OrderItem`:

```
var orderOne = new OrderItem {
        Name = "50-80mm Scottish Cobbles",
        Description = "These rounded stones are
          frequently used for edging paths and to add
            interest to gardens",
        QuantityOrdered = 4,
        UnitPrice = 199
};
```

   As you can see, the quantity that's being ordered is `4`.

4. Add `orderTwo` with the same values but with an `OrderQuantity` of `7`.

5. Then, add `orderThree` with the same values, but with an `OrderQuantity` of `31`. We will demonstrate simple pattern matching in the `GetDiscount` method:

```
static int GetDiscount(object order) =>
    order switch
    {
        OrderItem o when o.QuantityOrdered == 0 =>
            throw
          new ArgumentException("Quantity must be
              greater than zero."),
        OrderItem o when o.QuantityOrdered > 20 => 30,
        OrderItem o when o.QuantityOrdered < 5 => 10,
```

```
        OrderItem => 20,
        _ => throw new ArgumentException("Not a known
            OrderItem!", nameof(order))
    };
```

6. Our `GetDiscount` method receives an order. `QuantityOrdered` is then evaluated. Argument exceptions are thrown if the order quantity is `0` and if the object type that's been passed in is not of the `OrderItem` type. Otherwise, a discount of the `int` type is returned for the quantity ordered. Notice that we use the type without using the discard operator on the line for the 20% discount.
7. Finally, we must add the following lines to the end of the `Program` class:

```
Console.WriteLine($"The discount for Order One is
    {GetDiscount(orderOne)}%.");
Console.WriteLine($"The discount for Order Two is
    {GetDiscount(orderTwo)}%.");
Console.WriteLine($"The discount for Order Three is
    {GetDiscount(orderThree)}%.");
```

8. These lines print the discount received for each of the orders to the console window. Now, let's modify our code so that it uses relational pattern matching. Add the following method to the `Program` class:

```
static int GetDiscountRelational(OrderItem orderItem)
    => orderItem.QuantityOrdered switch
    {
        < 1 => throw new ArgumentException("Quantity
            must be greater than zero."),
        > 20 => 30,
        < 5 => 10,
        _ => 20
    };
```

9. Using relational pattern matching, we have received the same outcome as with simple pattern matching, but with less code. It is also very readable, which makes it easy to maintain. Add the following three lines of code to the end of the `Program` class:

```
Console.WriteLine($"The discount for Order One is
    {GetDiscountRelational(orderOne)}%.");
Console.WriteLine($"The discount for Order Two is
    {GetDiscountRelational(orderTwo)}%.");
Console.WriteLine($"The discount for Order Three is
    {GetDiscountRelational(orderThree)}%.");
```

10. In these three lines, we simply output the discount for each order to the console window. Run the program; you will see the following output:

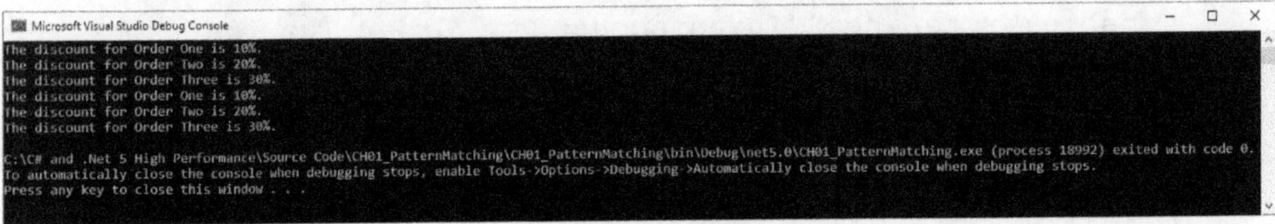

Figure 1.9 – Simple and relational pattern matching output showing the same results

From the preceding screenshot, you can see that the same outcome has been received for both discount methods.

11. The logical `AND`, `OR`, and `NOT` methods can be used in logical pattern matching. Let's add the following method:

```
static int GetDiscountLogical(OrderItem orderItem) =>
    orderItem.QuantityOrdered switch
    {
        < 1 => throw new ArgumentException("Quantity
            must be greater than zero."),
        > 0 and < 5 => 10,
        > 4 and < 21 => 20,
        > 20 => 30
    };
```

12. In the `GetDiscountLogical` method, we employ the logical AND operator to check whether a value falls in that range. Add the following three lines to the end of the `Program` class:

```
Console.WriteLine($"The discount for Order One is
    {GetDiscountLogical(orderOne)}%.");
Console.WriteLine($"The discount for Order Two is
    {GetDiscountLogical(orderTwo)}%.");
Console.WriteLine($"The discount for Order Three is
    {GetDiscountLogical(orderThree)}%.");
```

13. In those three lines of code, we output the discount value for the order to the console window. Run the code; you will see the following output:

```
Microsoft Visual Studio Debug Console
The discount for Order One is 10%.
The discount for Order Two is 20%.
The discount for Order Three is 30%.
The discount for Order One is 10%.
The discount for Order Two is 20%.
The discount for Order Three is 30%.
The discount for Order One is 10%.
The discount for Order Two is 20%.
The discount for Order Three is 30%.

C:\C# and .Net 5 High Performance\Source Code\CH01_PatternMatching\CH01_PatternMatching\bin\Debug\net5.0\CH01_PatternMatching.exe (process 21756) exited with code 0.
To automatically close the console when debugging stops, enable Tools->Options->Debugging->Automatically close the console when debugging stops.
Press any key to close this window . . .
```

Figure 1.10 – Simple, relational, and logical pattern matching showing the same results

The output for the logical pattern matching is the same as for simple and relational pattern matching. Now, let's learn how to use new expressions with targeted types.

Using new expressions with targeted types

You can omit the type of object being instantiated. But to do so, the declared type must be explicit and not use the `var` keyword. If you attempt to do this with the ternary operator, you will be greeted with an exception:

1. Create a new console application and add the `Student` record:

```
public record Student
{
    private readonly string _firstName;
    private readonly string _lastName;

    public Student(string firstName, string
        lastName)
    {
```

```
            _firstName = firstName;
            _lastName = lastName;
        }

        public void Deconstruct(out string firstName,
            out string lastName)
            => (firstName, lastName) = (_firstName,
                _lastName);
    }
```

2. Our `Student` record stores the first and last name values, which have been set via the constructor. These values are obtained via the `out` parameters of the `Deconstruct` method. Add the following code to the `Program` class:

```
Student jenniferAlbright = new ("Jennifer",
    "Albright");
var studentList = new List<Student>
{
    new ("Jennifer", "Albright"),
    new ("Kelly", "Charmichael"),
    new ("Lydia", "Braithwait")
};
var (firstName, lastName) = jenniferAlbright;
Console.WriteLine($"Student: {lastName}, {firstName}");
(firstName, lastName) = studentList.Last();
Console.WriteLine($"Student: {lastName}, {firstName}");
```

3. First, we instantiate a new `Student` without declaring the type in the `new` statement. Then, we instantiate a new `List` and add new students to the list while omitting the `Student` type. The fields are then defined for `firstName` and `lastName` and assigned their values through the assignment of the named student. The student's name is then printed out on the console window. Next, we take those fields and reassign them with the name of the last student on the list. Then, we output the student's name to the console window. Run the program; you will see the following output:

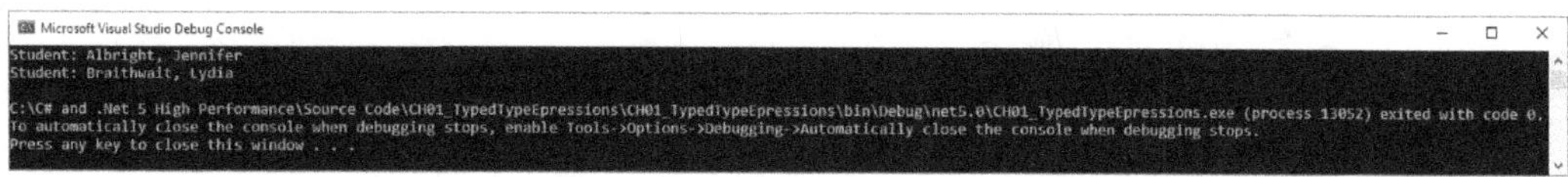

Figure 1.11 – Using targeted types with new expressions

From the preceding screenshot, you can see that we have the correct student names printed. Now, let's look at covariant returns.

Using covariant returns

With covariant returns, base class methods with less specific return types can be overridden with methods that return more specific types. Have a look at the following array declaration:

```
object[] covariantArray = new string[] { "alpha", "beta",
    "gamma", "delta" };
```

Here, we declared an `object` array. Then, we assigned a `string` array to it. This is an example of covariance. The `object` array is the least specific array type, while the `string` array is the more specific array type.

In this example, we will instantiate covariant types and pass them into a method that accepts less and more specific types. Add the following class and interface declarations to the `Program` class:

```
public interface ICovariant<out T> { }
public class Covariant<T> : ICovariant<T> { }
public class Person { }
public class Teacher : Person { }
public class Student : Person { }
```

Here, we have a covariant class that implements a covariant interface. We declared a general type of `Person` that is inherited by the specific `Teacher` and `Student` types. Add `CovarianceClass`, as shown here:

```
public class CovarianceExample
{
public void CovariantMethod(ICovariant<Person> person)
{
      Console.WriteLine($"The type of person passed in is
```

```
            of type {person.GetType()}.");
    }
}
```

In the `CovarianceExample` class, we have a `CovariantMethod` with a parameter that can accept objects of the `ICovariant<Person>` type. Now, let's put covariance to work by adding the `CovarianceAtWork` method to the `CovarianceExample` class:

```
public void CovarianceAtWork()
{
ICovariant<Person> person = new Covariant<Person>();
ICovariant<Teacher> teacher = new Covariant<Teacher>();
ICovariant<Student> student = new Covariant<Student>();
CovariantMethod(person);
CovariantMethod(teacher);
CovariantMethod(student);
}
```

In this method, we have the general `Person` type and the more specific `Teacher` and `Student` types. We must pass each into `CovariantMethod`. This method can take the less specific `Person` type and the more specific `Teacher` and `Student` types.

To run the `CovarianceAtWork` method, place the following code after the `using` statement and before the `covariantArray` example:

```
CovarianceExample.CovarianceAtWork();
```

Now, let's look at native compilation.

Native compilation

When .NET code is compiled, it is compiled into **Microsoft Intermediate Language** (**MSIL**). MSIL gets interpreted by a JIT compiler when it is needed. The JIT compiler then compiles the necessary MSIL code into native binary code. Subsequent calls to the same code call the binary version of the code, not the MSIL version of the code. This means that MSIL code is always slower than native code, as it is compiled to native on the first run.

JIT code has the advantage of being cross-platform code at the expense of longer startup times. The code of an MSIL assembly that runs is compiled to native code by the JIT compiler. The native code is optimized by the JIT compiler for the target hardware it is running on.

By default, UWP applications are compiled to native code using .NET Native, while iOS applications are compiled to native code via Xamarin/Xamarin.Forms. Microsoft .NET Core can also be compiled into native code.

Performing native compilation of .NET Core applications

When using `dotnet` to compile an assembly to native code, you will need to specify a target framework. For a list of supported target frameworks, please refer to `https://docs.microsoft.com/en-us/dotnet/standard/frameworks`. You will also need to specify a **Runtime Identifier** (**RID**). For a list of supported RIDs, please refer to `https://docs.microsoft.com/en-us/dotnet/core/rid-catalog`.

> **Note**
>
> At the time of writing, native compilation against .NET 5.0 does have its issues. So, to keep things simple, we will demonstrate native compilation into a single executable against netcoreapp3.1 and win10-x64.

To demonstrate the compilation of Microsoft .NET Core applications into natively compiled single executables, we will write a simple demonstration application that traverses a directory structure and converts audio files from one format into another:

1. Start a new console application and target .NET 6.
2. Visit `https://ffmpeg.org/download.html` and download `ffmpeg` for your operating system. Mine is Windows 10.
3. On Windows 10, extract the `ffmpeg` files into the `C:\Tools\ffmpeg` folder. Add the following `using` statements to the top of the `Program.cs` file:

```
using System;
using System.Diagnostics;
using System.IO;
```

4. We will be batch processing audio files in a folder hierarchy on our local systems. Here, the `using` statements listed will help us debug our code and perform I/O on the filesystem. Now, at the top of the `Program` class, add the following three fields:

```
private static string _baseDirectory = string.Empty;
private static string _sourceExtension = string.Empty;
private static string _destinationExtension = string
    .Empty;
```

5. The `BaseDirectory` member holds the starting directory that will be processed. `sourceExtension` holds the extension of the file type, such as `.wav`, we are after converting to, while `destinationExtension` holds the extension, such as `.ogg`, of the file type we are after converting to. Update your `Main` method so that it looks as follows:

```
static void Main(string[] args)
{
Console.Write("Enter Source Directory: ");
_baseDirectory = Console.ReadLine();
Console.Write("Enter Source Extension: ");
_sourceExtension = Console.ReadLine();
Console.Write("Enter Destination Extension: ");
_destinationExtension = Console.ReadLine();
new Program().BatchConvert();
}
```

6. In our `Main` method, we have requested that the user enters the source directory, source extension, and destination extension. Then, we set out member variables and called the `BatchConvert` method. Let's add our `BatchConvert` method:

```
private void BatchConvert()
{
var directory = new DirectoryInfo(_baseDirectory);
ProcessFolder(directory);
}
```

7. The `BatchConvert` method creates a new `DirectoryInfo` object called `directory` and then passes the `directory` object into the `ProcessFolder` method. Let's add this method now:

```
private void ProcessFolder(DirectoryInfo
    directoryInfo)
{
Console.WriteLine($"Processing Directory:
    {directoryInfo.FullName}");
var fileInfos = directoryInfo.EnumerateFiles();
```

```
var directorieInfos = directoryInfo.
    EnumerateDirectories();

        foreach (var fileInfo in fileInfos)
                if (fileInfo.Extension.Replace(".", "")
                    == sourceExtension)
                        ConvertFile(fileInfo);

foreach (var dirInfo in directorieInfos)
                ProcessFolder(dirInfo);
}
```

8. The `ProcessFolder` method outputs a message to the screen so that the user knows what folder is being processed. Then, it obtains an enumeration of the `FileInfo` and `DirectoryInfo` objects from the `directoryInfo` parameter. After this, it converts all the files in that folder that have the required source file extension. Once all the files have been processed, each of the `DirectoryInfo` objects is processed by calling the `ProcessFolder` method recursively. Finally, let's add our `ConvertFile` method:

```
private void ConvertFile(FileInfo fileInfo)
{
}
```

9. Our `ConvertFile` method takes a `FileInfo` parameter. This parameter contains the file that is to undergo conversion. The remaining code will be added to this `ConvertFile` method. Add the following three variables:

```
var timeout = 10000;
var source = $"\"{fileInfo.FullName}\"";
var destination = $"\"{fileInfo.FullName.Replace
      (_sourceExtension, _destinationExtension)}\"";
```

10. The `timeout` variable is set to 10 seconds. This gives the process 10 seconds to process each file. The `source` variable contains the full name of the file to be converted, while the `destination` variable contains the full path of the newly converted file. Now, add the check to see if the converted file exists:

```
if (File.Exists(fileInfo.FullName.Replace
      (_sourceExtension, _destinationExtension)))
```

```
{
Console.WriteLine($"Unprocessed: {fileInfo.FullName}");
        return;
}
```

11. If the destination file exists, then the conversion has already taken place, so we do not need to process the file. So, let's output a message to the user to inform them that the file is unprocessed, and then return from the method. Let's add the code to perform the conversion:

```
Console.WriteLine($"Converting file: {fileInfo.FullName}
     from {_sourceExtension} to {_destination
          Extension}.");

using var ffmpeg = new Process
{
StartInfo = {
                FileName = @"C:\Tools\ffmpeg\bin
                   \ffmpeg.exe",
                 Arguments = $"-i {source}
                   {destination}",
                 UseShellExecute = false,
                 RedirectStandardOutput = true,
                 RedirectStandardError = true,
                 CreateNoWindow = true
}
};
ffmpeg.EnableRaisingEvents = false;
ffmpeg.OutputDataReceived += (s, e) => Debug.WriteLine
      ($"Debug: e.Data");
ffmpeg.ErrorDataReceived += (s, e) => Debug.WriteLine
      ($@"Error: {e.Data}");
ffmpeg.Start();
ffmpeg.BeginOutputReadLine();
ffmpeg.BeginErrorReadLine();
ffmpeg.WaitForExit(timeout);
```

12. Here, we output a message to the window informing the user of the file being processed. Then, we instantiate a new process that executes `ffmpeg.exe` and converts an audio file from one format into another, as specified by the user. The converted file is then saved in the same directory as the original file.
13. With that, we have completed our sample project. So, let's see it running. On an external hard disk, I have some Ghosthack audio samples that I own. The files are in `.wav` file format. However, they need to be transformed into `.ogg` files to be used in an Android program that I use. You can use your own audio file or music folders.

> **Note**
>
> If you don't have any audio files to hand to test this small program, you can download some royalty-free sounds from `https://www.bensound.com`. You can check the following page for links to various public music domains: `https://www.lifewire.com/public-domain-music-3482603`.

14. Fill out the questions and press *Enter*:

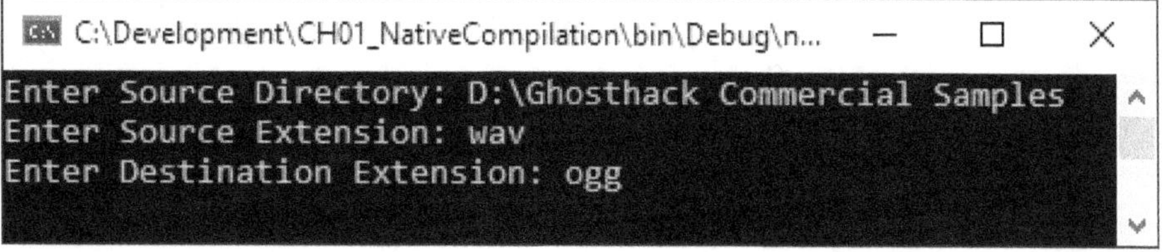

Figure 1.12 – Our file converter showing the directory and file conversion formats

The program will now process all files and folders under the specified parent folder and process them.

The program is working as expected in its MSIL form. However, we can see the delay in performing the file conversions. Let's compile our file converter into a single native executable, and then see if it is visibly any faster:

1. Open the Visual Studio Developer Command Prompt as an administrator and navigate to the folder that contains your solution and project file. When publishing the file, it is worth noting that the `TargetFramework` property of the project should also be updated to netcoreapp3.1; otherwise, this may not work – that is, if it is set to `net5.0`. Type the following command and then press *Enter*:

```
dotnet publish --framework netcoreapp3.1 -
    p:PublishSingleFile=true --runtime win10-x64
```

2. When the command has finished running, your command window should look as follows:

```
Administrator: Developer Command Prompt for VS 2019 Preview

C:\Development\CH01_NativeCompilation>dir
 Volume in drive C has no label.
 Volume Serial Number is D8E3-8785

 Directory of C:\Development\CH01_NativeCompilation

31/07/2020  22:17    <DIR>          .
31/07/2020  22:17    <DIR>          ..
31/07/2020  20:00    <DIR>          bin
31/07/2020  20:01               178 CH01_NativeCompilation.csproj
31/07/2020  20:00             1,149 CH01_NativeCompilation.sln
31/07/2020  20:01    <DIR>          obj
31/07/2020  22:17             2,995 Program.cs
               3 File(s)          4,322 bytes
               4 Dir(s)  29,551,661,056 bytes free

C:\Development\CH01_NativeCompilation>dotnet publish --framework netcoreapp3.1 -p:PublishSingleFile=true --runtime win10-x64
Microsoft (R) Build Engine version 16.8.0-preview-20371-01+e8338f5ad for .NET
Copyright (C) Microsoft Corporation. All rights reserved.

  Determining projects to restore...
  Restored C:\Development\CH01_NativeCompilation\CH01_NativeCompilation.csproj (in 294 ms).
  You are using a preview version of .NET. See: https://aka.ms/dotnet-core-preview
  CH01_NativeCompilation -> C:\Development\CH01_NativeCompilation\bin\Debug\netcoreapp3.1\win10-x64\CH01_NativeCompilation.dll
  CH01_NativeCompilation -> C:\Development\CH01_NativeCompilation\bin\Debug\netcoreapp3.1\win10-x64\publish\

C:\Development\CH01_NativeCompilation>_
```

Figure 1.13 – The Developer Command Prompt in administrative mode showing the native compilation output

3. If you navigate to the publish directory, you will see the following output:

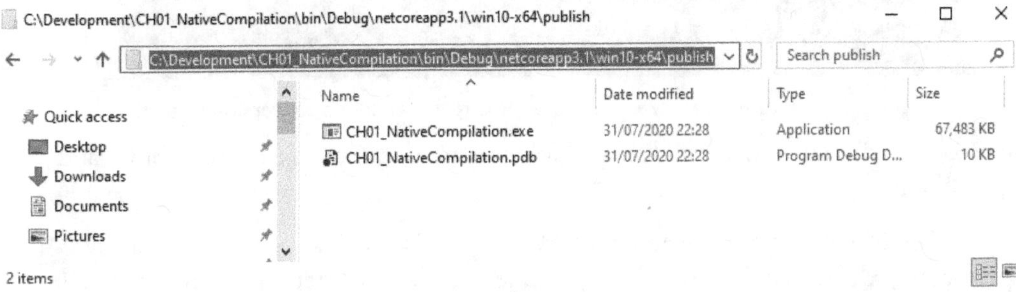

Figure 1.14 – Windows Explorer displaying the output files resulting from native compilation

4. Run the `CH01_NativeCompilation.exe` file. You will see that `.wav` files are processed into `.ogg` files much quicker.

In this section, we learned how to write a console app. We compile the console app to MSIL and then compile the console app into a single native executable file. Visually, from the user's perspective, the file processes batch audio files much quicker in native form than in MSIL form.

Now, let's learn how to improve Windows Store applications.

Improving Windows Store performance

Here are some basic tips for improving the performance of Windows Store applications:

- **Perform the Microsoft Store app performance assessment**: For information on how to do this, visit `https://docs.microsoft.com/en-us/windows-hardware/test/assessments/microsoft-store-app-performance`.
- **Understand the Microsoft Store app performance assessment's Results**: To help you understand the results of the Windows Store App Performance Assessment, visit `https://docs.microsoft.com/en-us/windows-hardware/test/assessments/results-for-the-microsoft-store-app-performance-assessment`
- **Address the issues highlighted in the Microsoft Store app performance assessment results**: The main areas to focus on are any that have issues highlighted in dark purple, followed by issues marked in medium purple. The primary metrics will be on Launch:Warm, Launch:Cold, Post Launch, Idle, and Suspend. You also need to pay attention to processor and storage usage, as well as processor and storage I/O delays, registry flushes, time accounting, missing symbols, long-running **Deferred Procedure Calls** (**DPCs**), and **Interrupt Service Routines** (**ISRs**) that can be perceived by the end user as performance issues.

In the next section, we'll learn how to improve performance with ASP.NET.

Improving ASP.NET performance

Here are some basic tips for improving the performance of web applications and APIs:

- **Perform baseline measurements**: Before making changes to the performance of your web application or API, take a baseline reading of your program's performance. This way, you can measure any adjustments to see if they improve performance or slow things down.
- **Begin by optimizing the code with the largest impact**: When you have completed your baseline measurements, start performance tuning on the piece of code that is the least performant and that has the biggest impact on your program's performance. This will provide you with your biggest win.

- **Enable HTTP compression**: To reduce the size of transmitted files over HTTP/HTTPS and improve network performance, enable compression. There are two types of compression. GZIP compression has been around for many years and is the de facto compression mechanism; it can reduce a file's size by one-third. An alternative compression mechanism is Brotli. Most major browsers have had support for this compression mechanism since 2016/2017.
- **Reduce TCP/IP connection overheads**: Reducing HTTP requests seriously improves HTTP communication performance. Each request uses network and hardware resources. When a hardware and software-specific number of connections is established, performance will start to show signs of degrading. This can be mitigated by reducing the number of HTTP requests.
- **Use HTTP/2 over SSL**: HTTP/2 over SSL provides various performance improvements of using HTTP. Multiplexed streams provide bi-directional sequences of text format frames. Server push enables a server to push cacheable data to the client in anticipation that the client may use it. Binary protocols have a lower overhead when it comes to parsing data and they are less prone to errors. Binary protocols offer more security and have better network utilization There are many more optimizations that you gain when you switch to HTTP/2 over SSL.
- **Employ minification**: Minification is the process of eliminating whitespace and comments in an HTML, CSS, or JavaScript web file. By making the size of the file smaller and by enabling compression, you can seriously speed up the network transmission of files, especially over poor Wi-Fi.
- **Place CSS in the head so that it loads first**: To efficiently render a web page, it is best to load the complete CSS before rendering to prevent reflows.
- **Place JavaScript at the end of HTML files**: For vanilla HTML, CSS, and JavaScript applications, the preferred location for JavaScript files is at the bottom of HTML files, before the closing `body` tag. For heavy framework-based applications, bootstrapping will be beneficial as only the JavaScript that is needed is loaded. An alternative is isomorphic JavaScript for rendering pages on both the client and the server. Isomorphic applications improve SEO, performance, and maintainability.
- **Reduce image size**: Images can vary greatly in size. Reduce the size of the images that are used on a page. When used with minification and compression, this technique can help fancy-looking web pages load fast.

You can find out more about other techniques for improving ASP.NET performance in the *Further reading* section. Now, let's summarize what we have learned in this chapter.

Summary

At the start of this chapter, you downloaded the latest source for the C# programming language. Then, you restored it, built it, and ran various tests. After that, you built a Hello, World! program that demonstrated C# 9.0 features.

Then, you learned what's new in .NET 5. This section covered topics on garbage collection, JIT compilation, text-based processing, threading and asynchronous operations, collections, LINQ, networking, and Blazor. We also covered the new performance-based APIs and analyzers. From what was covered, you now have a high-level appreciation of the many performance improvements made by Microsoft and third parties to the new version of the .NET programming language. These performance improvements are a solid reason to move to .NET 5. But another compelling reason is also the move to .NET for true cross-platform development from a single code base.

After reviewing the performance improvements and additions to .NET 5, we looked at the new C#10.0 features. You learned how to write a program with just one line of code using top-level statements. Then, you learned how to implement init-only properties, records, new pattern-matching features, new expressions with targeted types, and covariant returns. From reviewing the new additions to the C# 9.0 language, you learned how to compile and run code in MSIL, and then compile and run native code in a single executable file. Visually, the end user experience was shown to be better when using the native binary over the MSIL assembly. For the example, we used a simple audio file format converter.

You were then provided with some guidance on how to improve Windows Store app performance. Links to the official Microsoft documentation were presented to you to help you generate performance reports, along with how to understand the results of the performance assessment. This guidance also highlighted the main metrics to pay attention to. Finally, we considered some ways in which you can improve the performance of your ASP.NET websites and APIs. In the *Further reading* section, you will find a link to the official Microsoft ASP.NET documentation. This documentation will help you architect and build quality websites.

Furthermore, in the *Further reading* section, you will find some links to documentation and the GitHub repository for .NET MAUI, which is due to be released in 2021 in concert with .NET 6. This user interface technology is an evolution of Xamarin.Forms with evolutionary changes based on customer research. It does look rather promising.

In the next chapter, we will be looking at .NET interoperability. But before that, work through this chapter's questions to see how well everything has sunk in.

Questions and exercises

Answer the following questions regarding this chapter:

1. What areas of .NET are being improved by .NET 6?
2. What is new to C# 10.0?
3. What tools are available for native compilation in .NET?
4. How can you improve the Windows Store app's performance?
5. How can you speed up ASP.NET?
6. Investigate the state of .NET MAUI, the future of frontend desktop and mobile development that is still undergoing development.
7. Write some console applications and practice using the new features of .NET 6 and C# 10.0.
8. Use Benchmark.NET to benchmark one of your small applications, and then upgrade it to use .NET 6 and C# 10.0. Measure its performance without making any changes if possible, and then measure its performance again. See if you notice any performance improvements by simply upgrading to C# 10.0 and .NET 6.

> **Note**
>
> The answers to questions 4 and 5 can be found in the external reference sources provided in their respective sections.

Further reading

To learn more about the topics that were covered in this chapter, take a look at the following resources:

- Download .NET 6: `https://dotnet.microsoft.com/download/dotnet/6.0`.
- Download Visual Studio Preview: `https://visualstudio.microsoft.com/vs/preview/`.
- *Introducing the .NET multi-platform app UI*: `https://devblogs.microsoft.com/dotnet/introducing-net-multi-platform-app-ui/`.
- .NET MAUI GitHub page: `https://github.com/dotnet/maui`.
- Learn from Microsoft how to build quality Windows 10 apps that reflect your brand: `https://docs.microsoft.com/en-us/windows-hardware/get-started/`.

- Learn from Microsoft how to architect and build quality websites using Microsoft technology: `https://dotnet.microsoft.com/apps/aspnet`.
- C#9.0 early review: `https://medium.com/dev-genius/c-9-early-review-5bcd88296c54#:~:text=Relax%20ordering%20of%20ref%20and%20partial%20modifiers%20Currently%2C,is%20a%20ref%20struct%2C%20ref%20must%20appear%20`.
- *File I/O Improvements in .NET 6*: `https://devblogs.microsoft.com/dotnet/file-io-improvements-in-dotnet-6/`.

2 Implementing C# Interoperability

This chapter is an optional chapter for those who would like to or need to use C# to interoperate with Excel, Python, C++, and **Visual Basic 6** (**VB6**).

Python has become a very popular programming language in recent months and is now a very big player in data science and machine learning. Since big data employs various technologies that are required to work with each other under various business scenarios, in this chapter, you will learn how to execute Python scripts and code from C#. You can also use IronPython.NET on the .NET platform, but since this book is for C# programmers, we will not be considering IronPython.NET in this chapter.

There are times when it is necessary to access libraries written in C++ – especially when performance is an issue, and you need that extra performance in advanced games.

In this chapter, you will learn about Microsoft .NET interoperability. It is advantageous to move your complete code base to a single code base that uses a familiar language that your whole development team is comfortable with using. But sometimes, to do this in one move is often not practical or cost-effective, or even safe. And that is where interoperability comes in.

In this chapter, you will learn how to interact with managed and unmanaged code. You will be looking at using unsafe code, unmanaged code with **Platform Invoke** (**P/Invoke**), COM interoperability, and disposing of unsafe code.

> **Note**
>
> Using unmanaged code in C# does not always improve performance. Sometimes, it degrades it. But the logic of including this chapter within this book on high performance is to provide the knowledge and tools you will need to gradually replace your unmanaged code base with a managed code base. By doing so, all your developers only work with a single language and its supporting languages (in this case, C#). Your software can use the high-performing and highly scalable features of Azure or any other .NET cloud provider to build world-class cloud-based systems. The other advantage of doing this is that it makes code management and maintenance much easier.

In this chapter, we will be covering the following topics:

- **Using unsafe code**: C# does a good job of shielding programmers from having to deal with pointers. But sometimes, it is necessary to use pointers to improve performance. Due to this, in this section, we will be looking at what unsafe code is and how to implement them.
- **Exposing static entry points using Platform Invoke**: P/Invoke allows you to access code in unmanaged libraries from your managed C# code. In this section, we will learn how to access code that hasn't been built using .NET.
- **Performing COM interoperability**: In this section, we will learn how to make COM components and libraries visible for C# projects to use. We will also look at how to make our components and libraries visible to COM components to use.
- **Safely disposing of unsafe code**: C# does a very good job of performing garbage collection to free up resources when code is finished with, but when you're dealing with unmanaged code, you are responsible for cleaning up unmanaged resources. So, in this section, you will be shown how to do this.

After completing this chapter, you will be able to do the following:

- Understand the use of unsafe code in C#
- Call native code from managed code
- Use COM libraries and components in managed and unmanaged code
- Release unsafe resources when they're no longer needed

Technical requirements

In this chapter, some of the code includes interoperability between C# managed assemblies and COM-based ActiveX UserControls, DLLs, and executables.

To write the code and build the projects in this chapter, you will need the following:

- Visual Studio 2022
- The latest x86 preview of .NET 6
- The latest x64 preview of .NET 6
- Optional: Visual C++
- Optional: Visual Studio Tools for Microsoft Office
- Optional: Visual Basic 6

The code files for this chapter can be found in this book's GitHub repository: `https://github.com/PacktPublishing/High-Performance-Programming-in-CSharp-and-.NET/tree/master/CH02`.

> **Note**
>
> Although Visual Basic 6 is obsolete and no longer supported by Microsoft, it is still heavily used in production code within various businesses and sectors, such as automotive software providers and the education sector. Interoping with VB6 and .NET enables phased migrations from VB6 to .NET. By modernizing applications built with old technology, you can make them highly scalable across time zones using various cloud providers, such as Azure.

We will start this chapter by looking at unsafe code.

Using Platform Invocation (P/Invoke)

P/Invoke is a **Common Language Infrastructure** (**CLI**) feature that enables native code to be called by managed applications. Native code is not managed by the **Common Language Runtime** (**CLR**), so, the code's safety is firmly placed in the hands of the programmer.

In managed code, the garbage collector automatically cleans up objects in memory and is responsible for assigning generations to objects. We will cover the garbage collector in more detail in *Chapter 4, Memory Management*. A new object always starts life as generation zero when it is less than 80,000 bytes in size and will be placed on the small object heap. Objects equal to or greater than 80,000 bytes in size are placed on the large object heap. Objects that survive generation zero get promoted by the garbage collector to generation one. Finally, objects that survive generation one get promoted to generation two.

> **Note**
> Instantiated objects equal to or greater than 80,000 bytes may start as generation zero but be promoted, so they would not be seen as generation zero.

When an object is promoted from one generation to another by the garbage collector, its memory address changes. This breaks any pointers that refer to that address. To prevent the address from being modified by the garbage collector, the pointer code must be declared using the `fixed` keyword.

Now, let's look at using the `unsafe` and `fixed` keywords.

Using unsafe and fixed code

To remind the programmer of their responsibility for ensuring code safety, unmanaged code is wrapped in a code block marked as unsafe using the `unsafe` keyword. Unsafe code makes use of pointers to refer to locations in memory.

Unsafe code provides programmers with access to pointer types in C#, which can be necessary when they're working with the underlying operating system, system drivers, or working on time-critical code that needs to be executed in the smallest amount of time.

Even though we say the code that deals with pointers is unsafe code, it is safe to work with. Such code is marked with the `unsafe` keyword. Despite being called unsafe, such code is safe to use in managed code – it is just not verified by the CLR. Therefore, it is possible to introduce security risks and/or pointer errors. You can have an unsafe `pointer_type`, `value_type`, or `reference_type`.

> **Note**
> The topic of unsafe code is deep, so if you wish to learn more, please view the language specification that discusses unsafe code at https://docs.microsoft.com/dotnet/csharp/language-reference/language-specification/unsafe-code.

In this section, we will write a console application that puts the various unsafe code mechanisms to work. You can view the project's source code at https://github.com/PacktPublishing/C-9-and-.NET-5-High-Performance/tree/master/CH02/CH02_UnsafeCode.

Consider the following computer program:

```
namespace CH02_UnsafeCode
{
```

```
    using System;

    class Program
    {
        static void Main(string[] args)
        {
            int[] array = new int[5] { 5, 4, 3, 2, 1 };
            Console.WriteLine(array[4]);

            unsafe
            {
                int* pointer = stackalloc int[5];
                int* cpointer = pointer;
                cpointer += 50;
                Console.WriteLine(*cpointer);
            }
        }
    }
}
```

In the preceding code, you can see that we allocate memory space for an array of five `int` values using the `new` keyword. We can do the same thing using unsafe code. But instead of using the `new` keyword, we can use `stackalloc` and wrap the code in a code block marked as `unsafe`.

When dealing with unsafe code such as array pointers, it is necessary to use the `fixed` keyword. To understand why the `fixed` keyword is important, you need to understand garbage collection.

When objects are created, they are generation-zero objects. The garbage collector will remove any unreferenced generation one objects. If the space for allocating generation zero objects becomes full, the garbage collector moves the generation zero objects to generation one. Then, new objects can be added to generation zero. If the generation one and generation two objects become full, and all the objects are in use, then the garbage collector moves the generation one objects to generation two. This, in turn, moves the generation zero objects to generation one.

New objects are then added to generation zero. At this point, if the generation two, generation one, and generation zero storage spaces are full, which means that no new objects can be added, then you end up with an out-of-memory exception. The following diagram shows this:

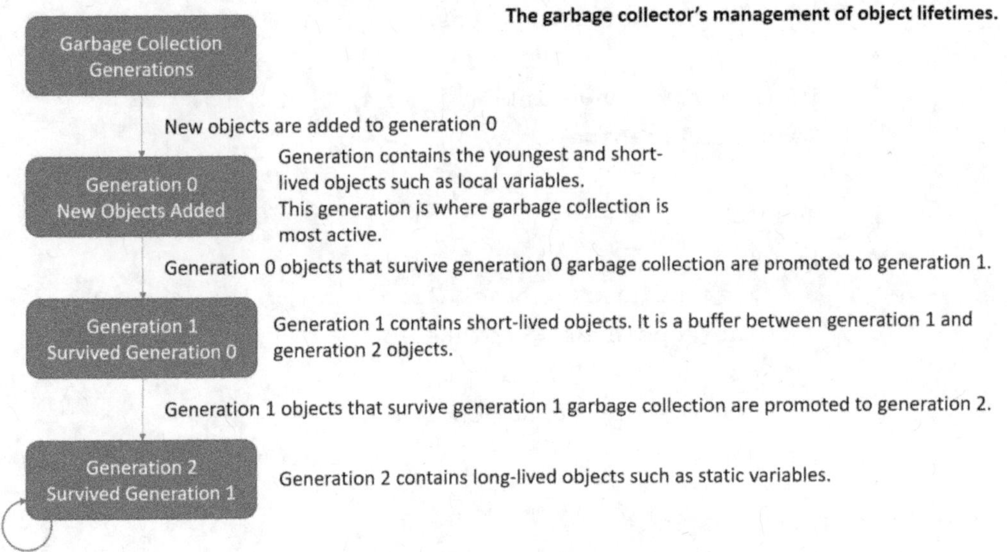

Figure 2.1 – Garbage collection management of object generations

Since the garbage collector is moving the items from one generation to another, the memory locations change. However, the pointers to those objects in your code do not change. Therefore, when retrieving the information from the pointer address, the data will be incorrect.

To prevent this from happening, we can use the `fixed` keyword. The `fixed` keyword tells the garbage collector to leave the address space that `arrayPointer` is pointing to alone. This means that we can ensure that the pointer will be pointing to the correct address space and data. The following code shows the `unsafe` and `fixed` keywords being used to deal with an array:

```
unsafe
{
        fixed (int* arrayPointer = array)
        {
        // Code omitted.
}
}
```

In the preceding code, because we are using unsafe code, we used an `unsafe` code block. Since we don't want the array to be affected by the garbage collector, we kept the object at its current generation by using the `fixed` code block.

One caveat you need to be aware of when using unsafe code is the effect of accessing an array that's out of bounds. When you access an array that's out of bounds in managed code, you are presented with `IndexOutOfBoundsException`. You do not have that luxury with unmanaged code. You are responsible for ensuring that the correct indexes are accessed. If you happen to access an index that is outside the bounds of the array, then you will not have `IndexOutOfBoundsException` thrown. Instead, you will have whatever is at that memory address returned to you. In that case, you may or may not end up with some type of exception being thrown. The following code demonstrates this:

```
int* pointerToArray = stackalloc int[100];
Console.WriteLine(pointerToArray[99]);
Console.WriteLine(pointerToArray[100]);
```

Here, the array is added to the stack. The value of the array at position `99` is correct, but the array position of `100` is out of bounds, so an incorrect value is returned. This means that `IndexOutOfBoundsException` is thrown. That is why you must be careful with unmanaged code when dealing with indexes.

> **Note**
>
> The reason for the `unsafe` keyword is to alert the programmer to their responsibility for code safety. When dealing with pointers, runtime exceptions aren't raised. Instead, whatever is at that memory location is returned. That's why you must take extra care when programming unsafe code. You must also use the `fixed` keyword when you can't afford for the garbage collector to switch the generations of your objects and move them.

In C#, you can only use structs and primitives with unsafe and fixed code. Classes and strings that access the heap are not allowed. This means that nothing that will be garbage collected can be referenced using unsafe code. So, when using C# pointers, you can use value types, but you cannot use reference types.

For example, the following code will not compile:

```
unsafe
{
        fixed (TestObject* testObject = new TestObject()) { }
```

```
        fixed (string* text = "Hello, World!") { }
}
```

The `testObject` variable is a reference type pointer, so the compiler throws an exception if you build the code. This code returns the following exception:

- `CS0208`: Cannot take the address of, get the size of, or declare a pointer to a managed type (`'TestObject'`)

The `text` variable is a string pointer, and the compiler throws an exception if you build the code. This code returns the following exception:

- `CS0208`: Cannot take the address of, get the size of, or declare a pointer to a managed type (`'string'`)

> **Note**
> Using fixed objects can result in memory fragmentation. So, avoid using the `fixed` keyword until you need to, and only use it for as long as you need it.

Now, let's look at exposing static entry points using P/Invoke.

Exposing static entry points using P/Invoke

P/Invoke allows you to make static entry points available to other applications. If you have ever used WinAPI, then you have accessed code in DLLs via their public static entry points. These access points would have been made available using P/Invoke.

To use P/Invoke, you will need to import the `System.Runtime.InteropServices` namespace. Then, you must make the static entry call using `DllImportAttribute`:

> **Note**
> To identify the static entry points of a file, you can use the `dumpbin.exe` file that's located in the `C:\Program Files (x86)\Microsoft Visual Studio\2019\Preview\VC\Tools\MSVC\14.28.29115\bin\Hostx64\x64` folder. This version of 14.28.29.115 was correct at the time of writing. When you come to execute the following code, this version will have changed. Use the latest version that you have installed on your computer.

Now, let's learn how to use `dumpbin` to see what methods and properties the `User32.dll` system library exports using the command line:

1. Open the command line or developer command prompt. Then, enter the following command (note that there might be a different version on your computer – use the latest version number you have):

   ```
   " C:\Program Files (x86)\Microsoft Visual
       Studio\2019\Preview\VC\Tools\MSVC\14.28.29304
         \bin\Hostx64\x64\dumpbin.exe /exports User32.dll
   ```

   You should see something like the following:

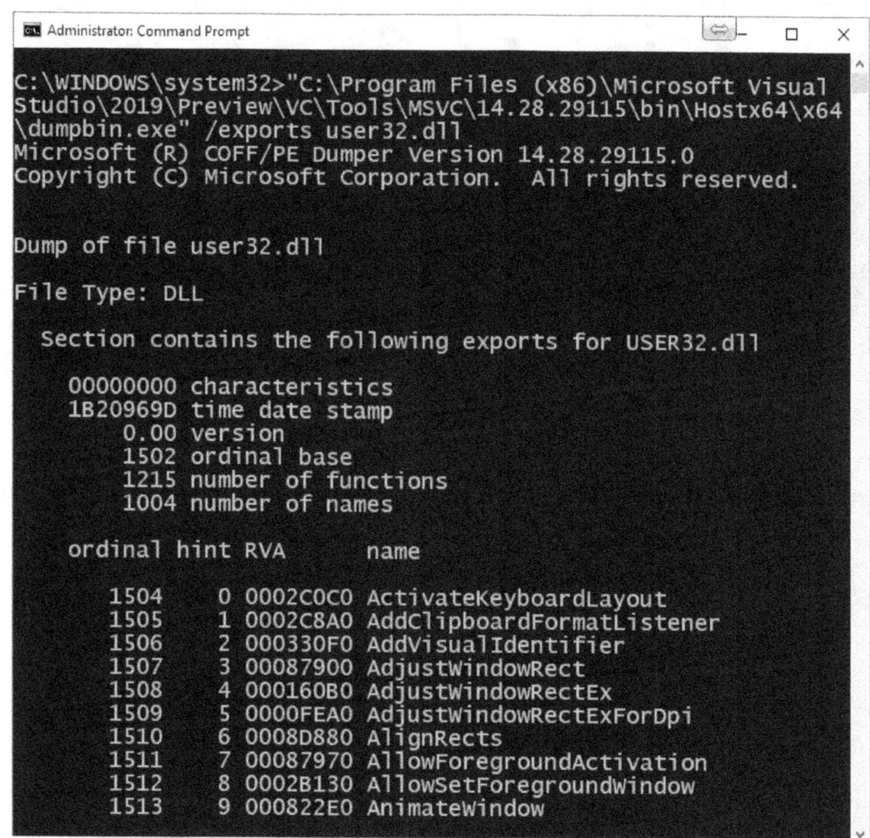

Figure 2.2 – Command line showing the outcome from executing dumpbin on User32.dll

2. Let's write a C++ library and call it from C# using P/Invoke. First, we must create a new empty C++ project, as shown in the following screenshot:

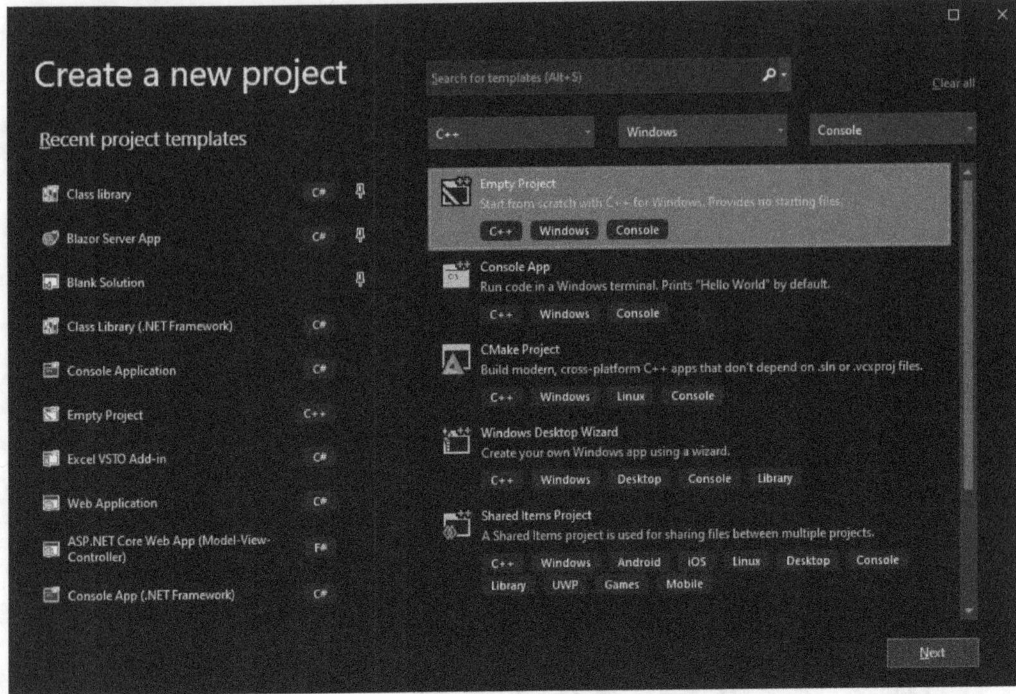

Figure 2.3 – Creating a new empty C++ project

3. Delete the `Header Files`, `Resource File`, and `Source Files` folders. Add a new class called `Product`. Delete the header file that has the `.h` file extension.

4. Modify the `Product.cpp` file so that it contains the following code:

```
#include <string>
#include <iostream>
#include <comdef.h>
struct Product {
int Id;
      BSTR Name;
      void BuyProduct() {
            std::wcout << "Product.BuyProduct(" <<
              Name << ");\n";
            std::cout << "Id: " << Id;
            std::cout << "\n";
      }
```

```
};
extern "C" __declspec(dllexport) Product
    CreateProduct() {
      Product product = Product();
      product.Id = 1;
      product.Name = SysAllocString(L"New Product");
      return product;
}
extern "C" __declspec(dllexport) void
    BuyProduct(Product product) {
      product.BuyProduct();
}
```

5. Now, we must import three libraries: `string`, `iostream`, and `comdef.h`. Then, we must declare a struct with `Id` and `Name` values. In C++, strings are typically defined using `std::string`, but when it comes to.NET, we declare strings as the BSTR type for OLE/automation by convention. The BSTR APIs use the `CoTask*` memory allocator, which is the implied interop contract for native on Windows. On non-Windows systems, .NET 5 uses `malloc/free`. We also have a void method called `BuyProduct()` that prints the `Id` and `Name` values as well as a newline, to the console's output window.
6. The next thing we must do is export two methods called `CreateProduct()` and `BuyProduct(Product product)`. Now, `CreateProduct()` creates a new `Product` and returns it to the caller, while `BuyProduct(Product product)` calls the `BuyProduct()` method on the passed-in `Product` struct.
7. Add a new class called `Greeting`. Delete the `Greeting.h` file. Update the `Greeting.cpp` file so that it contains the following source code:

```
#include <iostream>
#include <comdef.h>

extern "C" __declspec(dllexport) void SendGreeting();
extern "C" __declspec(dllexport) int Add(int, int);
extern "C" __declspec(dllexport) bool
    IsLengthGreaterThan5(const char*);
extern "C" __declspec(dllexport) BSTR GetName();
void SendGreeting() {
      std::cout << "Dear C#, C++ says hello!\n";
```

```
}
int Add(int x, int y) {
        return x + y;
}
bool IsLengthGreaterThan5(const char* value) {
        return strlen(value) > 5;
}
BSTR GetName() {
        return SysAllocString(L"Packt Publishing");
}
```

Here, we have included `iostream` and `comdef.h`. We have four methods called `SendGreeting()`, `Add(int x, int y)`, `IsLengthGreaterThan5(const char* value)`, and `GetName()`. We expose these methods to external callers.

`SendGreeting()` takes no parameters and outputs a string to the standard output window. `Add(int x, int y)` adds to integers passed in by the caller and returns the result. `IsLengthGreaterThan5(const char* value)` checks if the length of the string that's been passed in by the caller is greater than `5`. If it is, then `true` is returned. Otherwise, `false` is returned. `GetName()` returns a string. The return type for a string must be `BSTR`. To return a string in a method, you must call `SysAllocString(L"the string you want returning")`. This correctly initializes the string to a wide-character array and initializes the count.

That is all there is to our C++ library. Now, we just need to configure it. But before we do that, we will write our C# client, which will consume the C++ library. The reason for doing this is that once we have the build folder for our C# client, we will get our C++ library to output the DLL to the C# build folder. Follow these steps:

1. Add a new .NET Core 3.1 console application project to your solution, and then set it as the startup project. Add a class called `Product`. Update the contents of the `Product.cs` file, as follows:

```
using System.Runtime.InteropServices;
[StructLayout(LayoutKind.Sequential)]
public struct Product
{
        public int Id;
```

```
[MarshalAs(UnmanagedType.BStr)]
public string Name;
}
```

Here, we have created a mirror of the C++ struct in our C# client and included the `System.Runtime.InteropServices` library. Our C# struct has the same two fields as our C++ struct and they are in the same order. The struct itself is annotated with `[StructLayout(LayoutKind.Sequential)]`, which states that the field order must be processed sequentially. This ensures a match between the fields in the C++ library and the fields in the C# library. Additionally, the `Name` property is a string, so it needs to be annotated with the `[MarshalAs(UnmanagedType.Bstr)]` annotation. This tells the compiler that the C# string is to be treated as a C++ BSTR.

2. Modify the `Program.cs` file, as follows:

```
namespace CH02_Pinvoke {
    using System;
    using System.Runtime.InteropServices;
    class Program {
        static void Main(string[] _) {
        }
    }
}
```

Here, we imported the `System` and `System.Runtime.InteropServices` libraries, and then modified the `Main(string[] args)` method by replacing the `args` parameter's name with the default operator.

3. Set the build configuration to x64.
4. Append the following line to the `PropertyGroup` section of your C++ project file:

```
<AppendTargetFrameworkToPath>false</AppendTargetFrame
    workToPath>
```

5. Build the project. This will produce our output folder where we will place our compiled C++ library.

6. Right-click on the C++ project and select **Properties**. You should see the **CH02_NativeLibrary Property Pages** dialog box:

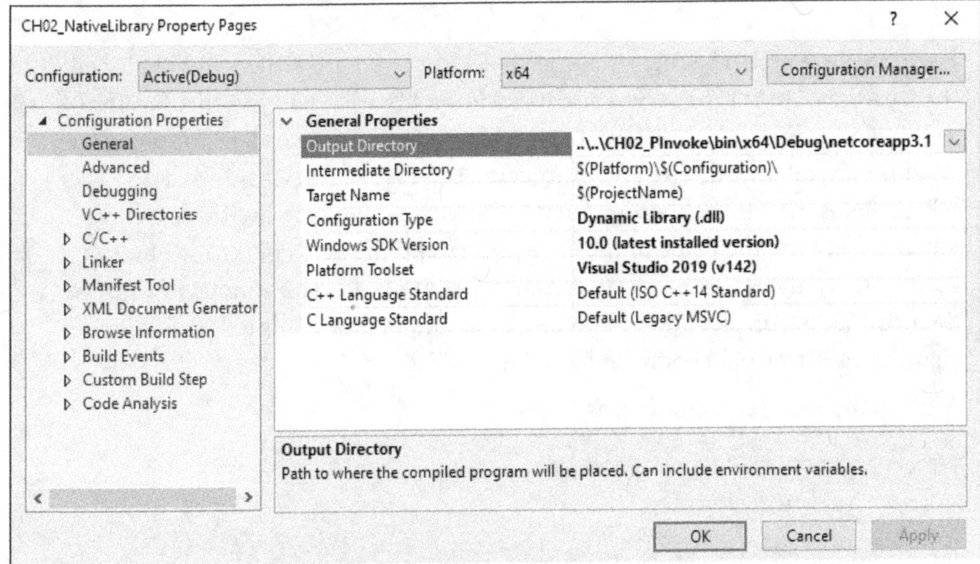

Figure 2.4 – CH02_NativeLibrary Property Pages

7. Change **Output Directory** to your C# project's output directory. Then, change **Configuration Type** to **Dynamic Library (.dll)**. Build the C++ library.
8. Back in your C# project, add the COM library by browsing for it in your C# build folder.
9. Add the following DLL imports to the `Program` class, above the `Main` method:

```
[DllImport("CH02_NativeLibrary.dll",
      CallingConvention = CallingConvention.StdCall
)]
[DllImport("CH02_NativeLibrary.dll", EntryPoint =
    "Add",CallingConvention = Calling
        Convention.StdCall
)]
public static extern int AddIntegers(int x, int y);
[DllImport("CH02_NativeLibrary.dll",
      CallingConvention = CallingConvention.StdCall
)]
```

```
public static extern bool IsLengthGreaterThan5(string
    value);
[DllImport("CH02 _ NativeLibrary.dll",
        CallingConvention = CallingConvention.StdCall
)]
[return: MarshalAs(UnmanagedType.BStr)]
public static extern string GetName();
[DllImport("CH02 _ NativeLibrary.dll",
        CallingConvention = CallingConvention.StdCall
)]
public static extern void BuyProduct(Product product);
[DllImport("CH02 _ NativeLibrary.dll")]
public static extern Product CreateProduct();
```

10. These `DllImport` statements make our `CH02_NativeLibrary.dll` methods available to C#. Update the `Main` method, as follows:

```
static void Main(string[] _)
{
SendGreeting();
    Console.WriteLine($"1 + 2 = {AddIntegers(1, 2)}");
     var answer = IsLengthGreaterThan5("C# is
       awesome!") ? "Yes." : "No.";
     Console.WriteLine($"Is \"C# is awesome!\" > than
       5? {answer}");
     Console.WriteLine($"Publisher Name: {GetName()}");
     var product = CreateProduct();
     Console.WriteLine($"Product: {product.Name}");
     BuyProduct(product);
     Console.ReadKey();
}
```

Our `Main` method calls the methods that were imported from our `CH02_NativeLibrary.dll` binary. We pass values in and receive values and structures back.

Now that you know what unsafe and fixed code is, let's learn how to interact with Python code in C#.

Interacting with Python code

Python is one of the world's top programming languages and is a favorite of data scientists and programmers working in the field of artificial intelligence and machine learning. Automation of day-to-day mundane infrastructure tasks has been carried out by infrastructure professionals using the Python programming language.

Python code has been designed in such a way that programmers can code tasks quicker than they can in C#. So, the programming writing experience in Python can be quicker than in C#. Some programmers state that Python can be more readable than C#, although I find C# easier to read and understand when compared to Python. This means that readability is rather subjective, but more programmers create programs in Python than they do in C#.

C# beats Python when it comes to compiled code performance. Python can be quicker to write but requires a lot of testing and its garbage collector and interpreter can affect the performance of Python applications. C# uses JIT, AOT, and Ngen, which are also available to VB.NET, C#, F#, and other .NET languages, to perform various types of compilation. The result is that C# produces native code on the target machine, thus providing much faster-executing code than Python. And with the advent of further performance improvements being added to .NET 5 and C# 9.0 by Microsoft, C# will be even faster than it was in its previous versions.

With so much good work being accomplished in the Python arena, it is good for C# programmers to be able to capitalize on Python by using Python code from C#. At the same time, some companies are striving to have all their code in a single code base, so they want to move away from languages such as Java and Python and become fully C#-oriented. Another advantage of moving the existing Python code over to C# is that the same tasks will be much faster in C# than they are in Python. The first step in being able to move away from Python to C# is to be able to use the existing Python code within the C# programming language.

In this section, you will learn how to execute Python code inside C#. You will also learn how to call and execute an external Python script. Follow these steps:

1. First, make sure you add the Python payload from within Visual Studio Installer and add Python to your `PATH` environment variable.
2. Start a new .NET Core 3.1 console application. Then, add the `IronPython` NuGet package. This will only work with Python 2.x code. If you require Python 3.x support, then use Python.NET, which is available at `http//pythonnet.github.io`. You will need the following `using` statements:

```
using System;
using IronPython.Hosting;
```

We need `System` because we will be outputting text to the console window. The `IronPython.Hosting` library is needed to host and execute Python code in C#.

3. Add a file called `welcome.py` to the project, set it to `Copy` always, and add the following code:

```
print("Welcome to the world of Python integration with
    C#!")
```

4. This Python code will print out the text to our console window. Add the following code to the `Main` method:

```
Console.WriteLine("Enter a string to be printed from
    Python: ");
var input = Console.ReadLine();
     var python = Python.CreateEngine();
     try
{
python.Execute("print('From Python: " + input + "')");
python.ExecuteFile("welcome.py");
}
catch (Exception ex)
{
Console.WriteLine(ex.Message);
}
finally
{
Console.ReadKey();
}
```

Here, we are prompting the user to input some text. Then, we read the line of text the user enters. A variable is created that can be used to execute Python code. A `try/catch/finally` block is then used to execute the Python code. First, we execute pure Python code directly from within C#. Then, we execute the code that was executed in our Python script. Any exceptions are caught with the exception message that was written to the console window. Finally, we wait for the user to press any key before we exit.

And that is all there is to executing Python code directly within C# and via external Python scripts. Now, let's learn the COM interface.

Performing Component Object Model (COM) interoperability

The **Component Object Model** (**COM**) is an interface standard that was introduced by Microsoft in 1993. It enables components written in the same or different languages to communicate with each other, and COM components can pass data between each other. Communication is accomplished through **Inter-Process Communication** (**IPC**) and dynamic object creation. COM is not a programming language; it provides a software architecture that consists of binary and network standards.

Many business employees use spreadsheets because they are an easy way to combine and manipulate data for various reasons. Spreadsheets are also the perfect tools for statistical analysis. Many companies expand the power of spreadsheets by building useful add-ons using C# and other languages. But spreadsheets are also useful for ingesting data into databases for day-to-day operations and reporting purposes. In this section, you will learn how to create and manipulate spreadsheets in C#, as well as write C# plugins for Excel.

> **Note**
>
> **Visual Studio Tools for Office** (**VSTO**) is only available in .NET 4.8 and below. It will not work in C# 9 and .NET 5.0. Due to this, we will perform C# interoperability using .NET 4.8. Microsoft has moved away from VSTO and the COM model to focus on the cross-platform extensibility of Excel using JavaScript. Since this book is on C#, we will focus on VSTO in .NET 4.8. To find out more about Microsoft Office extensibility using the JavaScript API, please read the following documentation: `https://docs.microsoft.com/office/dev/add-ins/develop/understanding-the-javascript-api-for-office`.

In this section, we will provide two demonstrations. The first demonstration will read data from an existing spreadsheet. It is useful to know how to do this as there is often a business need for programmers to work with spreadsheet data. After that, we will add an Excel VSTO add-in for Excel. It can be very useful to provide add-ins to end users that make their work more expedient and enjoyable.

Reading data from an Excel spreadsheet

In this section, we are going to write a small program to read an Excel file, count the number of lines, and then update the Excel spreadsheet with the used line count from within C#. Follow these steps:

1. Add a folder called `C:\Temp`. Then, create a new spreadsheet in it called `LineCount.xlsx`. Add 10 rows of text in the first column. Save and close the spreadsheet.
2. Add a new .NET 4.8 console application. Add the following reference using the NuGet package manager to install the latest versions:

```
Microsoft.Office.Interop.Excel
Microsoft.VisualStudio.Tools.Applications.Runtime
```

3. Add the following namespaces to the `Program` class:

```
using System;
using Microsoft.Office.Interop.Excel;
```

4. With that, we can interact with Excel from C#. Now, modify the `Main` method, as follows:

```
var excel = new Application();
var workbook = excel.Workbooks.Open
     ("C:\\Temp\\LineCount.xlsx");
var worksheet = excel.ActiveSheet as Worksheet;
Range userRange = worksheet.UsedRange;
int countRecords = userRange.Rows.Count;
int add = countRecords + 1;
worksheet.Cells[add, 1] = $"Total Rows: {countRecords}";
workbook.Close(true, Type.Missing, Type.Missing);
excel.Quit();
```

The preceding code creates a new Excel application. The workbook we created and modified earlier on is opened. At this point, we can obtain the actively used range on the active sheet and the count of how many rows there are. The count is then saved on a new row, after which we can close the workbook and quit Excel.

5. Run the code as many times as you like and then open the spreadsheet. You should see something similar to the following:

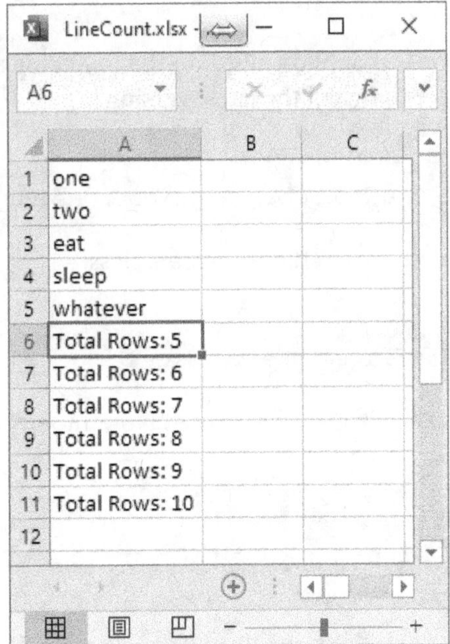

Figure 2.5 – Excel showing rows added by C#

As you can see, working with Excel files is straightforward.

> **Tip**
> The most performant way to populate an Excel spreadsheet from a database result set is to use `Worksheet.Range.CopyFromRecordset(Object, Object, Object)`. See the official Microsoft documentation at `https://docs.microsoft.com/dotnet/api/microsoft.office.interop.excel.range.copyfromrecordset?view=excel-pia`.

Now, let's create an Excel add-in.

Creating an Excel add-in

What does creating an Excel add-in have to do with.NET high performance? Well, VSTO performance can be improved by implementing the following strategies:

- Load VSTO add-ins on demand.
- Publish Office solutions by using Windows Installer.
- Bypass Ribbon reflection.
- Perform expensive operations in a separate thread.

In this section, we are going to write an Excel add-in that will appear on the **Add-ins** tab within Excel. When the button is clicked, it will read the text in the currently selected cell and display the contents in a message box. Follow these steps:

1. Create a new Excel VSTO add-in project. This will target .NET 4.8. You cannot use VSTO with .NET 5.0.
2. Add a new Ribbon (Visual Designer) and call it `CsRibbonExtension`.
3. Rename `group1` to `CsGroup` and change the label to `C# Group`.
4. Add a button to `CsGroup`.
5. Change the button's name to `GetCellValueButton` and change its label to `Get Cell Value`.
6. Double-click the button to generate the click event. Update the click event like so:

```
private void GetCellValueButton _ Click(object sender,
    RibbonControlEventArgs e)
{
CultureInfo originalLanguage = Thread.CurrentThread
    .CurrentCulture;
        Thread.CurrentThread.CurrentCulture = new
            CultureInfo("en-US");
        var activeCell = Globals.ThisAddIn.Application
            .ActiveCell;
        if (activeCell.Value2 != null)
            MessageBox.Show(activeCell.Value2
                .ToString());
            Thread.CurrentThread.CurrentCulture =
                originalLanguage;
}
```

7. In our click event, we save the current language and then change it to American English. Then, we obtain the active cell. The `Value2` property is a dynamic type. We check if the value for the active cell is null. If the cell is not null, then we display the active cell's value in a message box. Finally, we return the language to its original language.
8. Build the project.
9. Then, press F5 to deploy the solution.
10. Open Excel and start a blank workbook.
11. On the ribbon, if the **Add-ins** tab is not visible, click on **Customize Quick Access Toolbar** and then **More Commands…** to bring up the **Excel Options** dialog, as shown in the following screenshot:

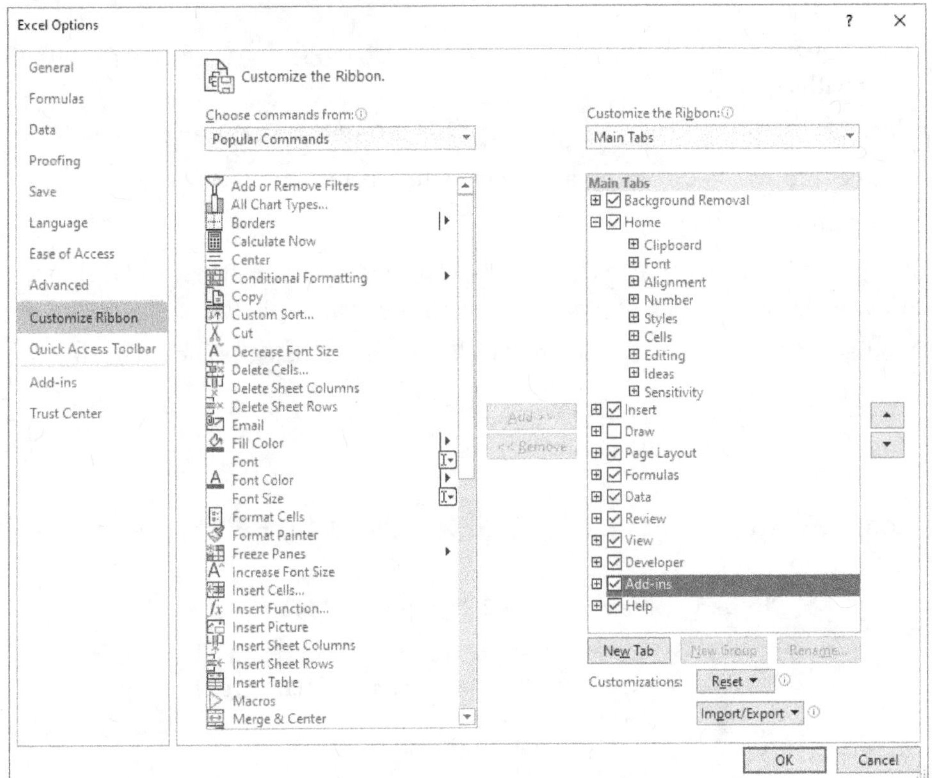

Figure 2.6 – The Excel Options dialog

12. Make sure that the **Add-ins** option is ticked, as shown in the preceding screenshot.

13. Click on **OK** to close the dialog. Type anything you like in a cell and then click on the **Add-ins** tab. You should see something similar to the following:

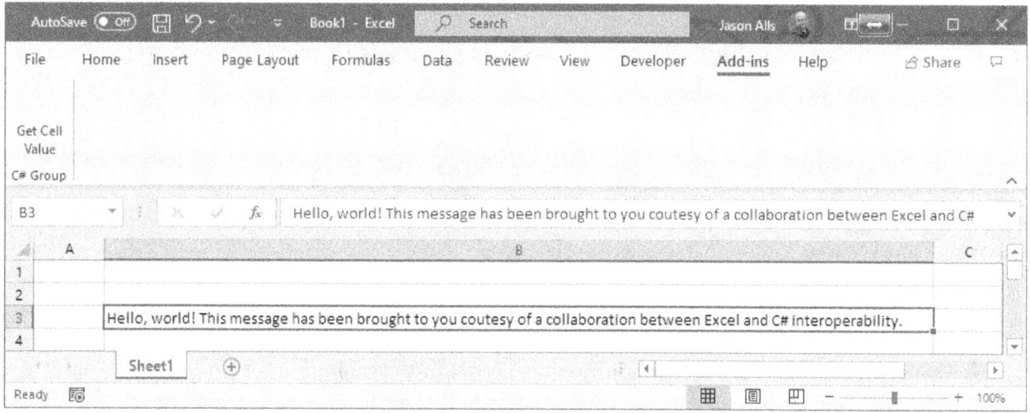

Figure 2.7 – Excel showing the Add-ins tab

14. Make sure that your text cell is selected. Then, click on the **Get Cell Value** ribbon item. You should see a message similar to the following:

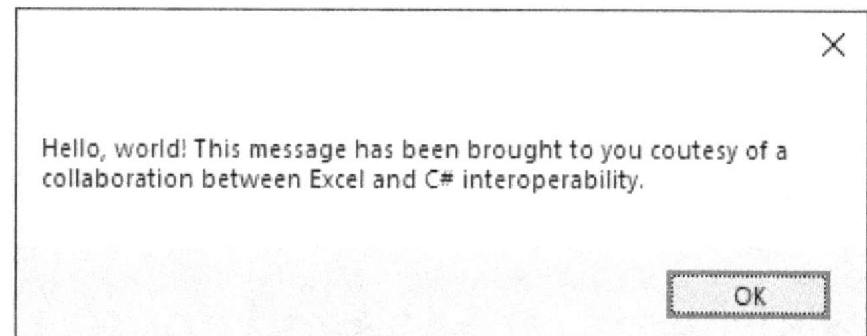

Figure 2.8 – Excel message displaying the text in the active cell

Loading our VSTO add-in on demand

Now, let's add a performance improvement to our Excel add-in by only loading it when the customer demands it instead of at startup. Follow these steps:

1. Right-click on the Excel add-in project and select **Properties**.
2. Then, select the **Publish** page.
3. On the **Publish** page, click on the **Options** button.
4. On the **Publish Options** dialog, select **Office Settings**.
5. Select the **Load on Demand** option and click on the **OK** button.

Bypassing Ribbon reflection

You can bypass Ribbon reflection by overriding `Microsoft.Office.Core.IRibbonExtensibility.CreateRibbonExtensibleObject()`. Instead of letting VSTO reflect what Ribbon object to load, you must use a conditional statement to explicitly load the correct Ribbon.

Executing expensive operations in a separate thread of execution

Any time-consuming tasks such as database operations and transferring objects over a network should be carried out in separate threads.

> **Note**
>
> You must execute calls to the Office object model in the main thread.

Further performance improvements

For further guidance on performance improvements that you can make to VSTO add-ins, check out the official Microsoft documentation: `https://docs.microsoft.com/en-us/visualstudio/vsto/improving-the-performance-of-a-vsto-add-in?view=vs-2019`.

So far, we have looked at various methods of interacting with other programs and programming languages. Now, let's learn how to safely dispose of unmanaged code.

Safely disposing of unmanaged code

When working with unmanaged resources, you must explicitly dispose of them yourself to free up resources. If you do not, then you may end up with exceptions being raised or, worse, your application completely crashing. You must make sure that your applications don't continue running and supplying wrong data when exceptions are encountered. Should exceptions be encountered where the data would become invalid if the application were to continue, then it is better to exit the program. You must also make sure that if your application encounters a catastrophic exception that it is unable to recover from, either a message is displayed or some kind of logging takes place before it shuts down.

In C#, there are two ways to dispose of unmanaged resources: using the disposable pattern and using finalizers. We will discuss both methods in this section via code examples.

Understanding C# finalization

A **finalizer** is a destructor in C# and is used to perform any necessary final cleanup that needs to be performed manually. You can use finalizers in classes, but you cannot use them in structs. A class can have one finalizer, but a class cannot inherit or overload finalizers. You cannot call finalizers as they are invoked automatically when the class is destroyed. Also, modifiers do not accept modifiers or have any parameters.

> **Note**
>
> You have no control over when a finalizer runs. If the GC was to run too infrequently, then you could experience `OutOfMemory` exceptions. Instead of relying on finalizers, you should implement the Dispose design pattern best practice, which will call the finalizer as a last resort. Consider finalizer code running as a bug when you're disposing of managed and unmanaged objects.

There are two syntactic ways to write finalizers in C#. The first is the classic method, as shown here:

```
public class Third : Second
{
      ~Third() // Destructor/Finalizer
{
      // Clean-up code goes here ...
}
}
```

The second way to write a finalizer is as follows:

```
public class Third : Second
{
      ~Third() => Console.WriteLine("Clean-up goes
          here ...");
}
```

As a programmer, you must know that, despite using finalizers to clean up code, you have no control over whether or when the garbage collector will call them.

> **Note**
>
> As a rule of thumb, most of your code is managed code. This means that there should never be a need for you to touch finalizers. Only use them if you need to when cleaning up unmanaged objects.

Using the disposable pattern to release managed and unmanaged resources

When you're dealing with managed and unmanaged objects, it is necessary to implement the disposable design pattern. The disposable pattern implements the **IDisposable** interface and makes use of finalizers. This is an aspect of the disposable pattern, not a requirement. You can write an abstract base class that implements the disposable design pattern, and then inherit from that class and override the `Dispose(bool disposing)` method, as shown in the source code for the `CH02_ObjectCleanup` project on GitHub. This is what we will do in this demonstration. Follow these steps:

1. Start a new .NET console application. Then, add a class called `DisposableBase`, as follows:

```
public abstract class DisposableBase : IDisposable
{
protected bool _disposed = false;
}
```

2. Here, we declared the class abstract and implemented the `IDisposable` interface. Our `_disposed` Boolean value will be accessed by subclasses, so we need to declare that it is protected. Add the `Dispose()` method, as follows:

```
public void Dispose()
{
	Dispose(true);
	GC.SuppressFinalize(this);
}
```

3. This method calls the `Dispose(bool disposing)` method, which cleans up both managed and unmanaged resources. Then, it stops the finalizer from being executed. Let's add the finalizer:

```
~DisposableBase()
{
```

```
    Dispose(false);
}
```

4. Should our finalizer run – and it is not guaranteed to run – it will call the `Dispose(bool disposing)` method when the programmer fails to call the `Dispose()` method. Now, let's add the final part of our `DisposableBase` class – that is, the `Disposable(bool disposing)` method:

```
protected virtual void Dispose(bool disposing)
{
if (_disposed)
            return;
if (disposing)
{
            // Free up any managed objects here.
}
// Free up any unmanaged objects here.
// Set large fields to null.
_disposed = true;
}
```

5. If our class has already been disposed of, then we can exit the method. If the class has not been disposed of, then we must free up managed resources. Once the managed resources have been cleaned up, we can clean up the unmanaged objects and set large fields to null. Finally, we must set the `_disposed` Boolean to `true`.

When a class inherits our abstract class, its finalizer will call `Dispose(false)`. The subclass will override the `Dispose(bool disposing)` method.

To create an object and destroy it, you can use the following code:

```
var objectThree = new ObjectThree();
objectThree.Dispose();
```

Here, the `ObjectThree` class is instantiated and then disposed of by calling the `Dispose()` method.

That brings us to the end of this chapter on C# interoperability. Let's summarize what we have learned.

Summary

In this chapter, we started by looking into P/Invoke regarding C# interoperability using pointer code. We looked at unsafe and fixed code. Unsafe code is code that is not managed by the .NET platform, while mixed code is objects fixed in memory that are not promoted by the garbage collector because they are accessed using pointers.

Then, we learned how to call methods in a C++ DLL, including passing parameters and returning structs.

Next, we learned how to interact with Python code. We learned how to install Python and then add the IronPython NuGet package. This allows us to execute Python 2.x code directly in a C# class and execute Python code that resides in a Python script. The ironPython 2.7.10 library only supports Python 2.x versions.

Then, we learned how to perform COM interoperability by reading data from an Excel spreadsheet. We also built an Excel add-in that was able to read the data of the active cell and display a message box.

Finally, we learned how to safely dispose of managed and unmanaged objects. We built a reusable abstract class called `DisposableBase`. At this point, you know to call `Disposable(false)` in subclass finalizers if `Dispose()` is not called, as well as how to override `Disposable(bool disposing)` in your base classes.

Now, it is time for you to answer some questions to reinforce your learning before moving on to the *Further reading* section. In the next chapter, we will learn about primitives and object types.

Questions

Answer the following questions to test your knowledge of this chapter:

1. What is P/Invoke short for?
2. Explain what P/Invoke is.
3. What is the `unsafe` keyword used for?
4. Explain object generations.
5. What is the `fixed` keyword used for?
6. What is the C++ type for a string?
7. What NuGet package do you have to import to work with Python code?

8. What pattern do you use to safely dispose of managed and unmanaged objects?
9. How do you dispose of large fields?

Further reading

To learn more about the topics that were covered in this chapter, take a look at the following resources:

- *Unsafe code language specification*: `https://docs.microsoft.com/en-us/dotnet/csharp/language-reference/language-specification/unsafe-code`.
- *C# tutorial for beginners: What is Unsafe Code?* `https://www.youtube.com/watch?v=oIqEBMw_Syk`.
- *Interoperating with unmanaged code*: `https://docs.microsoft.com/en-us/dotnet/framework/interop/`.
- *Interop Marshaling*: `https://docs.microsoft.com/en-us/dotnet/framework/interop/interop-marshaling`.
- *Marshalling Data with Platform Invoke*: `https://docs.microsoft.com/en-us/dotnet/framework/interop/marshaling-data-with-platform-invoke`.
- *P/Invoke Tips*: `http://benbowen.blog/post/pinvoke_tips/`.
- *Debugging Finalizers*: `https://docs.microsoft.com/en-us/archive/msdn-magazine/2007/november/net-matters-debugging-finalizers`.
- *Destructors in C#*: `https://www.geeksforgeeks.org/destructors-in-c-sharp/`.
- .NET Memory Performance Analysis: `https://github.com/Maoni0/mem-doc/blob/master/doc/.NETMemoryPerformanceAnalysis.md#The-effect-of-a-generational-GC`.
- *Improving the performance of a VSTO add-in*: `https://docs.microsoft.com/en-us/visualstudio/vsto/improving-the-performance-of-a-vsto-add-in?view=vs-2019`.
- *When everything you know is wrong, part one*: `https://ericlippert.com/2015/05/18/when-everything-you-know-is-wrong-part-one/`.

- *.NET Memory Performance Analysis*: `https://github.com/Maoni0/mem-doc/blob/master/doc/.NETMemoryPerformanceAnalysis.md`.
- *OLE/Automation BSTR (String Manipulation Functions)*: `https://docs.microsoft.com/previous-versions/windows/desktop/automat/string-manipulation-functions`
- *How to pass arrays of objects from C# to C++*: `https://alekdavis.blogspot.com/2012/07/how-to-pass-arrays-of-objects-from-c-to.html`.

3
Predefined Data Types and Memory Allocations

In this chapter, you will learn about **C#** predefined (that is, *built-in*) data types and C# object types, along with the different types of **memory allocations**.

The most basic requirement for improving the performance of your application is to understand the predefined data types and their sizes. There may be times when the memory usage of your applications is critical. Knowing the size of data types and the values they hold can help you make accurate memory usage estimates, as do memory profiling tools such as **dotTrace** and **dotMemory**, which are developed by **JetBrains**. We will be discussing the use of dotTrace and dotMemory in the next chapter. It also makes sense to know the different types of memory allocations and how they affect your code performance. Here, we will be benchmarking the performance of various operations using **BenchmarkDotNet**.

In this chapter, we will be covering the following topics:

- **Understanding the predefined .NET data types**: In this section, we will perform a review of the C# value and object types that are built into the C# programming language. Understanding these types and their size in bytes is useful when you need to provide memory usage estimates.
- **Understanding the various types of memory used in C#**: In this section, we delve into the different types of memory used in C#, including the *stack*, *heap*, *small object heap*, and *large object heap*. It is useful to know what data gets stored in memory and how it gets stored. This can have a big effect on the performance of your applications. For instance, did you know that value types do not always get stored on the stack?
- **Passing by value and passing by reference**: In this section, we will cover the differences between passing values by value and by reference, and the effects this has on the original variables. You will also understand how passing by value and by reference work in memory.
- **Boxing and unboxing:** In this section, we will discuss what happens in memory when we *box* and *unbox* a variable, and we will explore how boxing and unboxing negatively impact the performance of programs. You will use the disassembler to view the intermediate language commands that perform the boxing and unboxing.

By the end of this chapter, you will have the skills to do the following:

- You will understand the different value type sizes.
- You will understand the different reference types.
- You will understand the different types of memory and how they are allocated.
- You will understand the difference between passing by values and passing by references.
- You will understand how boxing and unboxing negatively impact performance and why.

We will first look at the technical requirements for following along with this chapter, then, we will move on to look at the various predefined C# data types.

Technical requirements

- Required: **Microsoft Visual Studio 2022**, latest version – preview
- Required: BenchmarkDotNet

The code files for this chapter can be found in this book's GitHub repository: `https://github.com/PacktPublishing/High-Performance-Programming-in-CSharp-and-.NET/tree/master/CH03`

You will need to clone the git repository and do a release build. The compiled executable will be found under `C:\Development\perfview\src\PerfView\bin\Release\net45`.

Understanding the predefined .NET data types

There are two types of predefined data types:

- **Reference types**
- **Value types**

The reference types are objects and strings. The value types consist of enumeration and struct types. Struct types are aggregated of simple types. Simple types consist of Boolean, char, and numeric types.

There are three main numeric types: decimal types, floating-point types, and integer types. Floating-point types consist of decimals, doubles, and floats. The integer types consist of bytes shorts, integers, longs, value tuples, and characters.

We are going to mention the stack and the heap in more detail later in the chapter. But for now, we should understand that the stack is *unmanaged* memory, and the heap is *managed* memory.

Value types live on the stack. Value types in arrays live on the heap. And reference types live on the heap, with their pointers living on the stack.

> **Note**
>
> Even if arrays are not ideal for some scenarios, in most cases, arrays will often perform faster than lists and other data structures. Array contents are placed contiguously on the heap. The variable for the array will be placed on the stack, and its contents on the stack will be a pointer to the memory address of the array on the heap.

The stack and the heap are the two main types of memory in **.NET**, and as mentioned, we will be covering them later in this chapter.

Now, let's look at the predefined value types in C#.

Understanding the predefined value types in C#

In this section, we will describe each predefined value type and its size in bytes. This is important for being able to choose the right data type to improve the memory performance of your applications. For those who are new to C#, you should know that *signed* data types are those data types that can have *positive* and *negative* values, whereas *unsigned* data types are those that can have only *positive* values.

Table 3.1 describes the different value types, their memory size, whether they are nullable, and their default, minimum, and maximum values, as well as providing notes where applicable:

Name	Size in bytes	Nullable	Default Value	Min Value	Max Value	Notes
`bool`	1	Yes	false			true/false
`byte`	1	Yes	0	0	255	
`char`	2	Yes	0 (\u0000)	0 (0\uFFFF)	65535 (\uFFFF)	
`DateTime`	8	Yes	01/01/0001 00:00:00	01/01/0001 00:00:00	31/12/9999 23:59:59 (3155378975599999999 ticks)	
`decimal`	16	Yes	0	79228162514264337593543950335	79228162514264337593543950335	
`double`	8	Yes	0	-1.7976931348623157E+308	1.7976931348623157E+308	
`enum`	4	Yes	0			Grows
`float`	4	Yes	0	-3.4028235E+38	3.4028235E+38	
`int`	4	Yes	0	-2147483648	2147483647	
`long`	8	Yes	0	-9223372036854775808	9223372036854775807	
`sbyte`	1	Yes	0	-128	127	
`short`	2	Yes	0	-32768	32767	
`struct`	Variable					
`value tuple`	1					Grows
`uint`	4	Yes	0	0	4294967295	
`ulong`	8	Yes	0	0	18446744073709551615	

Table 3.1 – The predefined value data types in C#

> **Note**
>
> The `enum` data type is 4 bytes (that is, 32 bits) in size, nullable, and has a minimum value of `0`. You can measure the size of a value type using `sizeof(Type type)`. Custom structs can be measured using `Marshal.SizeOf(typeof(NameOfCustomStruct))`. The `ValueTuple` data type is 1 byte (8 bits) in size and grows with each type parameter. For example, `ValueTuple<double, double, double>` is 24 bytes (192 bytes) in size.

We will now look at understanding the predefined reference types in C#.

Understanding the predefined reference types in C#

A **reference type** is a type that is placed in managed memory called the **managed heap**. The four predefined reference types in C# are the object type, string type, delegate type, and dynamic type.

> **Note**
>
> Unfortunately, with reference types, you cannot use `sizeof` (which is of the object type) to get the size of a reference type, and the `BinaryFormatter` class has been made obsolete. That means that you cannot serialize an object into binary, save it into a memory stream, and get its size from the memory stream's position.
>
> We are, however, recommended to serialize and deserialize objects using **JSON**. We can then assign the JSON to a memory stream, and in doing so, the length of the memory stream will give us the size of our object in memory.

Let's look at each of these in turn in terms of memory usage.

Describing the object reference type

The .NET `System.Object` type is aliased as object in C#. All types in C# either directly or indirectly inherit from `System.Object`. This includes predefined and user types (such as classes, enums, and structs), reference types, and value types. Objects can be nullable.

To obtain the memory size of your objects programmatically, serialize them to **XML** or JSON and load them into a memory stream, and the length of the memory stream will give you your object size in bytes. Alternatively, you can profile the memory of your application using a tool such as dotMemory to profile your application's memory usage.

Describing the string reference type

A `string` type uses 2 bytes (16 bits) for each character. So, our famous little `string`, *Hello, World!*, which uses 13 characters, is 13 x 2 bytes long, which equates to 26 bytes (208 bits) of memory. Strings can be nullable, and they can be empty.

Strings are immutable in .NET. But what do we mean by this?

When you create a `string` type, it is added to the heap. A variable is added to the stack that has an address pointer to the string's location on the heap. If you add the `string` type to another variable, that variable will be placed on the stack, and it will hold a copy of the address of the same string on the heap. But if you append an existing `string` type with another `string` type, a new `string` type is created in memory to hold the existing `string` type, plus the `string` type to be appended. The address pointer for the `string` type is updated on the stack to point to this new location.

Building an immutable string example program

We are going to write a simple **.NET 6** console application that demonstrates the immutability of strings. Start by creating a new .NET 6 console application called `CH03_StringsAreImmutable`. Then, update the `Main(string[] _)` method as follows:

```
static void Main(string[] _)
{
Console.WriteLine("Chapter 3: Strings are immutable");
var greeting1 = "Hello, world!";
var greeting2 = greeting1;
Console.WriteLine($"greeting1={greeting1}");
Console.WriteLine($"greeting2={greeting2}");
greeting1 += " Isn't life grand!";
Console.WriteLine($"greeting1={greeting1}");
Console.WriteLine($"greeting1={greeting2}");
}
```

We output a header to the console, and then we set the `greeting1 string` type to `"Hello, world!"`. Then, we assign `greeting1` to the `string greeting2` type. The contents of both `string` variables are output to the console window. We then amend `greeting1` by appending `" Isn't life grand!"` to the end of it. Next, we output the contents of both `greeting1` and `greeting2`. Run the program, and you should see the following:

Figure 3.1 – The immutable strings example

As you can see, although we assigned `greeting1` to `greeting2` and then updated `greeting1`, `greeting2` remains unchanged. So, we now have two strings on the heap. We have `"Hello, world!"`, and we have `"Hello, world! Isn't life grand!"`. And so, from our little example, we can see that strings are indeed immutable. And now, we will describe the `delegate` reference type.

Describing the delegate reference type

A **delegate** reference type points to methods with specific parameters and returns types. Methods referred to by the `delegate` type must have the same signature and return type. When you compile code that uses delegates, a private sealed class is created for the delegate that inherits from `System.MulticastDelegate`.

> **Note**
>
> Please check *section I.8.9.3* in the following link for more information on delegates: `https://www.ecma-international.org/publications/files/ECMA-ST/ECMA-335.pdf`.

We will now describe the `dynamic` reference type.

Describing the dynamic reference type

Type checking is performed at compile time. This ensures type safety when your applications are executed at runtime. Type safety aims to prevent erroneous or undesirable program behavior that is caused by discrepancies between types.

Types that are defined as `dynamic` bypass type checking at compile time, as they and the members are resolved at runtime. The advantage of the `dynamic` type is that it simplifies our access to COM APIs (such as the **Office Automation** API) to dynamic APIs (such as the **IronPython** libraries) and to the HTML **Document Object Model** (**DOM**).

Dynamic types are compiled as objects and exist as objects at runtime. A `dynamic` type only exists at compile time and not at runtime. When a `dynamic` type is compiled, it becomes an `object` type. Later in this section, and after we have written and built our console application, we will use ILDASM to show the IL type of a compiled dynamic variable.

When the object runs for the first time, it is correctly resolved by the runtime. This resolution incurs a performance penalty that can be considerable depending upon the type being resolved. Since `dynamic` is compiled into an object, boxing and unboxing take place. And as you know, boxing costs processor cycles.

Let's demonstrate the performance difference when using different variations of `var` and `dynamic` when we are declaring variables and assigning values to them, compared to using the correct types and assigning them without having to use casting.

Start a new .NET 6 console application called `CH03_DynamicPerformance`. You will need the following references:

```
using System;
using System.Diagnostics;
using System.Security.Cryptography;
using BenchmarkDotNet.Attributes;
using BenchmarkDotNet.Running;
```

Add a new member variable at the top of the `Program` class:

```
dynamic _dynamicType;
```

This variable declaration will be investigated by using ILDASM after we have run our benchmarks. Next, update the `Main(string[] _)` method as follows:

```
static void Main(string[] _)
{
        BenchmarkRunner.Run<BenchmarkTests>();
}
```

We are running the benchmarking tests in a class called `BenchmarkTests`. Add a new class called `BenchmarkTests` by using the same statements as the preceding example. Then, add the `MeasureVarUsage()` method:

```
[Benchmark]
public void MeasureVarUsage()
{
        var x = 3.14159;
}
```

This method assigns a `double` object to the `x` variable of a type that will be resolved at runtime. Next, add the `MeasureVarDynamicUsage()` method:

```
[Benchmark]
public void MeasureVarDynamicUsage()
{
        var x = (dynamic)3.14159;
}
```

Here, we are still assigning a number to the `x` variable of a type that will be resolved at runtime. But this time, we prefix the number with the `(dynamic)` cast. Remember that the `dynamic` keyword only exists at compile time. When compiled, `dynamic` types become the `object` type. Now, add the `MeasureTypeDynamicUsage()` method:

```
[Benchmark]
public void MeasureTypeDynamicUsage()
{
        double x = (dynamic)3.14159;
}
```

This time, we declare the variable as `double` and cast the assigned number as `(dynamic)`. At runtime, this number will be boxed in an `object` type, and so it will need to be unboxed. And for our final method, add the `MeasureTypeTypeUsage()` method:

```
[Benchmark]
public void MeasureTypeTypeUsage()
{
        double x = 3.14159;
}
```

In this method, we declare a `double` type and assign a `double` type. Compile the project in Release mode. Then, open a command line and navigate to your release folder. Type the name of the executable and press *Enter*. This will cause BenchmarkDotNet to detect the benchmarks within the project and sequentially run through them. You should see a summary similar to the following, albeit with different mean times:

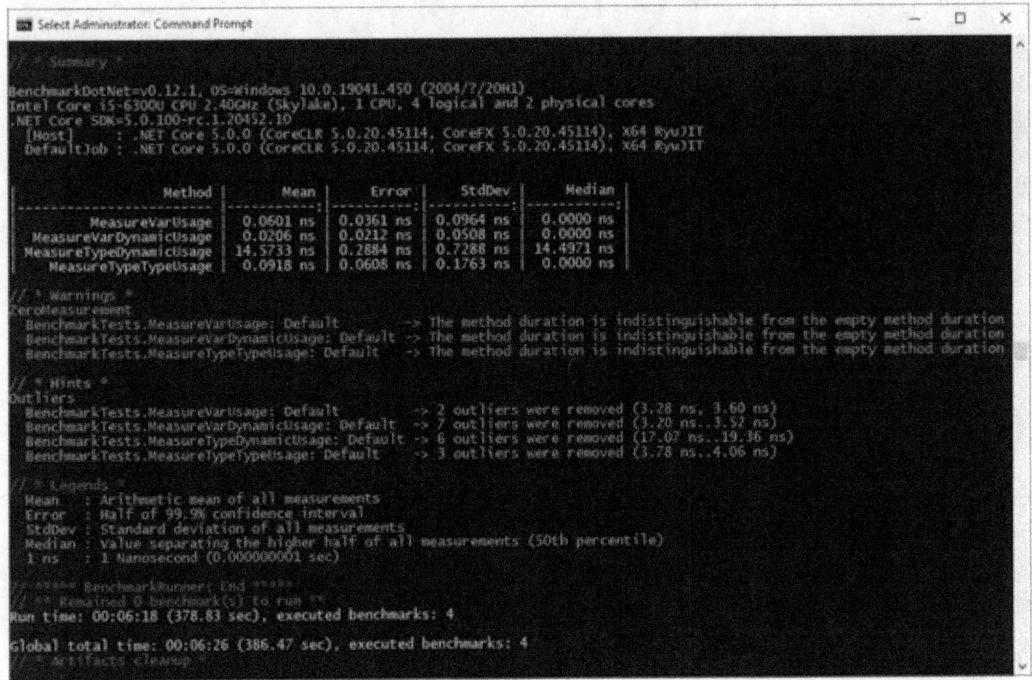

Figure 3.2 – The variable type declaration and the assignment's benchmarked mean timings

Figure 3.2 shows us that there are differences in performance when we declare variables and assign values depending on the methods we use. The fastest combination of declaration and assignment is `var variableName = (dynamic)value`.

Well, we have run our benchmark tests. So, let's view the IL code for the dynamic variable. Open the developer command prompt, then type `ildasm.exe` and press *Enter*. This will start the ILDASM application.

> **Note**
>
> **.NET Core** and **.NET 6** applications are compiled differently from previous versions of the **.NET Framework**. Previously, ILDASM would open the compiled executable. But .NET Core and .NET 6 applications get compiled into a **dynamic-link library** (**DLL**), and a native executable is produced to run the code in the resulting DLL.

Open your compiled DLL. Expand the `CH03_DynamicPerformance` node and then expand the `CH03_DynamicPerformance.Program` node. Then, locate the `_dynamicType : private object` line call, as shown in *Figure 3.3*:

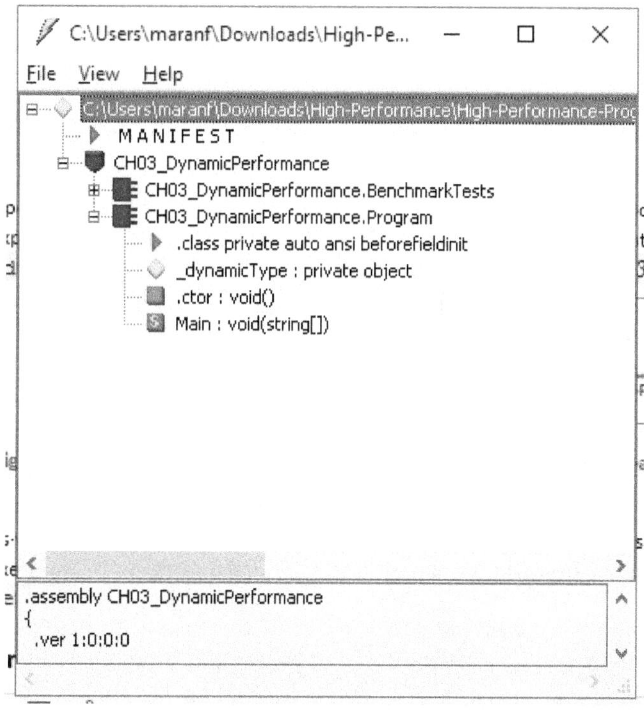

Figure 3.3 – ILDASM showing us that the compiler converts a dynamic type into an object type at compile time

As you can see, our `dynamic` type gets compiled into an `object` type. As a little exercise, play about with the ILDASM settings and view the code for the `BenchmarkTests` class for yourself. Now, let's look at static types.

Understanding static types

In .NET versions earlier than .NET Core and **.NET 5.0**, when you compile and run your applications, they run in their own application domains. If you run your applications multiple times, each running instance of your application will have its own app domain. In **ASP.NET**, you use multiple app domains for a single application. This becomes important when using static types in ASP.NET applications. In a single app domain, there will only be one instance of a static type. The runtime must create an instance of the static type before it can be used.

The `AppDomain` object has its own static heap. Static value and reference types will be placed on the static heap and managed by the app domain. Static types are considered by the garbage collector, but they are never collected. The reason the garbage collector considers them is that they may have references to objects on other heaps. Static types and variables in other app domains are isolated from each other.

In **.NET Native** and .NET 5.0, application domains have been discontinued as they require expensive runtime support. Developers use application domains for various purposes, including code isolation. It is recommended by Microsoft to replace the use of application domains with processes and/or containers. Microsoft also recommends the new `AssemblyLoadContext` class for the dynamic loading of assemblies. By *processes and/or containers*, Microsoft means that you should split your single applications/modules into separate, interacting applications/modules/processes/containers. So, you are encouraged by Microsoft to refactor code using microservices so that you no longer need to use application domains.

The `System.Runtime.Loader.AssemblyLoadContext` object represents a load context. A *load context* creates a scope for loading, resolving, and unloading assemblies. For more information on the `AssemblyLoadContext` class, see the official Microsoft documentation at `https://docs.microsoft.com/dotnet/api/system.runtime.loader.assemblyloadcontext?view=net-5.0`.

Static classes are instantiated only once by the runtime. You cannot instantiate a static class yourself. Static constructors are executed at the time the class is loaded into memory. If a non-static class has a static constructor and an instance constructor, the static constructor will be called before the instance constructor. Static constructors are parameterless, and there can only be one static constructor per class. Static constructors do not have access modifiers. Memory is allocated for static variables when a class loads and deallocated when a class is unloaded. Variables, constructors, and methods belong to the class and not to instantiated objects. So, modifying variables will modify the variable across all instances of a class.

On the call stack, static methods tend to be faster to call than instance methods. The compiler emits a nonvirtual call sites static members. Nonvirtual call sites prevent runtime checks that ensure the current object pointer is non-null. Although you may not see any visual performance improvements, performance gains can be measured for performance-sensitive code.

Now that we have covered the various predefined C# data types, it is time to look at C# memory and how it works.

Understanding the various types of memory used in C#

There are two main types of memory in C#: the stack and the heap. The heap is further broken down into the *small object heap* and the *large object heap*. In terms of physical memory, there is no difference between the stack or heap, as they are both stored in physical memory. Their differences are in their implementations.

When your application starts up, it is allocated a portion of memory. A pointer will be assigned to your application that will be your application's memory starting point. Above the pointer will be the stack, and below the pointer will be the heap. The heap will grow downwards, and the stack will grow upwards, as shown in *Figure 3.4*:

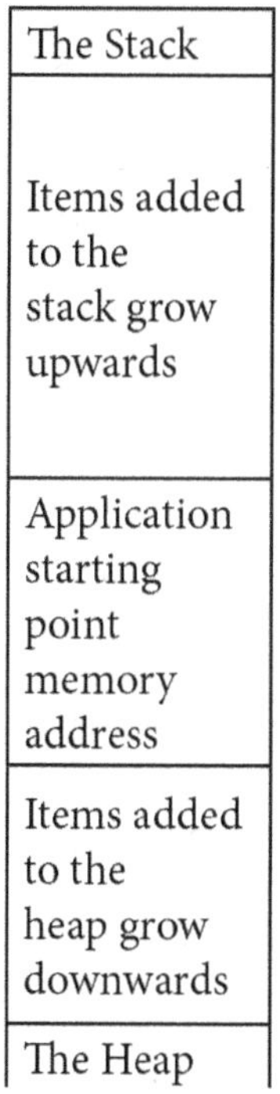

The Stack
Items added to the stack grow upwards
Application starting point memory address
Items added to the heap grow downwards
The Heap

Figure 3.4 – The stack, heap, and application starting point memory address

The following diagram visually represents the stack and heap for a simple program:

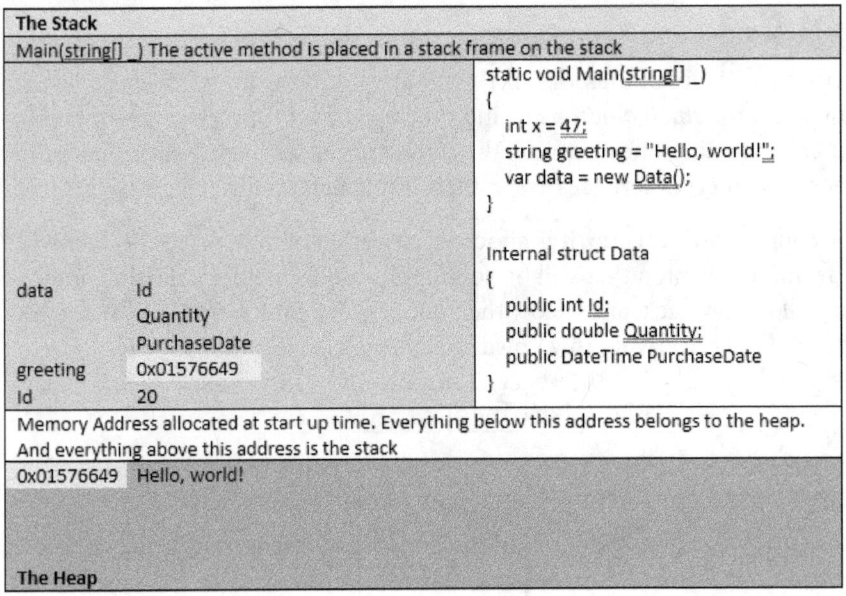

The Stack

Main(string[] _) The active method is placed in a stack frame on the stack

```
static void Main(string[] _)
{
   int x = 47;
   string greeting = "Hello, world!";
   var data = new Data();
}

Internal struct Data
{
   public int Id;
   public double Quantity;
   public DateTime PurchaseDate
}
```

data	Id
	Quantity
	PurchaseDate
greeting	0x01576649
Id	20

Memory Address allocated at start up time. Everything below this address belongs to the heap. And everything above this address is the stack

0x01576649	Hello, world!

The Heap

Figure 3.5 – The stack and heap at work

To understand the different types of memory in C#, first, we'll look at the stack and how it operates.

The stack

The *stack* is used to store value types and pointers to memory locations on the heap. When you call a method, it is added to a stack frame on the stack. Then, within that frame, the value types are added to the stack. If there are any reference types in the method, these are placed on the heap, and a variable is placed on the stack and assigned a pointer to a memory address for the reference type on the heap.

> **Note**
>
> Even though we can state that value types are added to the stack, this is not always true. For example, if you have an array of integers, the array – by virtue of being a reference type – will be added to the heap, and each of the integers that belong to the array will be added contiguously to the heap.

If a `struct` object has a reference type, the struct is placed on the stack, the reference type is placed on the heap, and a pointer to the address of the reference type on the heap is stored in the variable on the heap.

The stack is faster than the heap. It is arranged like a stack data structure. When you execute a method, the method is added to the stack in a stack frame. The local variables are then added to the stack frame on top of each other. When the method has completed execution, the memory is reclaimed immediately. The heap, however, must keep track of memory allocations, pointers, and reference counters, whereas the stack does not have to manage itself in this way.

> **Tip**
>
> With the stack, you can simply pop things on and off the stack. To increase the performance of your applications, look for heap usage in your applications. Measure the performance when using the stack and using the heap. If the stack is faster, then replace heap usage with stack usage.

Keep in mind that the cost of using memory is not at the time of allocation but at the point of deallocation. The deallocation of items on the stack is more predictable than the deallocation of items on the heap. In some cases, the garbage collector is doing similar pointer arithmetic when *freeing* memory in generation 0 or generation 1.

Memory calls are also expensive because they are placed on the stack but may also reference the heap. Method performance is affected by code that does not execute. Therefore, you should refactor your methods to be as small as possible and remove any code that will not be executed, such as dead code that is no longer used. This will reduce the number of local variables in use and thereby reduce the stack size. And so, you will eliminate performance loss.

The heap

The *heap* is used to store reference types. They are called reference types because they are reference-counted. To be reference-counted means that a count of variables referencing the allocated reference type is being kept by the runtime. When the reference count diminishes to zero, the reference type is deallocated by the garbage collector. For example, if I have a product object in memory and two variables on the stack pointing to that object, the product object has a reference count of two.

You may be surprised to learn that the allocation of objects in C# can sometimes be faster than in **C++**. The price is paid in C# when it comes to garbage collection. So, instantiating many objects does not cost us much at all, but the cleanup of those objects does. This means that the more objects you create, the harder the garbage collector must work, which negatively impacts your application's performance. Therefore, avoid using reference types if alternative value types can be used. Do not create objects if you do not need to.

When a new object is instantiated, it is placed on the heap. The variable is placed on the stack and is assigned a pointer to the address of the object on the heap.

Arrays of reference types are placed on the heap. The variable that references the array will be placed on the stack and it will be assigned to the memory address of the array on the heap. The array itself will contain a contiguous list of memory addresses, as shown in *Figure 3.5*:

The Stack			
Array:			
Object 1 Memory Address	Object 2 Memory Address	Object 3 Memory Address	Object 4 Memory Address
Object 1			
Object 2			
Object 3			
Object 4			

Figure 3.6 – The heap displaying objects on the heap and their memory addresses within an array

These memory addresses are pointers to the memory addresses of reference type address locations on the heap. This is because when an array is placed on the heap that contains reference types, each of the reference types in the array is assigned to its own area of memory. The memory addresses of the reference types are then placed inside the array.

> **Note**
> Array performance has been prioritized, followed by string performance. Arrays are often faster than lists and other data structures. But it is best to use benchmarks to decide which is better for your situation and choose the data structure that performs best for you.

When it comes to maximizing the performance of memory usage, you need to ensure that objects on the heap are placed as close to their reference pointers as possible. The reason for this is to reduce the required CPU cycles when locating the memory that is being referenced by the pointer. The rule of thumb for memory performance is that the further memory is from its pointer, the more it costs you in CPU performance. Although, it must be said that predictive memory access reduces this greatly, and memory usage can be dependent on the system page file setup.

> **Note**
>
> The order in which you instantiate arrays, instantiate objects, assign values to objects, and assign values and objects to arrays affects the performance of your applications. This will be down to the placement of those items within memory. Remember that items on the heap should be close to their memory pointers, which may be stored either on the heap or on the stack.

As already stated, object deallocation on the heap is slower than deallocation on the stack. The more objects you add to the heap, the slower your performance will be. The reason for this is that you give the garbage collector more work to do due to the frequent allocation and deallocation. It is this cycle of allocation and deallocation that causes the performance issues.

There are two heaps within the main heap:

- **Small object heap**: When a new object is instantiated, it is placed on the small object heap as generation 0 if it is less than 80,000 bytes in size.
- **Large object heap**: When a new object is instantiated that is 80,000 bytes or larger in size, it is added to the large object heap. Large objects are always allocated in generation 2 because they are only garbage collected during a generation 2 collection.

We will be looking at the heap in more detail when we look at garbage collection in *Chapter 4, Memory Management*.

Building a stack versus building a heap (example project)

Now, we will write a simple project that will get the number of ticks for object and struct instantiation with and without reference type properties. Start by adding a new .NET 6 console application called `CH03_StackAndHeap`. Then, add the `BenchmarkDotNet` `nuget` package. You will need to use the following `using` statements:

```
using System;
using System.Diagnostics;
```

```
using System.Security.Cryptography;
using BenchmarkDotNet.Attributes;
using BenchmarkDotNet.Running;
```

Then, update the `Main(string[] _)` method as shown:

```
static void Main(string[] _)
{
        BenchmarkRunner.Run<BenchmarkTests>();
}
```

In the method, we are calling the `BenchmarkTests` class that contains our benchmarks. Now, add the `ClassNoReference` class:

```
internal class ClassNoReferences
{
        public ClassNoReferences(
            int id,
            decimal price,
            DateTime purchaseDate
        )
        {
            Id = id;
            Price = price;
            PurchaseDate = purchaseDate;
        }
        public int Id { get; private set; }
        public decimal Price { get; private set; }
        public DateTime PurchaseDate { get; private set; }
}
```

This class has three value type properties and no reference type properties. Add the `ProcessClassNoReferences()` method in the `BenchmarkTests` class:

```
[Benchmark]
public void ProcessClassNoReferences()
{
        var _ = new ClassNoReferences()
        {
```

```
            1,
            1.50M
            DateTime.Now
        };
}
```

The `ProcessClassNoReferences()` method declares a new instance of the `ClassNoReferences` class. It will be used as a benchmarking method. Add the `StructNoReferences` class:

```
internal class StructNoReferences
{
    public StructNoReferences(
        int id,
        decimal price,
        DateTime purchaseDate
    )
    {
        Id = id;
        Price = price;
        PurchaseDate = purchaseDate;
    }
    public int Id { get; private set; }
    public decimal Price { get; private set; }
    public DateTime PurchaseDate { get; private set; }
}
```

This struct has three value type properties and no reference types. Let's add the `ProcessStructNoReferences()` method to the `BenchmarkTests` class:

```
[Benchmark]
public void ProcessStructNoReferences()
{
    var _ = new StructNoReferences()
    {
        1,
        1.50M,
        DateTime.Now
```

```
        };
}
```

The `ProcessStructNoReferences()` method will be used as a benchmark, and it creates a new `StructNoReferences` struct. Next, add the `ClassWithReferences` class:

```
class ClassWithReferences
{
    public ClassWithReferences(
        int id,
        string name,
        decimal price,
        DateTime purchaseDate,
        Dictionary<string, string> keyValueData
    )
    {
        Id = id;
        Name = name;
        Price = price;
        PurchaseDate = purchaseDate;
        KeyValueData = keyValueData;
    }
    public int Id { get; private set; }
    public string Name { get; private set; }
    public decimal Price { get; private set; }
    public DateTime PurchaseDate { get; private set; }
    public Dictionary<string, string> KeyValueData
        { get; private set; }
}
```

This class has value and reference type properties. Now, we will add the `ProcessClassWithReferences()` method:

```
[Benchmark]
public void ProcessClassWithReferences()
{
    var _ = new ClassWithReferences(
```

```
            Id = 1,
            "The quick brown fox jumped over the lazy dog.",
            1.50M,
            DateTime.Now,
    );
}
```

The `ProcessClassWithReferences()` method will be used as a benchmark, and it creates an instance of `ClassWithReferences`. Next, we will add the `StructWithReferences` struct:

```
internal struct StructWithReferences
{
    public StructWithReferences(
        int id,
        string name,
        decimal price,
        DateTime purchaseDate,
        Dictionary<string, string> keyValueData
    )
    {
        Id = id;
        Name = name;
        Price = price;
        PurchaseDate = purchaseDate;
        KeyValueData = keyValueData;
    }
    public int Id { get; private set; }
    public string Name { get; private set; }
    public decimal Price { get; private set; }
    public DateTime PurchaseDate { get; private set; }
    public Dictionary<string, string> KeyValueData
        { get; private set; }
}
```

This struct has value and reference types. And now, we will add our final method, `ProcessStructWithReferences()`:

```
[Benchmark]
public void ProcessStructWithReferences()
{
        var _ = new StructWithReferences()
        {
            Id = 1,
            Name = "Discard",
            Price = 1.50M
        };
}
```

The `ProcessStructWithReferences()` method will be used as a benchmark, and it creates a new `StructureWithReferences` struct.

Compile the code in release mode. Then, run the executable. Your code will then be benchmarked, and you will see the following benchmark report:

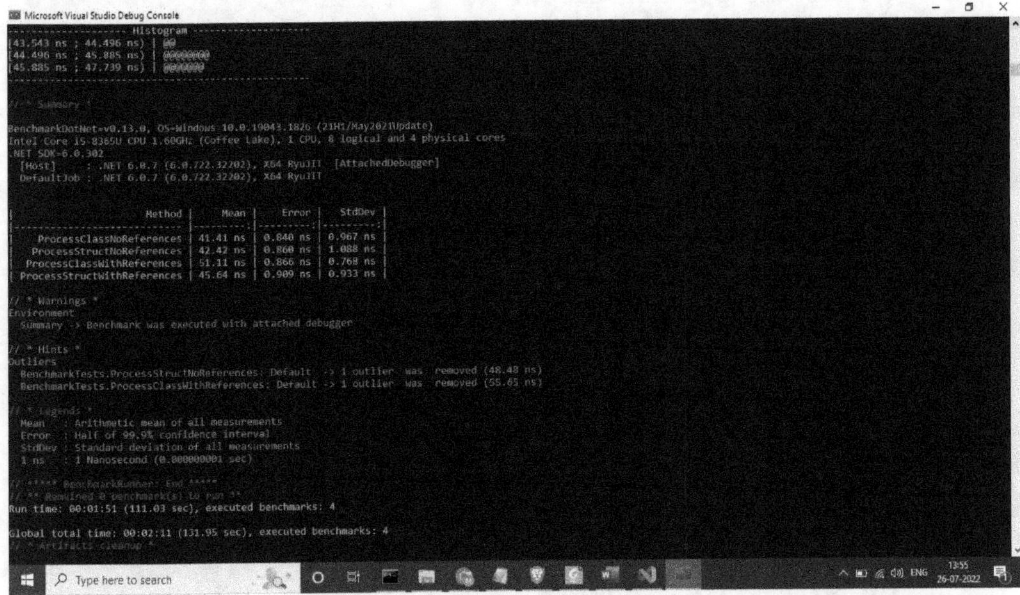

Figure 3.7 – The benchmark report comparing structs and classes with and without references

The benchmark results reveal the following insights:

- Processing a class with no references is faster than processing a struct with no references
- Processing a class with references is slower than processing a struct with references

As the benchmark results show, depending on the scenario, a struct can be faster than a class and vice versa. This is a good reason for benchmarking code, as you could be thinking your code is optimal when in fact it is slow.

So, how do you choose whether to use a struct or a class?

Choosing between a struct and a class

As a rule of thumb, Microsoft recommends that we define our types as classes. If a type is embedded in other objects or if it is short-lived, then consider using a struct. When defining a struct, it should have the following characteristics:

- Logically, the struct represents a single value.
- The struct instance size is under 16 bytes.
- The struct is immutable.
- The struct is not frequently boxed and unboxed.

A *struct* is a *value type*. Value types are allocated on the stack or inline inside containing types. A value type will be deallocated when the stack is unwound or during the deallocation of the containing type. Value types are not garbage collected. The allocation and deallocation of value types on the stack are considered cheap. However, when a value type is boxed, it is wrapped in a reference type or cast to an interface, and this causes a performance slowdown. A performance slowdown is also experienced when a value type is unwrapped from inside a reference type, which is known as *unboxing*. You should do your best to avoid boxing and unboxing value types for performance reasons. When you assign value types, a complete copy of the value is passed into the assignment. The assignment of large value types can be more expensive than the assignment of large reference types.

A *class* is a *reference type*. Reference types are objects allocated on the heap with a pointer to the memory location placed on the stack. When a reference type comes to the end of its life, it is garbage-collected. The allocation and deallocation of reference types on the heap are considered expensive when compared with the allocation and deallocation of value types on the stack. Unlike value types, no boxing occurs when casting reference types. When you assign a reference type, a copy of the reference is passed to the assigned variable. The assignment of large reference types can be cheaper than the assignment of large value types.

An array of reference types contains pointers to the actual types on the heap. An array of value types contains the actual values of those reference types. The allocation and deallocation of value type arrays are cheap, and they have better locality when compared to arrays of reference types, as the value type values are inline.

Let's move on to look at *passing by value* and *passing by reference*.

Passing by value and passing by reference

When passing values into a method or constructor, there are two ways to do this. They are *passing by value* and *passing by reference*:

- **Passing by value**: By default, all value types are passed by value into constructors and methods using *copy semantics*. This means that a copy is made of the value being passed in. The original value remains unchanged, and it is the copy that is used with the constructor or method.
- **Passing by reference**: When a reference type is passed into a constructor or method, a variable is made on the stack that points to the same object on the heap. So, both the variable that is passed in and the copied variable used inside the constructor or method operate on the same object in memory.

Now that we know what passing by value and passing by reference are, let's write a simple program that demonstrates what we have learned.

Building a pass-by-reference example program

We are going to write a very simple program that demonstrates the effects of passing by value and passing by reference. Add a new .NET 6 console application called `CH03_PassByValueAndReference`. Then, modify the `Main(string[] _)` method as follows:

```
static void Main(string[] args)
{
```

```
int x = 0;
Console.WriteLine("Chapter 3: Pass by value and reference");
Console.WriteLine($"=======================================");
Console.WriteLine($"int x = 0;");
AddByValue(x);
Console.WriteLine($"    AddByValue(x): {x}");
AddByReference(ref x);
Console.WriteLine($"AddByReference(x): {x}");
}
```

Here, we have declared an integer called x and assigned it a value of 0. Some text is output to the console window, and we call two methods and output the value of x after they have been called. Let's add the first method that is called – the `AddByValue(int x)` method:

```
static void AddByValue(int x)
{
    x++;
}
```

As you can see, it is a very simple method that increments the value for the variable passed in. Now, let's repeat the same process, but this time, we will pass the value by reference:

```
static void AddByReference(ref int x)
{
    x++;
}
```

Run the program, and you should see the following output:

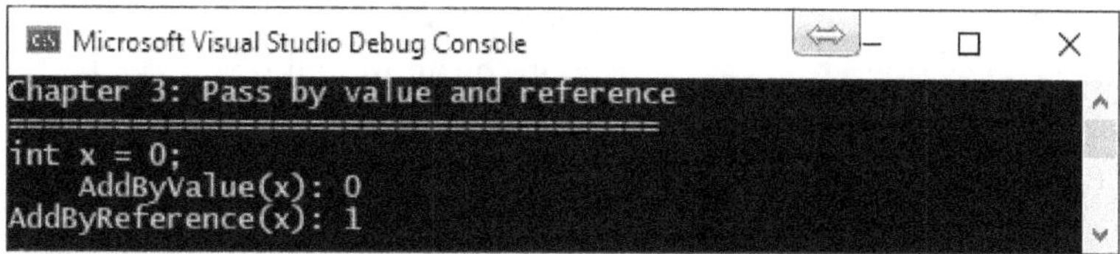

Figure 3.8 – The value of x after incrementing using pass by value and pass by reference

We can see that the original value is not updated when we pass by value. But it is updated when we pass by reference. We will now extend the application to cover the `in` parameter modifier.

Arguments passed with the `in` keyword are passed by reference. However, `in` arguments cannot be modified. Let's demonstrate this – add a new method called `InParameterModifier()`:

```
static void InParameterModifier()
{
        int argument = 13;
        InParameterModifier(argument);
        Console.WriteLine(argument);
}
```

In the `InParameterModifier()` method, we create an integer and assign to it a value of `13`. We then call a method of the same name and pass in the variable as an argument. Then, we print out the value to the console window. Now, we will write the `InParameterModifier(in int argument)` method:

```
static void InParameterModifier(in int argument)
{
        // Error CS8331: Cannot assign to variable 'in int'
        // because it is a readonly variable.
        // argument = 47;
}
```

The code is commented out because if we assign a value to the argument, we will get the compiler warning you see in the comment. Call the method from the `Main(string[] _)` object and run the program. You will see that the variable remains at `13`, as the compiler prevented us from being able to change it in the called method. Finally, in the next part of our program, we will look at the `out` keyword.

An `out` argument does not have to be initialized before being passed in. This is different from a ref value that must be initialized before it is passed in. All `out` parameters are passed by reference. Any operation carried out on the argument inside the method becomes available to the external code that can see the argument. An example will make this easier to understand.

We will be adding two methods to demonstrate how the `out` parameter works. Add a new method called `OutParameterModifier()` to the `Program` class:

```
static void OutParameterModifier()
{
```

```
        int x;
        OutParameterModifier(out x);
        Console.WriteLine($"The value of x is: {x}.");
    }
```

In the preceding code, we declare an integer variable. Then, we call a method that has an `out` parameter and we pass in our integer with its default value of `0`. Next, we print out the value of the integer once the method has returned. Now, add the `outParameter(out x)` method:

```
    static void OutParameterModifier(out int argument)
    {
        argument = 123;
    }
```

Here, we are simply setting the argument to `123` and exiting. Call the `OutParameterModifier()` method from `Main(string[] _)`. If you run the code, you will see that our integer was updated to the value of `123` inside the method that we called. This is shown in *Figure 3.9*:

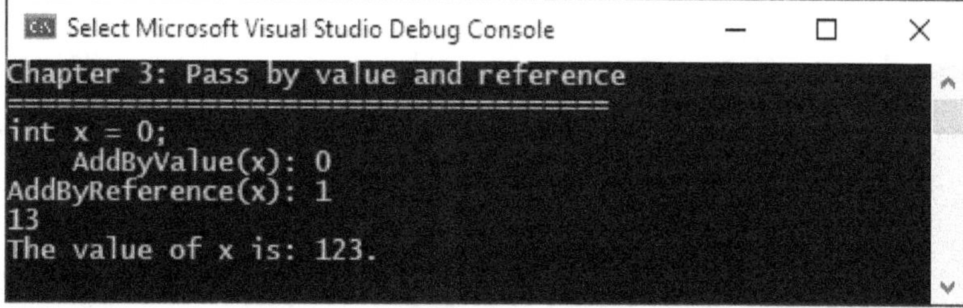

Figure 3.9 – Our integer has been updated inside the method we passed it into

In the following section, we will look at *boxing* and *unboxing*.

Boxing and unboxing

Boxing and *unboxing* variables negatively impact the performance of your applications. To improve your application's code, you should do your best to avoid boxing and unboxing – especially when your code is mission-critical. In this section, we will look at what happens when you package (that is, box) a type.

Performing boxing

When a variable is boxed, you are wrapping it in an object that gets stored on the heap. As you know, objects on the heap incur costs, as they must be managed by the runtime. On top of this, you also increase the memory used by the variable, as well as the number of CPU cycles needed to process the variable.

An empty `class` definition is 12 bytes on a 32-bit operating system and 24 bytes on a 64-bit operating system. This may not sound like a lot. But if a value type is boxed that does not need to be boxed, you will be wasting 12 or 24 bytes of memory unnecessarily.

Now, we will look at what happens when you unbox a variable

Performing unboxing

A variable is copied to the evaluation stack that references an object on the heap. The variable is then unboxed (that is, unpacked) and the variable is placed on the evaluation stack. Then, whatever needs to be done with the unboxed variable can be done. Once all the work has been done with the variable, it then must be boxed up again and placed on the heap. This will create a new object on the heap, and the variable on the stack will be updated with its memory location.

Building a boxing-and-unboxing example program

Now, we will write a simple .NET 6 console application that shows the time difference between not boxing and boxing/unboxing on performance using `BenchmarkDotNet`. First, start a new .NET 6 console application and call it `CH03_BoxingAndUnboxing`. You will need to add the `BenchmarkDotNet` package and the following two namespaces:

```
using System;
using System.Diagnostics;
using System.Security.Cryptography;
using BenchmarkDotNet.Attributes;
using BenchmarkDotNet.Running;
```

We need these namespaces to perform benchmarking. In the `Main(string[] _)` method, add the following line:

```
BenchmarkRunner.Run<BoxingAndUnboxingBenchmarkTests>();
```

This line of code starts the benchmarks running. Next, add a new class called `BoxingAndUnboxingBenchmarkTests`:

```
public class BoxingAndUnboxingBenchmarkTests { }
```

This class will hold two benchmarking methods called `NonBoxingUnboxingTest()` and `BoxingUnboxingTest()`. Add the `NonBoxingUnboxingTest()` method:

```
[Benchmark]
public void NonBoxingUnboxingTest()
{
        int z = 0, a = 4, b = 4;
        z = a + b;
}
```

In this method, we declare and assign three integers: `z = 0`, `a = 1`, and `b = 6`. We then add `a` and `b` together and assign the resulting value to `z`. Now, add the `BoxingUnboxingTest()` method:

```
[Benchmark]
public void BoxingUnboxingTest()
{
        object a = 4, b = 4;
        int z;
        z = (int)a + (int)b;
}
```

This time, we declare and assign two objects: `a = 4` and `b = 4`. We also declare an integer: `z`. Then, we cast `a` and `b` to integers, add them together, and assign the result to the `z` integer variable.

Perform a release build of your code. Then, open a command line and navigate to your executable. Run your executable from the command line, and you should see the following summary:

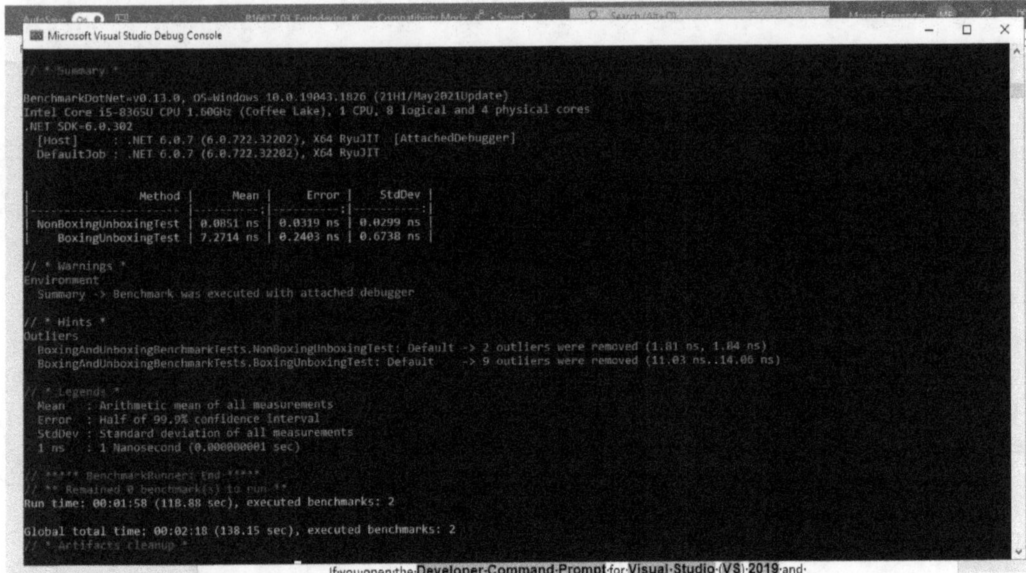

Figure 3.10 – The boxing-and-unboxing example project addition output

As you can see from the screenshot in *Figure 3.10*, unboxing does add overhead to the performance of your applications.

If you open the **Developer Command Prompt** for **Visual Studio (VS) 2019** and type `ILDASM`, this will load the intermediate language disassembler. Open the DLL file in your build folder, and expand the tree until you see the `Main : void(string[])` line, as shown in *Figure 3.11*:

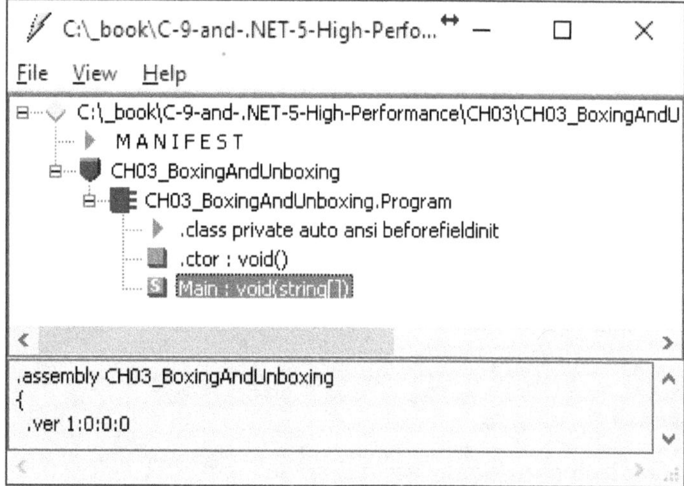

Figure 3.11 – The Intermediate Language Disassembler (ILDASM)

Double-click the `Main` method. This will bring up the window that shows the disassembled intermediate language for our `Main(string[] _)` method, as shown in *Figure 3.12*:

```
CH03_BoxingAndUnboxing.Program::Main : void(string[])
Find  Find Next
.method private hidebysig static void  Main(string[] args) cil managed
{
  .entrypoint
  // Code size       265 (0x109)
  .maxstack  5
  .locals init (int32 V_0,
           int32 V_1,
           object V_2,
           object V_3,
           int32 V_4,
           class [System.Runtime]System.Diagnostics.Stopwatch V_5,
           int32 V_6,
           bool V_7)
  IL_0000:  nop
  IL_0001:  ldstr      "Chapter 3: Boxing and unboxing example."
  IL_0006:  call       void [System.Console]System.Console::WriteLine(string)
  IL_000b:  nop
  IL_000c:  ldc.i4.1
  IL_000d:  stloc.0
  IL_000e:  ldc.i4.2
  IL_000f:  stloc.1
  IL_0010:  ldc.i4.4
  IL_0011:  box        [System.Runtime]System.Int32
  IL_0016:  stloc.2
  IL_0017:  ldc.i4.4
  IL_0018:  box        [System.Runtime]System.Int32
  IL_001d:  stloc.3
  IL_001e:  newobj     instance void [System.Runtime]System.Diagnostics.Stopwatc
  IL_0023:  stloc.s    V_5
```

Figure 3.12 – The disassembled intermediate language for our Main(string[] _) method

Study the disassembled code. When you see the `box` command, the value type is being wrapped inside of an object, which is a reference type that gets placed on the heap. And when you see the `unbox.any` command, the value type is being unwrapped from the object and assigned to an int value type that belongs on the stack.

You now understand why boxing and unboxing affect the performance of your applications, and now we have come to the end of the chapter. In the next chapter, we will be focusing on how the garbage collector works and what we can do to improve its performance. But first, let's summarize what we have learned. You are then encouraged to answer the questions that follow and further your reading on this subject.

Summary

We started the chapter by looking at the various predefined .NET data types. First, we described the various value types, and then we moved on to the predefined reference types. Then, we concluded our discussion of predefined .NET data types by exploring static types.

You learned that value types live on the stack. But if they are part of an array, they are placed on the heap with the array that happens to be a reference type. You also learned that reference types live on the heap and that they have pointers to them in the form of variables that live on the stack.

Next, we looked at the different types of memory used in C#. First, we looked at the stack. Then, we looked at the heap, which consists of the small object heap and the large object heap. After looking at the differences between the stack and the heap, we saw that the stack performs much faster than the heap. The reason for this is that the stack memory does not have to be managed by the runtime. It is simply popped onto the stack when it is needed and popped off the stack when it is not needed. In contrast, the heap must be managed by the runtime that allocates the objects – it keeps a reference count of all the variables that reference those objects, and then it deallocates the objects when they are no longer needed.

We then looked at passing by value and passing by reference. Values passed by value have a copy taken of them that is passed into the constructor or method. This copy is utilized, and the original value remains untouched. When passed by reference, a copy of a value is made and placed on the stack, and it is assigned the memory location of the object on the heap.

Finally, we looked at the boxing and unboxing of variables and why this negatively impacts your application's performance.

With all that you have learned in this chapter, you can reduce the amount of memory your applications use by using the right types, and you can reduce the number of ticks per operation by avoiding boxing and unboxing. And now that you know how memory allocations work, you can improve performance by keeping methods small and using the stack instead of the heap when it is practical to do so.

In the next chapter, we will be learning more about garbage collection.

Questions

1. List the predefined .NET value types.
2. List the predefined reference types.
3. What does the runtime have to do before a static type can be accessed and utilized?
4. Is there a physical difference in the memory that is used that makes the stack run faster than the heap?
5. Why is the stack faster than the heap?
6. Explain why strings are immutable.
7. What is the approximate size of objects placed on the small object heap?
8. What is the approximate size of objects placed on the large object heap?

Further reading

- *The C# type system*
- `https://docs.microsoft.com/en-us/dotnet/csharp/programming-guide/types/`
- *C# Different Types of Heap Memory*
- `https://vivekcek.wordpress.com/tag/stub-heap/`
- *Drill Into .NET Framework Internals to See How the CLR Creates Runtime Objects*
- `https://web.archive.org/web/20140724084944/http://msdn.microsoft.com/en-us/magazine/cc163791.aspx`
- *Passing Parameters (C# Programming Guide)*
- `https://docs.microsoft.com/en-us/dotnet/csharp/programming-guide/classes-and-structs/passing-parameters`

- *Boxing and Unboxing (C# Programming Guide)*
- `https://docs.microsoft.com/en-us/dotnet/csharp/programming-guide/types/boxing-and-unboxing`
- *The large object heap on Windows systems*
- `https://docs.microsoft.com/en-us/dotnet/standard/garbage-collection/large-object-heap`
- *.NET Memory Allocations and Performance*
- `https://www.youtube.com/watch?v=aylUPfOVM90`
- *Replacing AppDomain in .NET Core*
- `https://www.michael-whelan.net/replacing-appdomain-in-dotnet-core/`

4
Memory Management

In this chapter, we will be looking at object generations and how to avoid memory issues, followed by a discussion on strong and weak references. Then, we will look at finalization and how we can suppress finalization by implementing the `IDisposable` pattern to clean up managed and unmanaged resources. Finally, we will take a high-level look at ways to avoid memory leaks.

In this chapter, we will be covering the following topics:

- **Object generations and avoiding memory issues**: In this section, we learn about object generations and `System.OutOfMemoryException`. We learn how to predict out-of-memory errors before they happen by using the `System.Runtime.MemoryFailPoint` class.
- **Understanding long and short weak references**: In this section, we learn about long and short weak references and how they are affected by the garbage collector.
- **Finalization**: In this section, we look at how to use finalizers to clean up resources, and understand why we have no control over if and when they will run.

- **Implementing the IDisposable pattern**: In this section, we look at how we can have more control over the cleanup of managed and unmanaged resources by implementing the `IDisposable` pattern.
- **Preventing memory leaks**: In this section, we look at how the use of the **Component Object Model** (**COM**) and managed events can be sources that generate memory leaks and what we can do to avoid memory leaks from being generated. We will be using Microsoft Excel and JetBrains dotMemory in this section to see how leaks can be generated and to see how using a memory profiler can be very useful in identifying memory leaks and their sources.

By the end of this chapter, you will have gained skills in the following areas:

- Understanding object generations
- Understanding how objects are disposed
- Understanding why it is best to avoid finalizers and implement `IDisposable`
- Understanding how to prevent memory leaks arising from the use of unmanaged COM libraries and components and from using events
- Using anonymous methods, long weak references, and short weak references to improve garbage collection

Technical requirements

To complete the steps in this chapter, there are some technical requirements, as outlined here:

- Visual Studio 2022
- JetBrains dotMemory
- Source code: `https://github.com/PacktPublishing/High-Performance-Programming-in-CSharp-and-.NET/tree/master/CH04`

Object generations and avoiding memory issues

There are three object generations in the .NET runtime, as follows:

- Generation 0

- Generation 1
- Generation 2

Generation 0 is the youngest generation and holds short-lived objects. Objects that are less than 80,000 bytes are generation 0 objects that get placed on the **small object heap** (**SOH**) when they are instantiated. Objects that are 80,000 bytes or larger are usually generation 2 objects and live on the **large object heap** (**LOH**). Generation 1 objects are those objects that survived generation 0 garbage collection and received a promotion to generation 1.

Generation 0 is where most of the garbage collection takes place. Objects that do not get collected when they are generation 0 will get promoted to generation 1 to make room for more generation 0 objects to be added to the heap. If generation 0 and 1 become full, then generation 1 objects are promoted to generation 2, and generation 0 objects are promoted to generation 1. If generations 0, 1, and 2 become full so that no more objects can be added to the heap, you then end up with a `System.OutOfMemoryException`-type exception.

We are now going to write a very simple program that will throw a `System.OutOfMemoryException`-type exception. Follow these next steps:

1. Start a new .NET 6 console application project called `CH04_OutOfMemoryExceptions`. Add the following `using` statements to the `Program.cs` file:

```
using System.Text.RegularExpressions;
using System;
using System.Collections.Generic;
using System.IO;
using System.Runtime;
using System.Text;
```

2. Add the following method calls to the `Main` method:

```
DataExportToCsv();
ReadCsvBroken();
ReadCsvPredictive();
Console.ReadKey();
```

3. The `DataExportToCsv()` method builds up a very large data file. `ReadCsvBroken()` reads in the **comma-separated values** (**CSV**) file, but the string limit is blown for the imported data file when the whole file is read at once. This will generate a `System.OutOfMemoryException`-type exception. The exception is avoided in the `ReadCsvPredictive()` method, as the method instantiates the `MemoryFailPoint` class to ensure that the data read of the file will not generate an exception. If the operation does generate a `System.OutOfMemory exception`-type exception, then the `MemoryFailPoint` object will raise an `OutOfMemoryException`-type exception. This saves memory, time, **central processing unit** (**CPU**) usage, and power consumption. Finally, we wait for the user to press any key before exiting. Add the following member variable to the top of the `Program` class:

```
private static string _filename
= @"G:\Temp\SampleData.csv";
```

4. This will be the file we will write to and read from. Add the following `DataExportToCsv()` method:

```
private static void DataExportToCsv()
{
    int row = 0;
    try
    {
    File.Delete(_filename);
    using (FileStream fs = new FileStream(_filename,
    FileMode.OpenOrCreate))
    {
        fs.Write(Encoding.Unicode.GetBytes("Id,
            Name, Description\n"));
            for (int i = 0; i <= 491616373; i++)
            {
                row = i;
                Console.WriteLine($"Writing row {row} to
                    CSV data. There are {491616373-row}
                        rows remaining.");
                fs.Write(Encoding.Unicode.GetBytes
                  ($"{i}, Name {i}, Description {i}\n"));
    }
```

```
        }
    }
    catch (Exception ex)
        {
            Console.WriteLine($"DataExportToCsv:
                {ex.GetBaseException().Message}")
        }
    }
```

5. This code writes 491,616,373 lines of data to a CSV file. Add the following `ReadCsvBroken()` method:

```
private static void ReadCsvBroken()
{
    int row = 0;
    try
    {
        string csv = File.ReadAllText(_filename);
    }
    catch (OutOfMemoryException oomex)
    {
    Console.WriteLine($"ReadCsvBroken:
        {oomex.GetBaseException().Message}");
    }
}
```

6. The `ReadCsvBroken()` method tries to read the massive 44.2 **gigabytes** (**GB**) CSV file all at once. However, the file produces a string that is too big to be assigned to a `string` variable. This operation throws a `System.OutOfMemoryException`-type exception. Add the following `ReadCsvPredictive()` method:

```
private static void ReadCsvPredictive()
{
    int row = 0;
    try
    {
        string alphabet = "abcdefghijklmnopqrstuvwxyz";
        using (new MemoryFailPoint(alphabet.length))
```

```
            {
                string alpha = alphabet;
            }
            FileInfo fi = new FileInfo(_filename);
            Int length = unchecked((int)fi.length);
            using (new MemoryFailPoint(length))
            {
                string csv = File.ReadAllText(_filename);
            }
        }
        catch (OutOfMemoryException oomex)
        {
            Console.WriteLine($"ReadCsvPredictive:
                {oomex.GetBaseException().Message}");
        }
    }
```

7. This code uses predictive memory checking using the `MemoryFailPoint` class. We show it working for the `alphabet` string, and we show that it highlights an error and fails with an `OutOfMemoryException`-type exception when the length of the file contents is assigned to the `length` variable that is passed into the `MemoryFailPoint` constructor. We use the unchecked struct since the length of the file is a long value, and this value to too big to be assigned to an `int` data type. If we used the checked struct instead, we would have an `ArithmeticOverflowException`-type exception.
8. Building and running the code takes hours. I recommend you build the code in `Release` mode, and then run the executable from a command window. The code will successfully build up the CSV file and save it. When the file contents are read all at once, they will generate an `OutOfMemoryException`-type exception. Then, the program will do a precheck prior to loading the file and will fail before the file read is attempted with a more detailed `OutOfMemoryException`-type exception.

Predicting memory exceptions saves time and improves application performance, as you are not wasting CPU cycles and memory performing an operation that is ultimately going to fail.

We have seen how easy it is for an application to run out of memory and how we can predict and prevent memory exceptions. So, let's now move on to discuss strong and weak references.

Understanding long and short weak references

In the .NET runtime, there are two types of references: **long weak references** and **short weak references**. These are described in more detail here:

- **Long weak reference**: When the `Finalize()` method has been called on an object, a long weak reference is retained in memory. You specify `true` in the `WeakReference` constructor to define a long reference. A long weak reference can be recreated, although its state can be unpredictable. A short weak reference will be applied when an object's type does not have a `Finalize()` method. The weak reference will only remain until its target is collected sometime after the finalizer is run. You will need to cast the target property of a `WeakReference` constructor to the type of an object if you want to create a strong weak reference that will be reused. When the object is collected, the `Target` property will be `null`. If it is not `null`, then you can continue to use the object because the application has regained a strong reference to it.
- **Short weak reference**: A weak reference is a managed object that will be garbage-collected the same as any other managed object. The parameterless constructor for `WeakReference` is a short weak reference. When the garbage collector reclaims a short weak reference, its target becomes `null`.

A long weak reference protects referenced objects from garbage collection, and a short weak reference does not protect referenced objects from garbage collection. This means that when garbage collection executes, the long weak referenced objects will not be garbage-collected, but the short weak referenced objects will be garbage-collected. We will demonstrate this with a code example.

Our code example will show both long and short weak references at work. Follow these next steps:

1. Start by adding a new .NET 6 console application called `CH04_WeakReferences`. Add the following class called `ReferenceObject`:

```
internal class ReferenceObject
{
public int Id { get; set; }
public string Name { get; set; }
}
```

This class will be our reference object that we will be adding to two different object managers.

2. Add a new class called `LongWeakReferenceObjectManager`. Then, add the following list field:

```
private readonly List<ReferenceObject> Objects
= new List<ReferenceObject>();
```

3. Our read-only `Objects` list will contain several `ReferenceObject` types. Now, add the following method to add items to the list:

```
public void Add(ReferenceObject o)
{
Objects.Add(o);
}
```

4. This method adds a `ReferenceObject` object to the list of reference objects. Then, the next task is to add a method that will print a list of stored objects to the console, as follows:

```
public void ListObjects()
{
    Console.WriteLine("Long Weak Reference Objects: ");
    foreach (var reference in Objects)
        Console.WriteLine($"- {reference.Name}");
}
```

The `ListObjects()` method prints out the contents of the list to the console window. That concludes our `LongWeakReferenceObjectManager` class.

5. Now, add a class called `ShortWeakReferenceObjectManager`. At the top of the class, add the following list field:

```
private readonly List<WeakReference<ReferenceObject>>
  Objects
= new List<WeakReference<ReferenceObject>>();
```

Notice with the list that the `ReferenceObject` object is wrapped in a `WeakReference` object.

6. Now, add a method to add items to the list, as follows:

```
public void Add(ReferenceObject o)
{
Objects.Add(new WeakReference<ReferenceObject>(o));
}
```

This method wraps the passed-in `ReferenceObject` object in a `WeakReference` object and assigns it to the list.

7. We now add the `ListObjects()` method, as follows:

```
public void ListObjects()
{
Console.WriteLine("Short Weak Reference Objects: ");
foreach (var reference in Objects)
{
    reference.TryGetTarget(
        out ReferenceObject referenceObject
    );
    if (referenceObject != null)
        Console.WriteLine($"- {referenceObject.Name}");
}
}
```

The `ListObjects()` method prints out to the console window all the weak objects that are stored in the list. Our focus now moves to the `Program` class.

8. Add the following two fields to the top of the `Program` class:

```
private static readonly StrongReferenceObjectManager
  StrongReferences = new StrongReferenceObjectManager();
private static readonly WeakReferenceObjectManager
  WeakReferences = new WeakReferenceObjectManager();
```

These are our read-only strong and weak object managers that we will use to demonstrate strong and weak references in action, with regard to the garbage collector.

9. Update the `Main(string[] _)` method by adding the following three method calls:

```
TestLongWeakReferences();
TestStrongReferences();
TestShortWeakReferences();
ProcessReferences();
```

The `TestLongWeakreferences()`, `TestStrongReferences()`, and `TestWeakReferences()` methods build up our lists of strong referenced objects and weak referenced objects respectively.

10. Add the `TestStrongReferences()` method, as follows:

```
private static void TestStrongReferences()
{
var o1 = new ReferenceObject() {
    Id = 1, Name = "Object 1"
};
var o2 = new ReferenceObject() {
    Id = 2, Name = "Object 2"
};
var o3 = new ReferenceObject() {
    Id = 3, Name = "Object 3"
};
StrongReferences.Add(o1);
StrongReferences.Add(o2);
StrongReferences.Add(o3);
}
```

This method adds three `ReferenceObject` objects to the `StrongReferences` list.

11. Next, add the `TestWeakReferences()` method, as follows:

```
private static void TestWeakReferences()
{
var o1 = new ReferenceObject() {
    Id = 1, Name = "Object 4"
};
```

```
var o2 = new ReferenceObject() {
    Id = 2, Name = "Object 5"
};
var o3 = new ReferenceObject() {
    Id = 3, Name = "Object 6"
};
WeakReferences.Add(o1);
WeakReferences.Add(o2);
WeakReferences.Add(o3);
o1 = null;
o2 = null;
o3 = null;
}
```

This method adds three weak referenced objects to the `WeakReferences` list and then sets the objects it instantiated to `null` so that they will be garbage-collected.

12. Finally, add the `ProcessReferences()` method, as follows:

```
private static void ProcessReferences()
{
int x = 0;
while(x < 10)
{
    StrongReferences.ListObjects();
    WeakReferences.ListObjects();
    Thread.Sleep(2000);
    GC.Collect();
    x++;
}
}
```

The `ProcesseReferences()` method loops 10 times. During each iteration, the `ListObjects()` method is called on the `StrongReferences` and `WeakReferences` fields. The program sleeps for 2 seconds, and then the garbage collector is executed manually.

13. It is now time to run the program. When you run the program, you should see the following output:

```
Microsoft Visual Studio Debug Console
Reference Object 1: Reference Object 1
Reference Object 2: Reference Object 1
Reference Object 3a: Reference Object 1
Reference Object 3b: Reference Object 1
Long Weak Reference Objects:
- Object 1
- Object 2
- Object 3
Short Weak Reference Objects:
- Object 4
- Object 5
- Object 6
Long Weak Reference Objects:
- Object 1
- Object 2
- Object 3
Short Weak Reference Objects:
Long Weak Reference Objects:
- Object 1
- Object 2
- Object 3
Short Weak Reference Objects:
Long Weak Reference Objects:
- Object 1
- Object 2
- Object 3
Short Weak Reference Objects:
Long Weak Reference Objects:
- Object 1
- Object 2
```

Figure 4.1 – Weak references' project output

As you can see from *Figure 4.1*, on the first iteration of the loop, both strong and weak reference objects exist, and the names of those objects are printed in the console window. However, after garbage collection is called, the weak references are garbage-collected, and so, from the second iteration onward, only the strongly referenced objects remain in memory.

A weakly referenced object's lifespan is not extended as it is for strong references. This means that they can be garbage-collected once all strong references have gone out of scope.

Objects that are large but cheap to rehydrate on-demand benefit from weak references.

> **Note**
>
> To improve the performance of your applications, avoid using weak references on many small objects as they can take up more memory space than the objects they wrap, thus adding performance overhead. But if you are working with many large expensive objects, using cached weak references may help improve your application's performance.

That concludes our look at strong and weak references. Let's move our focus and attention to finalization in C#.

Finalization

In C#, there is no direct way of destroying an object. The nearest thing we have is **finalization**. A finalizer in C# is the C# equivalent of a destructor in C++. Except in C#, you have no control over if and when it will run this down to the garbage collector to make that decision.

> **Note**
>
> The terms *finalizer* and *destructor* are used interchangeably in C#. A finalizer is where the user-defined finalizer code is run. After the finalizer in an object is run, it is once again considered alive and the garbage collector will then finally collect the object. This means an object is actually marked "`collectable`" twice if it has a finalizer defined.

Finalization is used by an object to release resources and perform other housekeeping operations prior to the object being garbage-collected. Cleanup operations to release unmanaged resources held by an object can be performed by overriding the protected `Finalize()` method.

You have to override the `Finalize()` method for the garbage collector to mark types derived from `Object` for finalization. When you override the `Finalize()` method, an entry for the instance is placed in a finalization queue. Before reclaiming memory, the `Finalize()` method is called for each object instance in the finalization queue. Once an object's `Finalize()` method has been run, then its memory can be reclaimed by the garbage collector.

The `Finalize()` method is not called if `GC.SupressFinalize()` has been called during the disposing of the object's resources, but the `Finalize()` method will be called automatically when an object is discovered to be inaccessible, and during **application domain** (**AppDomain**) shutdown (even if the object is accessible).

> **Note**
>
> AppDomains isolate applications from one another, but their usage is very expensive. In .NET 5+, some AppDomain **application programming interface (API)** surface is exposed to help ease migration from older frameworks. Some functionality has been removed, and so will either do nothing or throw an exception. Microsoft has no plans to add support for adding extra AppDomains. The present advice from Microsoft to implement code isolation is to use separate processes or containers and use the `AssemblyLoadContext` class for dynamic assembly loading.

`Finalize()` methods only run once unless `GC.SuppressFinalize()` has not been called and `GC.ReRegisterForFinalize()` is called; then, the `Finalize()` method can be called again.

When overriding `Finalize()`, there are a few things to keep in mind, as follows:

- You have no control over when the `Finalize()` method will be called.
- To guarantee the release of managed and unmanaged resources within your instance, implement the `IDisposable.Dispose()` method using the `IDisposable` pattern. There is no guarantee of the order in which finalizers will run.
- Finalizers run on an unspecified thread, and they implicitly call the `Finalize()` method on the base class.

To avoid the need to override the `Finalize()` method and for us to ensure the cleanup of our managed and unmanaged resources, we will look at implementing the `IDisposable` pattern.

Using finalization

We are going to write a sample application that demonstrates the use of `Finalize()`. Then, we will modify the program to implement the `IDisposable` pattern and suppress the call to `Finalize()`, while ensuring the deterministic release of our managed and unmanaged resources. Follow these next steps:

1. Start a new .NET 6 console application called `CH04_Finalization`. Add a new internal class called `Product`. Then, add the following properties:

```
public int Id { get; set; }
public string Name { get; set; }
public string Description { get; set; }
public decimal UnitPrice { get; set; }
```

2. We have created four properties—`Id`, `Name`, `Description`, and `UnitPrice`. Now, add the constructor, as follows:

```
public Product()
{
Console.WriteLine("Product constructor.");
}
```

3. Our constructor writes a message to the console window so that we know we have entered the constructor. Next, add the finalizer, as follows:

```
~Product()
{
Console.WriteLine("Product finalizer.");
}
```

4. In our finalizer, we write a message to the console window so that we know our finalizer has been called. For the last bit of code in our `Product` class, we will override the `ToString()` method, as follows:

```
public override string ToString()
{
   return $"Id: {Id}, Name: {Name},
   Description: {Description}, Unit Price: {UnitPrice}";
}
```

5. Our `ToString()` method returns a string that outputs the values of each of the properties of the `Product` class. For now, unless stated otherwise, the following code is to be added to the `Program` class. Add the following variable:

```
private static Product _product;
```

6. The `_product` variable will be used to store an instance of our `Product` class. Update the `Main` method, as follows:

```
static void Main(string[] _)
{
InstantiateObject();
PrintObjectData();
RemoveObjectReference();
RunGarbageCollector();
InstantiateLocalObject();
```

```
RunGarbageCollector();
DisplayGeneration(_product);
RemoveObjectReference();
RunGarbageCollector();
}
```

7. As you can see, we have several methods that instantiate the object, print object data, remove object references, display object generations, and run the garbage collector. We will now add each of the methods in turn. Add the `InitiateObject()` method, as follows:

```
private static void InstantiateObject()
{
  Console.WriteLine("Instantiating Product.");
  _product = new Product()
{
    Id = 1,
    Name = "Polly Parrot",
    Description = "Cudly child's toy.",
    UnitPrice = 7.99M
};
}
```

8. In this method, we write a console window message, create a new product, and assign it to the `_product` member variable. Now, we will add the `PrintObjectData()` method, as follows:

```
private static void PrintObjectData()
{
Console.WriteLine(_product.ToString());
}
```

9. Here, we are printing the contents of the `Product` class to the console window. Next, we will write the `RemoveObjectReference()` method, as follows:

```
private static void RemoveObjectReference()
{
    _product = null;
}
```

10. We are setting the `Product` object to `null`. This removes references to the object and makes it eligible for garbage collection. We now add a method to call the garbage collection, as follows:

```
private static void RunGarbageCollector()
{
    GC.Collect();
}
```

11. In this method, we call the garbage collector, as follows:

```
private static void InstantiateLocalObject()
{
        var product = new Product()
        {
            Id = 2,
            Name = "Cute Kittie",
            Description = "Cudly child's toy.",
            UnitPrice = 5.75M
        };
        DisplayGeneration(product);
        _product = product;
        GC.Collect();
}
```

12. In this method, we create a local object. Then, we call the method to display the current generation. We then assign the local product to the member product, followed by a call to the garbage collector. Our final method, for now, is the `DisplayGeneration(Product product)` method, as illustrated in the following code snippet:

```
private static void DisplayGeneration(Product product)
{
        Console.WriteLine($"local product:
            generation {GC.GetGeneration(product)}");
}
```

13. This method prints out the generation of the product passed into it. Run the code. You should see the following output:

```
Instantiating Product.
Product constructor.
Id: 1, Name: Polly Parrot, Description: Cudly child's toy., Unit Price: 7.99
Product constructor.
Product destructor.
local product: generation 0
local product: generation 2
Product destructor.
```

Figure 4.2 – The finalization project output

As you can see, our code demonstrates construction and finalization. We have both generation 0 and generation 2 code, and both our constructor and finalizer methods do get called. Now, we will look at implementing `IDisposable` to make the cleanup of our code more deterministic so that `Finalize()` does not need to be called.

Implementing the IDisposable pattern

In this section, we will implement a reusable `IDisposable` pattern. We will have a base class that implements `IDisposable`. This base class will provide two methods that subclasses can override. One method will be for cleaning up managed resources, and the other method will be for disposing of unmanaged resources. For us to implement the `IDisposable` pattern, proceed as follows:

1. Add a new class called `DisposableBase` that implements `IDisposable`, as follows:

```
public class DisposableBase : IDisposable
{
      public void Dispose()
      {
         Dispose(true);
      }
      private void Dispose(bool disposing)
      {
          if (disposing)
          GC.SuppressFinalize(this);
          ReleaseManagedResources();
          ReleaseUnmanagedResources();
      }
```

```
protected virtual void ReleaseManagedResources(){}
protected virtual void ReleaseUnmanagedResources(){}
}
```

This class acts as a base class that can be inherited. It implements the `IDisposable` interface and calls two virtual methods called `ReleaseManagedResources()` and `ReleaseUnmanagedResources()` that will be overridden in the subclass.

2. Move the code from `Main` into a new method called `Finalization()`. Then, modify `Main`, as follows:

```
static void Main(string[] _)
{
        Finalization();
        Disposing();
}
```

We are calling two methods. The `Finalization()` method demonstrates using finalization to clean up resources that you have no control over when finalization will be called by the garbage collector. `Disposing()` demonstrates the determined disposing of managed and unmanaged resources, with finalization being suppressed so that it is not called by the garbage collector. Your `Finalization()` method should look like this:

```
private static void Finalization()
{
        Console.WriteLine("--- Finalization ---");
        InstantiateObject("Finalization");
        PrintObjectData();
        RemoveObjectReference();
        RunGarbageCollector();
        InstantiateLocalObject("Finalization");
        RunGarbageCollector();
        DisplayGeneration(_product);
        RemoveObjectReference();
        RunGarbageCollector();
}
```

We are passing "Finalization" into the InstantiateObject(string cleanUpMethod) and InstantiateLocalObject(string cleanUpMethod) methods so that we know the objects being finalized were instantiated in our Finalization() method.

3. Add a new method called Disposing(), as follows:

```
private static void Disposing()
{
Console.WriteLine("--- Disposing ---");
InstantiateObject("Disposing");
PrintObjectData();
DisposeOfObject();
InstantiateLocalObject("Disposing");
DisplayGeneration(_product);
DisposeOfObject();
RunGarbageCollector();
}
```

4. In the Disposing() method, we write a message to the console identifying that the Disposing() method is running. We then call InstantiateObject("Disposing"). Next, we print the object data and dispose of the object. Then, we instantiate a local object that will get assigned to the member variable. The generations of the local and member variables are printed to the console window, and then we dispose of the object and call garbage collection.

5. Add the DisposeofObject() method, as follows:

```
private static void DisposeOfObject()
{
        _product.Dispose();
}
```

6. The DisposeOfObject() method calls the Dispose() method on the _product object to free up resources. Update the Product class, as follows:

```
private string _cleanUpMethod;
public Product(string cleanUpMethod)
{
   Console.WriteLine("Product constructor.");
```

```
    _cleanUpMethod = cleanUpMethod;
}
~Product()
{
    Console.WriteLine($"Product destructor: {_
        cleanUpMethod}.");
}
```

7. We are storing the name of the cleanup method we are using so that when the finalizer is called, we will know the method of cleanup the object uses. Modify the `InstantiateObject()` method, as follows:

```
private static void InstantiateObject(string
        cleanUpMethod)
{
Console.WriteLine("Instantiating Product.");
_product = new Product(cleanUpMethod)
{
            Id = 1,
            Name = "Polly Parrot",
            Description = "Cudly child's toy.",
            UnitPrice = 7.99M
};
}
```

8. We are assigning the method of cleanup to the `Product` object. Do the same with the `InstantiateLocalObject()` method so that the code looks like this:

```
private static void InstantiateLocalObject(string
        cleanUpMethod)
{
var product = new Product(cleanUpMethod)
{
        Id = 2,
        Name = "Cute Kittie",
        Description = "Cudly child's toy.",
        UnitPrice = 5.75M
};
```

```
DisplayGeneration(product);
_product = product;
}
```

9. Again, we are assigning the method of cleanup to the `Product` object. Update `Product` to inherit from `DisposableBase`. Then, add the `ReleaseManagedResources()` method to the `Product` class, as follows:

```
protected override void ReleaseManagedResources()
{
base.ReleaseManagedResources();
Console.WriteLine("Releasing managed resources.");
}
```

10. This method will be used to release managed resources. Now, add the `ReleaseUnmanagedResources()` method to the `Product` class, as follows:

```
protected override void ReleaseUnmanagedResources()
{
base.ReleaseUnmanagedResources();
Console.WriteLine("Releasing unmanaged resources.");
}
```

    This method will be used for cleaning up unmanaged resources.

11. Run the code and you should see the output, as shown here:

```
Microsoft Visual Studio Debug Console
--- Finalization ---
Instantiating Product.
Product constructor.
Id: 1, Name: Polly Parrot, Description: Cudly child's toy., Unit Price: 7.99
Product constructor.
Product destructor: Finalization.
local product: generation 0
local product: generation 1
Product destructor: Finalization.
--- Disposing ---
Instantiating Product.
Product constructor.
Id: 1, Name: Polly Parrot, Description: Cudly child's toy., Unit Price: 7.99
Releasing managed resources.
Releasing unmanaged resources.
Product constructor.
local product: generation 0
local product: generation 0
Releasing managed resources.
Releasing unmanaged resources.
```

Figure 4.3 – The output of finalization and disposing code

As you can see, the finalization code calls the finalizer, but the methods used for releasing managed and unmanaged resources explicitly do not get called. Objects also survive the generation 0 garbage collection. Conversely, the disposing code explicitly releases the managed and unmanaged code, and finalization being suppressed is not called by the garbage collector. No objects in our example survive generation 0 garbage collection.

Another way to implicitly call `Dispose()` on disposable classes is to use a `using` statement. Here is an example, as can be seen in the `Program` class:

```
private static void UsingDispose()
{
        Console.WriteLine("--- UsingDispose() ---");
        using (var product = new Product("using")
            {
                Id = 2,
                Name = "Cute Kittie",
                Description = "Cudly child's toy.",
                UnitPrice = 5.75M
            }
        )
        {
            DisplayGeneration(product);
        }
}
```

The `using` statement is used with disposable objects. When the code block completes, the object is automatically disposed of. The object's generation is 0. Add a call to `UsingDispose()` in the `Main` method.

Well, you have seen how to use finalization and implement the `IDisposable` pattern in relation to the garbage collector. Now, let's look at how we can avoid memory leaks in C#.

Preventing memory leaks

In this section, we will understand the issues around COM objects and what can lead to memory leaks using COM objects. We will look at interoping with the Excel COM library for our example code. We will see how instances of Excel are kept alive after our code exits. By using Windows Task Manager, we will be able to see instances of Excel being generated. Our Excel code will be developed in such a way as to avoid memory leaks and ensure that every Excel instance is closed when our code has completed running so that no instances of Excel remain in memory.

We will then move on to look at how using events can be a common source of memory leaks at runtime and how we can avoid them. Using JetBrains dotMemory, we will profile a runtime build executable of our program code. As the code is running, we will generate snapshots. As the profiler runs, you will see the memory usage gradually climbing. Clicking on the snapshots will display detailed memory information for our running profile. We will also be able to see if we have any memory leaks, and will see that we have event-based memory leaks. In this section, we will also be looking at anonymous methods and weak references.

The outcome of this section will be that you understand how COM and the use of events, if not handled correctly, can introduce memory exceptions, and you will see how you can write your code so that no memory exceptions are generated.

Understanding the dangers of using Marshal.ReleaseComObject

The Visual Studio team ran into problems with Visual Studio 2010. Their problems arose due to rewriting native C++ components in managed C# code. The components that were rewritten as managed C# code were the window manager, command bars, and text editor.

With the release of Visual Studio 2010, there were two extension enablers—the existing extension mechanism that uses COM interfaces for older extensions, and a new managed programming model.

In order for the **Common Language Runtime** (**CLR**) to make COM objects appear as regular managed objects, COM objects are wrapped in an object called a `RuntimeCallableWrapper` or **RCW**. An RCW acts as a bridge between the worlds of COM and managed code.

All COM components must, at the very minimum, implement the `IUnknown` interface. When an object that implements the `IUnknown` interface enters the managed runtime, it is wrapped in an RCW. An RCW is, therefore, a regular managed object that references native code that implements the `IUnknown` interface.

There are two types of objects that can reference an RCW in a managed .NET computer program: COM objects and managed objects. This is the point at which issues can start to present themselves.

At this point, we will now consider a typical scenario that will result in memory issues between COM objects and managed objects.

The `DatabaseSearch` component begins the `Find` operation by asking the **global service provider** (**GSP**) for the `DatabaseManager` service. A valid instance of `IDatabaseManager` is returned to the `DatabaseSearch` component. The `DatabaseManager` component returned to the `DatabaseSearch` component is a native COM component. Because the `DatabaseManager` component is a native COM component, it is wrapped in an RCW by the runtime. The `DatabaseSearch` component does not know or care whether the `DatabaseManager` component is a native COM component or managed code component because all it sees is the `IDatabaseManager` interface. The `Find` operation continues with the `DatabaseSearch` component making various calls through `IDatabaseManager` to complete its task. Once the `Find` operation is completed, it is exited. Since `IDatabaseManager` is an RCW, it has the same lifetime semantics as managed objects. As a result, the `IDatabaseManager` component will be cleaned up when the garbage collector runs. The garbage collector may not run for a long time if there is not a lot of memory pressure, and there is the possibility that it may not even run. At this point, we end up with a native and managed memory clash because of the different ways in which they both manage system memory. The managed `DatabaseSearch` component is finished with the `DatabaseManager` component until it needs it again. If there are no references to the `DatabaseManager` component, then this would be a good time for the garbage collector to run and remove `DatabaseManager`. Any component written in native code would, as soon as the `Find` method is exited, call `Release` on `IDatabaseManager`. This would indicate that the reference to `IDatabaseManager` is no longer needed. Since the final `Release` is not being called until the next garbage collection, it appears that there is a memory leak with `IDatabaseManager`.
This is an example of non-deterministic finalization. The inability to determine when an object should be garbage-collected is known as non-deterministic finalization. The `Finalize()` method is executed on a special thread allocated by the garbage collector whenever the object it belongs to is being garbage-collected and finalization has not been suppressed when there are non-managed resources to be disposed of.

This scenario that we have looked at would result in expensive objects being reported as leaked objects, and this would be during application shutdown.

The natural solution would be to call `Marshal.ReleaseComObject(object)`. This call would be made as soon as the expensive object is no longer needed. In our scenario, it would be when `DatabaseManager` is no longer needed. This call causes the RCW to be released, and the internal reference count is decremented by one. At this point, the underlying COM object is usually released.

However, calling `Marshal.ReleaseComObject(object)` can be dangerous.

Consider that as part of a migration away from COM, `DatabaseManager` has been written in managed code. The `DatabaseSearch` managed component requests the `DatabaseManager` component via the GSP. An `IDatabaseManager` instance is returned to the `DatabaseSearch` component. The instance returned is an RCW that wraps a COM object. As a result, we have double wrapping that consists of an RCW wrapped around a **COM Callable Wrapper** (**CCW**). The CLR can easily deal with these scenarios, and so this is not a problem. It is when the `Find` operation exits that problems arise. The `DatabaseSearch` component still calls `Marshall.ReleaseComObject(object)` for the RCW of `DatabaseManager` when terminating.

This results in an `ArgumentException`-type exception being raised. The exception message generated is `"The object's type must be _ComObject or derived from _ComObject."` When this happens, remove the call to `Marshal.ReleaseComObject(object)`. An alternative is to call `Marshal.IsComObject` before `ReleaseComObject` is called.

Calling `Marshal.IsComObject` causes further problems. The `DatabaseManager` RCW has been declared as being no longer needed, but the problem is that the `DatabaseManager` RCW is still a valid object, meaning that it may still be reachable by managed objects. The next time the object is accessed, if reachable from managed code, an `InvalidComObjectException`-type exception will be raised by the CLR, stating: `"COM object that has been separated from its underlying RCW cannot be used."`

If the COM components used by our `DatabaseManager` RCW are cached by managed code instead of being returned to the GSP each time our `DatabaseManager` component is requested, our cached COM components will be checked first. This is done to avoid costly calls across the boundary between managed and unmanaged code. If several components then request the same COM component, they will each receive the same RCW.

The problem here is that the component calling the RCW that has had `ReleaseComObject` called will often be blamed as the component that generated the exception. But this is not the case—it is the component that called `ReleaseComObject` that is the component at fault, which in our scenario would be the `DatabaseSearch` component.

> **Note**
>
> It is recommended by Microsoft developers, especially those on the Visual Studio team, that unless you are 100% certain that there are no managed code items that have access to the RCW, you do not call `Marshal.ReleaseComObject`.

We will delve deeper into what we have just been discussing by looking at an Excel example.

Using the Microsoft Excel 16.0 Object Library in .NET 6

We are going to be looking at COM interoperability in .NET 6 in this section, by referencing the Microsoft Excel 16.0 Object Library. This library is a COM library. You will see how to use Excel to create a new application, modify it, and save it. When the first example is run a few times, you will see that your code does not fail. But in Task Manager, each time the method is run, another instance of Excel will remain open, as seen in Windows Task Manager. Then, we will move on to see how we can correctly dispose of COM objects so that instances of Excel are not kept open when our applications complete. Let's start by viewing what happens when we don't release Excel COM objects.

Investigating what happens when Excel COM objects are not released

In this section, we will create a spreadsheet, add data to it, and then save the file. This will reveal memory issues that arise from using Excel and not cleaning up properly after ourselves when we have finished using Excel. We will also see how to use Excel and clean up after ourselves so that we prevent memory issues through using Excel.

Add a COM reference to the `CH04_PreventingMemoryLeaks` project for the *Microsoft Excel 16.0 Object Library*.

> **Note**
>
> If you add a COM reference to your project, you will have IntelliSense available to you. But when you come to run your successfully compiled program, when it attempts to create an Excel application, it will raise a `FileNotFoundException`-type exception. Therefore, you need to set the values for `EmbedInteropTypes` and `Private` to `true`.

Since a `FileNotFoundException`-type exception is the last thing we need, edit your project file and then update the `COMReference` section, as follows:

```
<ItemGroup>
      <COMReference Include="Microsoft.Office.Excel.dll">
          <WrapperTool>tlbimp</WrapperTool>
          <VersionMinor>9</VersionMinor>
          <VersionMajor>1</VersionMajor>
          <Guid>00020813-0000-0000-c000-000000000046</Guid>
          <Lcid>0</Lcid>
```

```
            <Isolated>false</Isolated>
            <EmbedInteropTypes>True</EmbedInteropTypes>
            <Private>true</Private>
        </COMReference>
    </ItemGroup>
```

This will ensure that we don't experience the `FileNotFoundException`-type exception. Add a new `UsingExcel` class to the project, and then add the following `using` statements:

```
using Microsoft.Office.Interop.Excel;
using System;
using System.Diagnostics;
using System.IO;
using System.Runtime.InteropServices;
using Excel = Microsoft.Office.Interop.Excel;
```

Now, add the `RunExcelExamples()` method, as follows:

```
public void RunExcelExamples()
{
        for (int i = 0; i < 10; i++)
            NotReleasingExcelComObjects();
        for (int i = 0; i < 10; i++)
            ReleasingExcelComObjects();
}
```

This method calls two methods. It calls each of these methods 10 times and then exits. Let's add the `NotReleasingExcelComObjects()` method, as follows:

```
private static void NotReleasingExcelComObjects()
{
        string filename = @"C:\Temp\BucketList.xlsx";
        Excel.Application application = new Excel.Application();
        application.Visible = false;
        Excel.Workbook workbook = application.Workbooks.Add();
        Excel.Sheets sheets = workbook.Sheets;
        Excel.Worksheet worksheet =(Worksheet)sheets
        .Add(sheets[1], Type.Missing, Type.Missing,
```

```
            Type.Missing);
        worksheet.Range["A1"].Value = "Bucket List";
        worksheet.Range["A2"].Value = "Visit New Zealand";
        worksheet.Range["A1"].Value = "Visit Australia";
        if (File.Exists(filename))
            File.Delete(filename);
        workbook.SaveAs(filename);
        workbook.Close();
        application.Quit();
    }
```

This method declares a `filename` string. It then instantiates a new Excel application that is not visible. It then adds a column header called "`Bucket List`", and adds two items to that bucket list column in the rows below. It then checks if the file exists. If the file does exist, then it is deleted. The workbook is then saved and closed, and the Excel application is exited. Comment out the following lines from the `RunExcelExamples()` method:

```
        for (int i = 0; i < 10; i++)
            ReleasingExcelComObjects();
```

If you then save your project and run it, you will find that once the program exits, you are left with multiple Excel processes. Each of these processes takes up memory. The following screenshot shows Excel processes that remain in memory after our program exits:

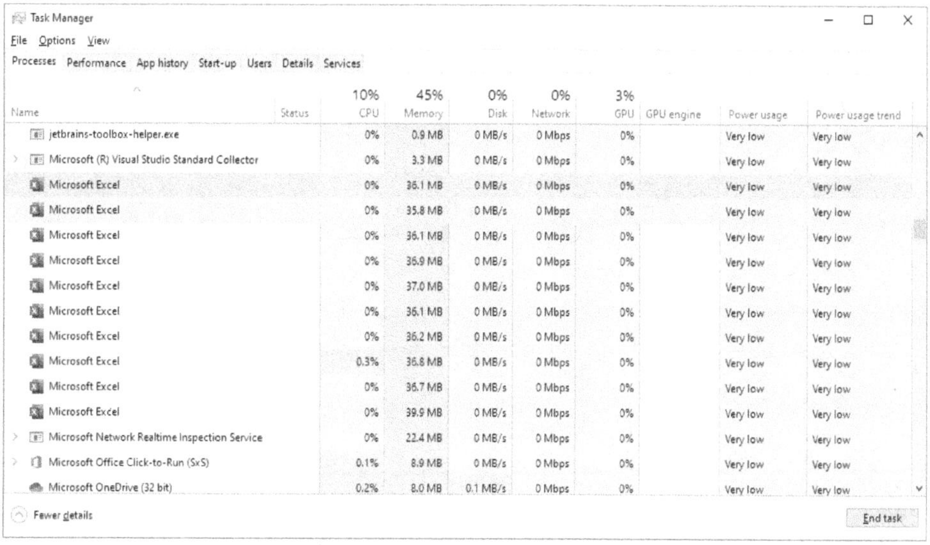

Task Manager

File Options View

Processes Performance App history Start-up Users Details Services

Name	Status	10% CPU	45% Memory	0% Disk	0% Network	3% GPU	GPU engine	Power usage	Power usage trend
jetbrains-toolbox-helper.exe		0%	0.9 MB	0 MB/s	0 Mbps	0%		Very low	Very low
Microsoft (R) Visual Studio Standard Collector		0%	3.3 MB	0 MB/s	0 Mbps	0%		Very low	Very low
Microsoft Excel		0%	36.1 MB	0 MB/s	0 Mbps	0%		Very low	Very low
Microsoft Excel		0%	35.8 MB	0 MB/s	0 Mbps	0%		Very low	Very low
Microsoft Excel		0%	36.1 MB	0 MB/s	0 Mbps	0%		Very low	Very low
Microsoft Excel		0%	36.9 MB	0 MB/s	0 Mbps	0%		Very low	Very low
Microsoft Excel		0%	37.0 MB	0 MB/s	0 Mbps	0%		Very low	Very low
Microsoft Excel		0%	36.1 MB	0 MB/s	0 Mbps	0%		Very low	Very low
Microsoft Excel		0%	36.2 MB	0 MB/s	0 Mbps	0%		Very low	Very low
Microsoft Excel		0.3%	36.8 MB	0 MB/s	0 Mbps	0%		Very low	Very low
Microsoft Excel		0%	36.7 MB	0 MB/s	0 Mbps	0%		Very low	Very low
Microsoft Excel		0%	39.9 MB	0 MB/s	0 Mbps	0%		Very low	Very low
Microsoft Network Realtime Inspection Service		0%	22.4 MB	0 MB/s	0 Mbps	0%		Very low	Very low
Microsoft Office Click-to-Run (SxS)		0.1%	8.9 MB	0 MB/s	0 Mbps	0%		Very low	Very low
Microsoft OneDrive (32 bit)		0.2%	8.0 MB	0.1 MB/s	0 Mbps	0%		Very low	Very low

Fewer details

End task

Figure 4.4 – Windows Task Manager displaying Excel processes no longer in use using up memory

As you can see, these Excel processes that remain in memory after our program finishes are using up 367.6 **megabytes** (**MB**) of RAM, which is the combined sum of all Excel processes' RAM. If this program in its current form were to be run multiple times, you would eventually run out of memory, as the Excel processes left running in memory constitute a memory leak. Each time the program runs, you are using up another 367 MB of RAM, or thereabouts. Eventually, the amount of memory available will not be enough, and you will end up with an out-of-memory exception.

The following screenshot shows the display in Task Manager after the program has been run once:

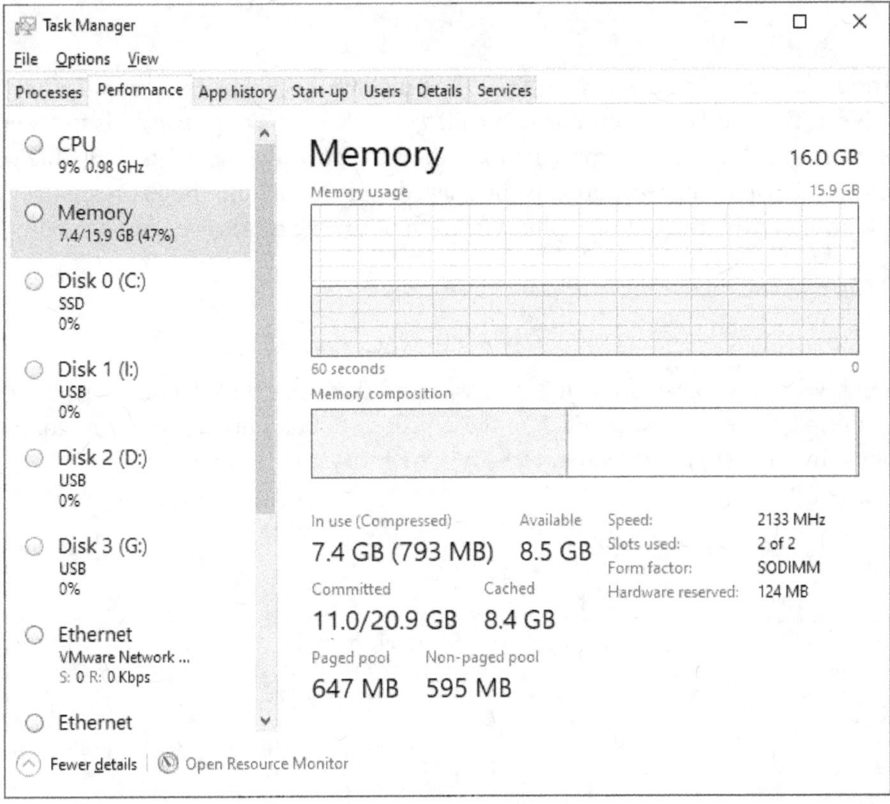

Figure 4.5 – Windows Task Manager after the program has been run once

From *Figure 4.5*, we can see that we are using 7.4 GB (793 MB), with 8.5 GB RAM still available to us. Run the program through a number of times continually. Each time the program is run, you will see the compressed memory rise and the available memory fall. At no point does the memory appear to be reclaimed, as shown in the following screenshot:

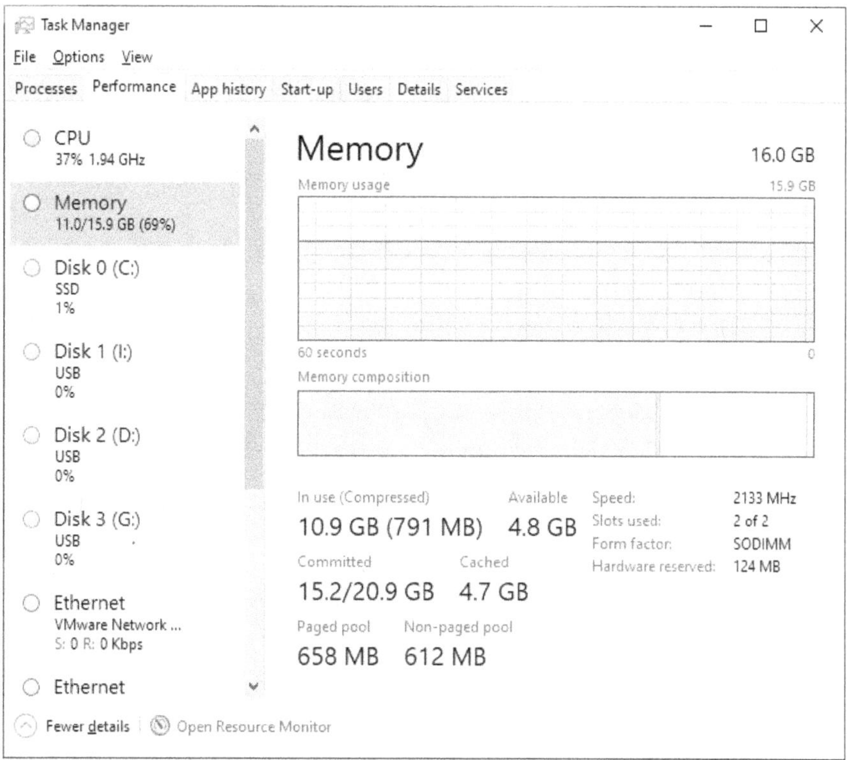

Figure 4.6 – Windows Task Manager displaying increased memory usage and diminished available memory after multiple program runs

After multiple continuous runs of our program, we can see that our **In use (Compressed)** memory has gone from 7.4 GB (793 MB) to 10.9 GB (799 MB) and our available memory has gone from 8.5 GB to 4.9 GB. This is clearly a problem that needs to be addressed, but how?

This is where the `ReleasingExcelComObjects()` method shown here comes in:

```
[System.Diagnostics.CodeAnalysis SuppressMessage
  ("Interoperability","CA1416:Validate platform compatibility",
    Justification = "Windows only code.")]
private static void ReleasingExcelComObjects()
{
      Excel.Application application = null;
      Excel.Workbooks workbooks = null;
      Excel.Workbook workbook = null;
      Excel.Sheets worksheets = null;
      Excel.Worksheet worksheet = null;
      Excel.Range range = null;
      Try
      {
          string filename = @"C:\Temp\BucketList.xlsx";
          application = new Excel.Application();
          application.Visible = false;
          workbooks = application.Workbooks;
          workbook = workbooks.Add();
          worksheets = workbook.Sheets;
          worksheet = (Worksheet)worksheets.Add(worksheets[1],
              Type.Missing, Type.Missing, Type.Missing);
          range = worksheet.Range["A1"];
          range.Value = "Bucket List";
          range = worksheet.Range["A2"];
          range.Value = "Visit New Zealand";
          range = worksheet.Range["A3"];
          range.Value = "Visit Australia";
          if (File.Exists(filename))
              File.Delete(filename);
          workbook.SaveAs(filename);
          workbook.Close();
          application.Quit();
      }
      Finally
      {
```

```
            if (range != null)
                Marshal.FinalReleaseComObject(range);
            if (worksheet != null)
                Marshal.FinalReleaseComObject(worksheet);
            if (worksheets != null)
                Marshal.FinalReleaseComObject(worksheets);
            if (workbook != null)
                Marshal.FinalReleaseComObject(workbook);
            if (workbooks != null)
                Marshal.FinalReleaseComObject(workbooks);
            if (application != null)
                Marshal.FinalReleaseComObject(application);
            range = null;
            worksheet = null;
            worksheets = null;
            workbook = null;
            worksheets = null;
            application = null;
            GC.Collect();
            GC.WaitForPendingFinalizers();
            Process[] processes =
                Process.GetProcessesByName("EXCEL");
            foreach (Process process in processes)
                process.Kill();
        }
    }
```

This rather lengthy method does what we need Excel to do—it releases the Excel COM objects, sets the managed objects to `null`, runs the garbage collector, and then terminates all running Excel processes. If you uncomment the code in the `RunExcelExamples()` method and then run the code once, you will see that we no longer have any Excel processes running in memory once our code has finished running. You will also see if you look at the **Performance** tab of Windows Task Manager that we have reclaimed our memory.

We have managed to fix our memory leak by terminating COM components and setting managed objects to `null` to remove managed references. Then, we killed all processes called `EXCEL`.

> **Note**
>
> Be careful when using the `process.Kill()` method to kill off all processes for a given name such as `EXCEL`. There may be other programs that also use that process that could be badly impacted by such termination. You should run such code in an isolated environment if doing batch processing on a server, or schedule such operations for a time when you can guarantee that other processes will not be affected by running such code.

It is now time to look at how using events can be a source of memory leaks.

How using events can be a source of memory leaks

In this section, we will look at how the use of events in your computer programs can be a source of memory leaks. We will demonstrate this using a very simple Windows Forms application that we will write. Then, we will analyze our memory usage using JetBrains dotMemory. There will be two methods employed to show events in use. One method will generate a memory leak, while the other won't generate a memory leak.

So, how can using events generate memory leaks?

Unless you are using anonymous methods, subscribing to an event holds a reference to the class that holds that event until such time as the event is unsubscribed from. Consider the following class:

```
internal class EventSubscriber
{
     public EventSubscriber(Control control)
     {
         Control.TextChanged += OnTextChanged
     }
     private void OnTextChanged(
         object sender,
         EventArgs eventArgs
     )
     {
         Text ((Control)sender).Text;
     }
}
```

If the control outlives the `EventSubscriber` class, then all instances of `EventSubscriber` will not be deallocated by the garbage collector. The end result is a memory leak. Here are some different ways to avoid event-based memory leaks:

1. Subscribe to anonymous methods.
2. Unsubscribe from events when you are finished with them.
3. Implement the weak-handler pattern.

Before we look at each of these ways of avoiding memory leaks, we will write our Windows Forms application that demonstrates a way to avoid memory leaks and a way to generate memory leaks. Follow these steps:

1. Start a new .NET Core Windows Forms project, and then change the target framework from .NET Core 3.1 to .NET 5 in the project settings.
2. Rename `Form1` to `MainForm`.
3. Add a label called `InformationLabel` with the text `"Information"`, a button called `RaiseEventsButton` with the text `"Raise Events"`, and another label called `ProgressLabel` with the text `"Progress:"`. You can lay the components out and style them according to your preference.
4. Double-click on the `RaiseEventsButton` button. This will generate a click event handler method.
5. Add a class to the project called `EventOne`. You will need the following `using` statements:

   ```
   using System;
   using System.Threading;
   ```

6. Add the following code to the top of the `EventOne` class:

   ```
   public event EventHandler OnEventRaised;
   private static int _count;
   public static int Count { get { return _count; } }
   ```

7. These elements are needed to handle the event and keep a count of how many instances are still being kept alive. Add the constructor, as follows:

   ```
   public EventOne()
   {
         Interlocked.Increment(ref _count);
   }
   ```

8. The constructor code increments the `_count` member variable in an atomic and thread-safe manner for each instance of the class. Add the `RaiseEvent(EventArgs e)` method, as follows:

```
public void RaiseEvent(EventArgs e)
{
        EventHandler eventHandler = OnEventRaised;
              if (eventHandler != null)
                    eventHandler(this, e);
}
```

9. This method is called by the clients and is responsible for firing the event upon request. Now, add the finalizer, as follows:

```
~EventOne()
{
        Interlocked.Decrement(ref _count);
}
```

10. The finalizer decrements the `_count` member variable in a thread-safe manner each time an instance of the class is terminated and collected by the garbage collector. Add a new `EventTwo` class to the project. You will need the following `using` statements:

```
using System;
using System.Threading;
using System.Windows.Forms;
```

11. Add the following code to the top of the `EventTwo` class:

```
private static int _count;
public static int Count { get { return _count; } }
public string Text { get; private set; }
```

12. The code stores the count of the number of alive instances and the current text of the subscribed control. Add the following constructor:

```
public EventTwo(Control control)
{
        Interlocked.Increment(ref _count);
        control.TextChanged += OnTextChanged;
}
```

13. The constructor takes a Windows Forms control as a parameter. It increments the `_count` member variable by one in a thread-safe manner. It then subscribes to the `TextChanged` event that is handled by the `OnTextChanged` method. Add the `OnTextChanged` method, as follows:

```
private void OnTextChanged(object sender, EventArgs
    eventArgs)
{
      Text = ((Control)sender).Text;
}
```

14. This method is fired when the `Text` property of the subscribed control is changed. It takes the `Text` content of the control and assigns it to the `Text` property of the `EventTwo` class. Add the `Finalizer()` method, as follows:

```
~EventTwo()
{
      Interlocked.Decrement(ref _count);
}
```

15. The finalizer decrements the `_count` member variable by one in a thread-safe manner each time an instance is garbage-collected. We now have in place the two classes that our form will use for raising events. Switch back to the `MainForm` class.
16. At the top of the `MainForm` class, add the following member variables:

```
private int _eventsGeneratedCount;
private int _eventSubscriberCount;
```

17. These two values will store the number of events that have been generated. Add the `SetTitleText()` method, as follows:

```
private void SetTitleText()
{
      Text = $"{_eventsGeneratedCount}/{EventOne.Count} -
         {_eventSubscriberCount}/{EventTwo.Count}";
}
```

18. This method sets the control's `Text` property for each method that raises events. The text displays the number of events raised and the number of events still alive for the non-memory leak method, and the same again for the memory leak method. Add the `SetInformationLabelText()` method, as follows:

```
private void SetInformationLabelText()
{
        StringBuilder sb = new StringBuilder();
        sb.AppendLine($"Raised Events (No Memory Leak):
            {_eventsGeneratedCount}, Alive Events:
              {EventOne.Count}");
        sb.AppendLine($"Raised Events (Memory Leak):
            {_eventSubscriberCount}, Alive Events:
              {EventTwo.Count}");
        InformationLabel.Text = sb.ToString();
}
```

19. The `SetInformationLabelText()` method updates the `InformationLabel` text to display the number of events raised in each method and the number of events remaining in memory once both methods have finished executing. Add the `RaiseEvent` method, as follows:

```
private void RaiseEvent(object sender, EventArgs e)
{
        ProgressLabel.Text = $"Event Raised:
            {DateTime.Now}";
        ProgressLabel.Invalidate();
        ProgressLabel.Update();
}
```

20. The `RaiseEvent` method updates the `ProgressLabel.Text` property, but so that it is updated in real time, it is necessary to call the `Invalidate()` and `Update()` methods. Now, add the `MemoryLeakMethod` method, as follows:

```
private void MemoryLeakMethod(EventArgs e)
{
        int count = 10000;
        for (int x = 0; x < count; x++)
```

```
        {
            var eventTwo = new EventTwo(this);
        }
        _eventTwoCount += count;
    }
```

21. This method declares a count of 10,000 items. It then loops through 10,000 iterations. A new `EventTwo` object is subscribed to with the reference to `MainForm` passed in. Once the loop completes, the `_eventTwoCount` variable is incremented by 10,000. Next, we will add the `NoMemoryLeakedMethod` method, as follows:

```
private void NoMemoryLeakMethod(EventArgs e)
{
        int count = 10000;
        for (int x = 0; x < count; x++)
        {
        EventOne eventOne = new EventOne();
        eventOne.OnEventRaised += RaiseEvent;
        eventOne.RaiseEvent(e);
        }
        _eventOneCount += count;
}
```

22. This method declares a count of 10,000. It iterates 10,000 times. During that 10,000 times, it instantiates a new `EventOne` object, adds an event handler called `RaisedEvent`, and then raises the event. Once, the loop has completed, the `_eventOneCount` variable is incremented by 10,000. Update the click event handler with the following code:

```
NoMemoryLeakMethod(e);
MemoryLeakMethod(e);
SetInformationLabelText();
SetTitleText();
```

23. Change the build mode to `Release` and build the project.

24. Open **JetBrains dotMemory**. Select **Local | .NET Core Application**, select an executable generated by the build process, then check the **Collect memory allocation and traffic from start** box. Your screen should look like this:

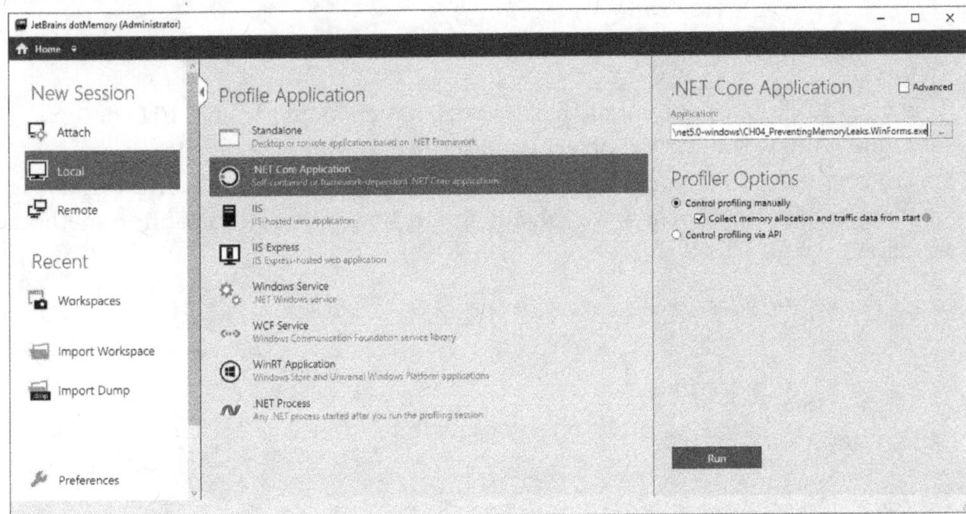

Figure 4.7 – The JetBrains dotMemory configuration screen

25. Click on the **Run** button. This will start your application and profiling session, as shown in the next two screenshots:

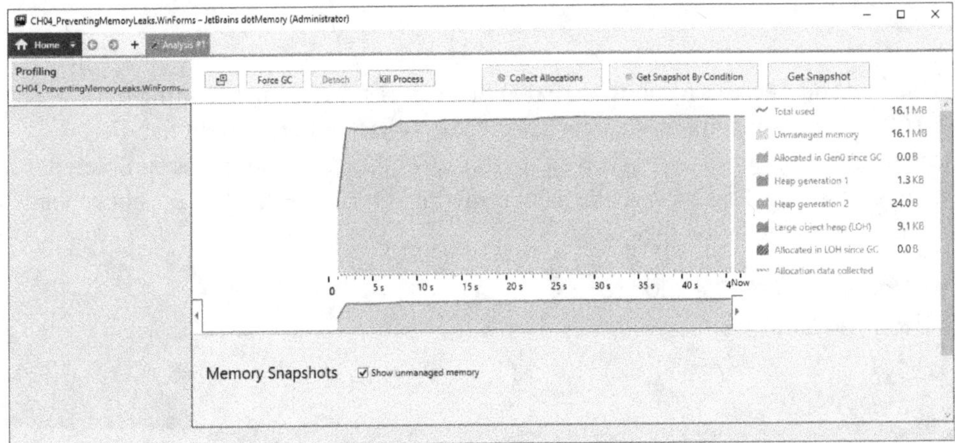

Figure 4.8 – JetBrains dotMemory profiling our Windows Forms application

Figure 4.9 – Our Windows Forms application before any events have been run

26. Click on the **Raise Events** button a few times. Each time you click on the button, the memory profile should change and the memory usage should increase, as shown in the following screenshot:

Figure 4.10 – Our Windows Forms application showing 50,000 alive events, indicating we have a memory leak

27. As you can see, we have a memory leak. Our `NoMemoryLeakMethod` method does not generate a memory leak. As you can see, after 50,000 raised events, the objects kept alive in memory is 0. But our `MemoryLeakMethod` method does produce a memory leak. Out of 50,000 raised events, 50,000 objects remain alive.

28. Run the program a few more times, and pay attention to what is going on in dotMemory. When you see a point of interest, click on the area and then click on **Get Snapshot**. This will take a snapshot of that moment in time that users can analyze to see if there are any issues. You should end up with something similar to this:

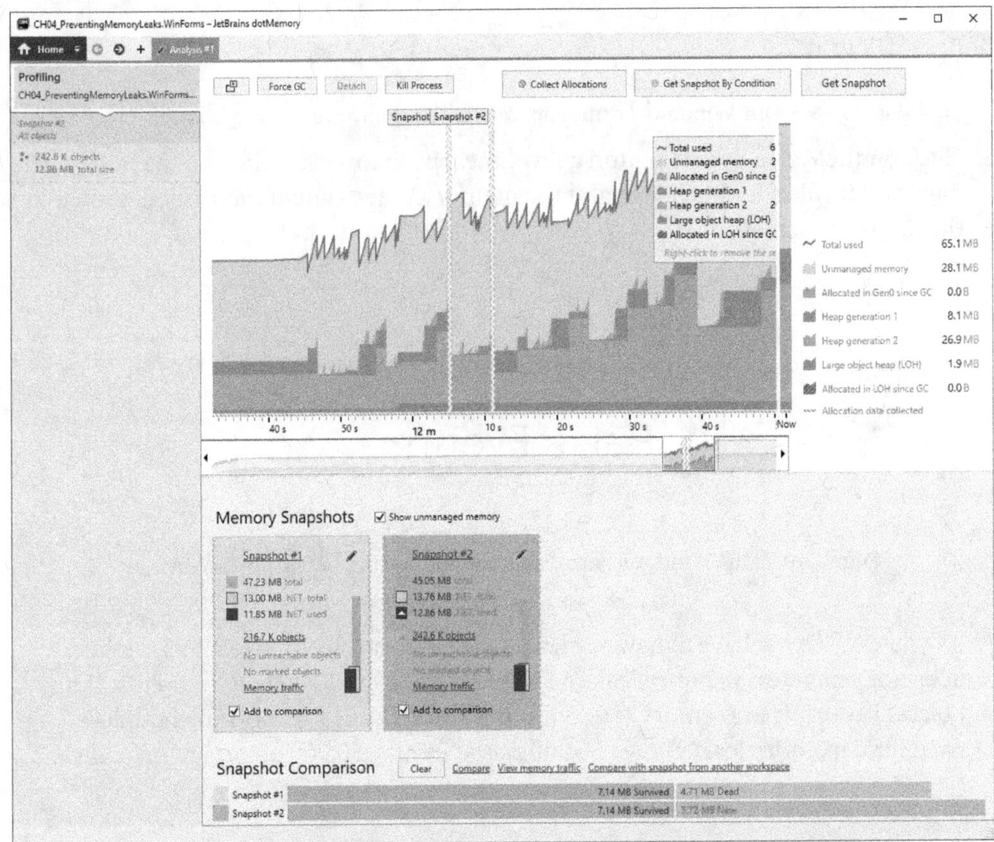

Figure 4.11 – JetBrains dotMemory profile of our Windows Forms application when events are raised and snapshots are taken

29. Click on any one of your snapshots. You should see an output like this:

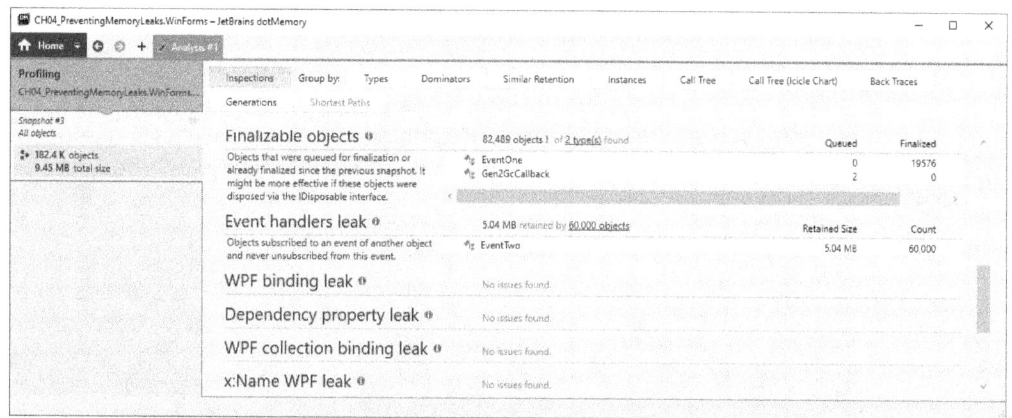

Figure 4.12 – A memory leak has been identified with the EventTwo class

30. JetBrains dotMemory has detected a memory leak in the `EventTwo` class. This is because the class subscribes to an event of another object, but never unsubscribes from it. However, you will see that all the objects for the `EventOne` class have been finalized.

You have seen how to use events in such a way that generates memory leaks and in such a way that all objects are finalized and a memory leak is prevented. Let's revisit the three ways to prevent memory leaks when using events, as follows:

1. Subscribe to anonymous methods.
2. Unsubscribe from events when you are finished with them.
3. Implement the weak-handler pattern.

Let's take a look at subscribing to anonymous methods and then unsubscribing

Using local methods

Prior to C# 7.0, you would use anonymous methods as a way of handling events such that you avoid introducing memory leaks. As of C# 7.0, you can use local methods. In this example, we will handle events using local methods. Follow these next steps:

1. Load the `CH04_PreventingMemoryLeaks` project.
2. Add a class called `Website`, as follows:

```
internal class Website
{
        public event EventHandler<EventArgs> Login;
```

```
        public event EventHandler<EventArgs> Logout;
}
```

3. This class has two events for logging in and logging out of a website. Add a new class called `AnonymousEventSubscription`. Add the `Login()` method, as follows:

```
public void Login()
{
        Website website = new Website();
        void LoginHandler(object sender, EventArgs args)
        {
            Debug.WriteLine("Anonymous login event handler
              using a local method.");
            website.Login -= LoginHandler;
        };
        website.Login += LoginHandler;
        LoginHandler(this, new EventArgs());
}
```

4. The `Login()` method instantiates a new `Website` object. It then has a local method called `LoginHandler` that writes a message to the debug window and then unsubscribes from the `Website.Login` event. Then, outside of the local method, it subscribes to the `Website.Login` event and raises the event. Let's add the `Logout()` method, as follows:

```
public void Logout()
{
        Website website = new Website();
        void LogoutHandler(object sender, EventArgs args)
        {
            Debug.WriteLine("Anonymous logout event handler
              using a local method.");
            website.Logout -= LogoutHandler;
        };
        website.Logout += LogoutHandler;
        LogoutHandler(this, new EventArgs());
}
```

5. The `Logout()` method instantiates a new `Website` object. It then has a local method called `LogoutHandler` that writes a message to the debug window and then unsubscribes from the `Website.Logout` event. Then, outside of the local method, it adds the event handler for the `Website.Logout` event, and then raises the event.

6. In the `Main` method, comment out the `RunExcelExamples()` line. Then, add the `UseAnonymousEventSubscription()` method call, as follows:

```
private static void UseAnonymousEventSubscriptions()
{
    for (int x = 0; x < 1000000; x++)
    {
        AnonymousEventSubscription aes = new
          AnonymousEventSubscription();
        aes.Login();
        aes.Logout();
    }
}
```

7. This code runs through 1,000,000 iterations. For each iteration, a new `AnonymousEventSubscription` is instantiated, with calls to `Login()` and `Logout()` made. These two calls will each have a subscription to an event, an event executed via a local method, and, as the local method is executed, the event it will be unsubscribed from.

8. If you build and run the code, you should see the following lines printed 1,000,000 times in your debug window:

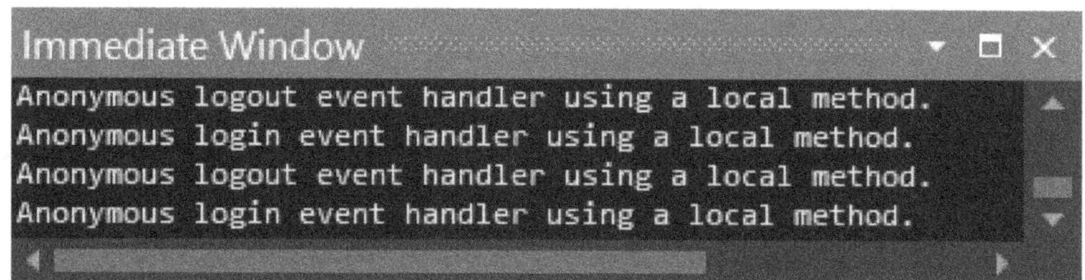

Figure 4.13 – The debug window showing events firing for Login and Logout

9. If you perform a release build and run dotMemory, you will see that we have no memory leak, considering we have just generated 2,000,000 event subscriptions and unsubscriptions—that is, 1,000,000 for `Login()` and 1,000,000 for `Logout()`.

We have seen how to effectively use anonymous events using local methods without causing memory leaks. Now, let's look at our final topic of the chapter—weak references.

Using weak reference events

We use the weak reference event pattern to allow an object to be garbage-collected if its only remaining link is an event handler. We will implement the weak reference event pattern in this section in the `CH04_PreventingMemoryLeaks` project. Follow these next steps:

1. In the Package Manager Console, type the following: `install-package WeakEventListener`. The `System.Windows.WeakEventManager` package only works with .NET 4.8 and older, which is why we install this package.
2. Add the following `SampleClass` class:

```
internal class SampleClass
{
        public event EventHandler<EventArgs> RaiseEvent;
        public void DoSomething()
        {
            OnRaiseEvent();
        }
        protected virtual void OnRaiseEvent()
        {
            RaiseEvent?.Invoke(this, EventArgs.Empty);
        }
}
```

3. In this class, we declare an event called `RaiseEvent`. The `DoSomething()` method calls the `OnRaiseEvent()` method. The `OnRaiseEvent()` method checks if the event is `null`; if it is not `null`, then the event is invoked. Add a new class called `UsingWeakreferences`. You will need the following references:

```
using System;
using System.Diagnostics;
using WeakEventListener;
```

4. Add the `RaiseWeakReferenceEvents()` method, as follows:

```
public void RaiseWeakReferenceEvents()
{
```

```
        bool isOnEventTriggered = false;
        bool isOnDetachTriggered = false;
        SampleClass sample = new SampleClass();
        WeakEventListener<SampleClass, object, EventArgs>
          weak = new WeakEventListener<SampleClass, object,
                  EventArgs>(sample);
        weak.OnEventAction = (instance, source, eventArgs)
          => { isOnEventTriggered = true; };
        weak.OnDetachAction = (listener) =>
          {isOnDetachTriggered = true; };
        sample.Raisevent += weak.OnEvent;
        sample.DoSomething();
        Debug.Assert(isOnEventTriggered);
        weak.Detach();
        Debug.Assert(isOnDetachTriggered);
  }
```

5. We have two variables that are `true` when an event has been triggered and when it has been detached. We instantiate a new `SampleClass` class instance. Then we declare a `WeakEventListener` package that references the `SampleClass` class. Anonymous methods are used to handle the `OnEventAction` and `OnDetachAction` methods. The `WeakReferenceListener.OnEvent` method is then assigned as the handler for the `SampleClass.RaiseEvent` event. We then call the `DoSomething()` method that raises the event. Then, we assert that the event has been triggered, detach the event, and then assert that the event has been detached.
6. Make sure the project is set to **Debug** mode, and then step through the code. It should work as expected, with the event being correctly triggered and detached.

Let's now summarize what we have learned in this chapter.

Summary

We looked at object generations and saw how easy it was to generate a `System.OutOfMemoryException`-type exception. We saw how we can use predictive out-of-memory exception checking to save time by preventing the running of code that will cause this exception.

Then, we moved on to discuss long weak references and short weak references. We learned that strong references are not garbage-collected, and weak references are garbage-collected.

We then looked at finalization and saw how the `Finalize()` method will be called on objects that are not disposed of, and that we have no control over when the `Finalize()` method will run. Then, we looked at how to implement the `IDisposable` pattern and suppress the need for garbage collection to call `Finalize()`.

Finally, we looked at the various ways to prevent memory leaks, such as properly disposing of managed resources and unmanaged resources. We also saw how to correctly handle events so that we do not cause memory leaks.

With what you have learned in this chapter, you will be able to overcome out-of-memory exceptions, improve memory performance, and improve garbage collection in your applications, and you will be to correctly use events and event handlers without generating memory leaks and will be able to effectively release COM objects and allocated memory. This will lead to better quality and more stable programs that make good use of memory.

In the next chapter, we will be looking at application profiling.

Questions

1. How many object generations are there?
2. Which sized objects get placed on the SOH?
3. Which sized objects get placed on the LOH?
4. What is a strong reference?
5. What is a weak reference?
6. How can we clean up objects without having to rely on finalization?
7. How do we avoid memory leaks when using events?
8. Which method do we use to release COM objects?
9. How do we prevent memory leaks when allocating memory?

Further reading

- Weak references: `https://www.youtube.com/watch?v=2WcDhh8lvJs`
- `ComWrappers` class: `https://docs.microsoft.com/ dotnet/api/system.runtime.interopservices.comwrappers?view=net-5.0`

- *Marshal.ReleaseComObject Considered Dangerous*: `https://devblogs.microsoft.com/visualstudio/marshal-releasecomobject-considered-dangerous/`
- *WeakEventManager Class:* `https://docs.microsoft.com /dotnet/api/system.windows.weakeventmanager?view=net-5.0`
- *Weak Event Patterns*: `https://docs.microsoft.com/en-us/dotnet/desktop/wpf/advanced/weak-event-patterns?view=netframeworkdesktop-4.8`
- *How to properly release Excel COM objects*: `https://www.add-in-express.com/creating-addins-blog/2013/11/05/release-excel-com-objects/`
- *Understanding and Avoiding Memory Leaks with Event Handlers and Event Aggregators*: `https://www.markheath.net/post/understanding-and-avoiding-memory-leaks`
- Why and how to avoid event handler memory leaks: `https://stackoverflow.com/questions/4526829/why-and-how-to-avoid-event-handler-memory-leaks`
- *.NET Framework technologies unavailable on .NET Core and .NET 5+:* `https://docs.microsoft.com/en-us/dotnet/core/porting/net-framework-tech-unavailable`

5
Application Profiling and Tracing

Application profiling is the internal examination of the inner workings of a computer program. We use application profiling to measure the performance of a program's internals. This helps us to identify any performance bottlenecks and memory issues. Then, we can use this information to refactor and improve the performance of the program.

Application tracing is used to monitor the internal performance of a computer program as it is running. You can trace the execution of your computer program during development, testing, and when released into production.

When used together, application profiling and application tracing can be very powerful and useful in identifying why computer programs are slow.

In this chapter, you will learn how to profile your applications to identify any poor areas of performance. You will come to understand code metrics and how to perform static code analysis. In your drive to write more performant code, you will learn how to make use of memory dumps, the loaded modules viewer, debugging, tracing, and `dotnet-counters`. By the time you have completed this chapter, you will have the necessary skills and experience you need to profile and trace your own applications.

In this chapter, we will be covering the following main topics:

- **Understanding code metrics**: In this section, we will be looking at what application, assembly, namespace, type, method, and field metrics various tools can offer us.
- **Performing static code analysis**: In this section, we will look at performing static code analysis with Visual Studio 2022. And we will be generating metrics for our software that consist of the maintainability index, cyclomatic complex, the depth of inheritance, class coupling, units of source code, and lines of executable code.
- **Generating and viewing memory dumps**: In this section, we will look at how to generate and view memory dumps when a breakpoint is hit in code or when an application is encountered.
- **Viewing loaded modules**: In this section, we will display the **Modules** window in Visual Studio so that we can view the modules that are loaded into memory by our application and view information about those modules.
- **Debugging your applications**: This section highlights the various debugging options that are available to us.
- **Using tracing and diagnostics tools**: In this section, we will introduce tools that can help us to perform tracing and diagnostics on our software applications. Specifically, we will consider Visual Studio 2022, JetBrains dotMemory, and JetBrains dotTrace.
- **Installing and using dotnet-counters**: In this section, we will install `dotnet-counters` and use them to list .NET processes that can be monitored, list the available counters that we can use to gather performance data, monitor a .NET process, and collect data for that process in a CSV file for post-processing analysis in Excel.
- **Tracking down and fixing a memory leak with dotMemory**: In this section, we will use dotMemory to hunt down a memory leak in a WPF application and fix it.
- **Finding the cause of a UI freeze with dotTrace**: In this section, we will use dotTrace to hunt down the cause of a UI freeze in a WPF application and fix it.
- **Optimizing application performance and memory traffic**: In this section, we will use dotTrace to identify opportunities to improve performance and memory traffic for a WPF application.

After completing this chapter, you will be skilled in the following things:

- Understanding code metrics and being able to use them to improve code quality and performance
- Performing static code analysis to improve code quality and performance
- Using loaded modules to identify what modules your code uses
- Effectively debugging software
- Effectively tracing software
- Using `dotnet-counters` to perform first-level performance investigations
- Using JetBrains dotMemory to track down memory leaks and fix them
- Using JetBrains dotTrace to track down the cause of UI freezes and fix them
- Using JetBrains dotTrace to track down performance and memory traffic issues and fix them

> **Note**
>
> Don't be alarmed if you are asked to access code from previous chapters for some of the examples. Due to the page limitation for chapters, adding code examples for those exercises would have exceeded the count limit for this chapter.

Technical requirements

The technical requirements to follow along with this chapter are as follows:

- Visual Studio 2022 or higher
- JetBrains dotMemory
- JetBrains dotTrace
- Source code: `https://github.com/PacktPublishing/High-Performance-Programming-in-CSharp-and-.NET/tree/master/CH05`
- Optional: Microsoft Excel or some other CSV file viewer

Understanding code metrics

In this section, we will be looking at the code metrics that can be gathered using various tools that are paid for, free, and open source. Source code metrics are extracted from source code and are used to measure the quality and performance of our source code.

> **Note**
> Different tools have different metrics that they can measure and calculate. Since each tool is different, it is a good idea for you to see what tools and metrics are available that satisfy your own project's requirements.

In the upcoming subsections, we will learn about the different code metrics that we can use to measure our code and improve performance.

Application metrics

Application metrics cover your application's complete source code across assemblies. They give you the big picture regarding how many lines of code your application has, along with how many lines are covered by tests.

In this section, we will cover, from a high level, the various metrics that certain tools such as the *ndepends* tool offer. As part of your own studies, identify different application metrics gathering tools. Then, see what metrics they offer. Choose the tool that best fits your needs. In the next section, the generation of code metrics will be demonstrated using Visual Studio's built-in static code analysis tool to generate the following metrics: the maintainability index, cyclomatic complexity, the depth of inheritance, class coupling, the lines of source code, and the lines of executing code. These and other metrics are described next.

Although metrics are different between tool vendors, available application metrics might include the following:

- **Lines of Code** (**LOC**): There are two types of LOC measurements. They include logical LOC and physical LOC. A logical LOC refers to those lines of code that can span one or more lines and are terminated by either a closing curly brace or a semicolon. A physical LOC refers to actual lines of code including comments and whitespace.
- **Lines of comment**: The number of lines used for comments.

- **Percentage comment**: This metric identifies the percentage of code that is made up of comments. It is calculated using this formula: *100 x Lines of Comment/(Lines of Comment + Lines of Code).*
- **IL instructions**: When your code compiles, it is converted into **Intermediate Language** (**IL**) code. Depending on how you code your C# code, this can lead to the generation of a large or small number of IL instructions. It makes sense to measure the number of IL instructions generated by your code. That's because even if the code is small, it could generate many IL instructions. And conversely, a method can be large but generate smaller lines of code compared to the smaller version of the code. The smaller number of IL instructions, the easier the method is to maintain.

> Note
>
> The company *ndepend* has a recommendation on their documentation code-metrics page that states methods that produce IL instructions higher than 100 are hard to understand and maintain. Additionally, they state that unless the methods are autogenerated by code generation tools, methods that produce 200 lines or more of IL instructions are extremely complex and should be split into smaller methods.

- **Application assemblies**: The application assembly count.
- **Application namespaces**: The application namespace count.
- **Application methods**: The application method count.
- **Application fields**: The application field count.
- **Lines of code covered**: The number of lines covered by tests.
- **Lines of code not covered**: The number of lines not covered by tests.

Now we will cover what assembly metrics are and what types of metrics can be gathered.

Assembly metrics

Assembly metrics are more focused on measuring the quality and stability of individual assemblies. Since an application can consist of many assemblies, problems can arise in any one or more of those assemblies. If multiple assemblies rely on one poorly performing assembly, then the whole application will be affected. Additionally, it is good to be able to reuse assemblies in different projects, so coupling should be kept to an absolute minimum.

Gathering assembly metrics enables you to understand how your assemblies are coupled together, and you can also see how abstract and stable or unstable they are. Additionally, you can determine whether they are reusable in their current form based on those metrics. The various metrics that are available to measure assembly source code include the following:

- **Afferent coupling**: This is the count of classes in other assemblies that rely on classes within the current assembly.
- **Efferent coupling**: This is the count of classes in the current assembly that depend upon classes in other packages.
- **Relational cohesion**: The average count of internal relationships per type within an assembly.
- **Instability**: The ratio of efferent coupling to total coupling.
- **Abstractness**: The ratio of internal abstract classes and interfaces to internal types.
- **Distance from the main sequence**: A number that indicates the balance between abstractness and stability.

Now, let's look at what namespace metrics are and what kind of metrics can be gathered.

Namespace metrics

Namespaces are an important part of any professional quality API. Correctly partitioning your code into relevantly named namespaces helps programmers understand your API and find what they are looking for more easily. Namespace metrics help you to understand whether you have dependency cycles and whether your assemblies are high-level, mid-level, or low-level.

The metrics that are available concerning the code quality of namespaces include the following:

- **Afferent coupling**: The count of namespaces that directly depend on the current namespace.
- **Efferent coupling**: The count of different namespaces that the current namespace depends on.
- **Level**: The level value of a namespace. This metric can help you identify dependency cycles. Additionally, it helps you objectively classify your assemblies, namespaces, methods, and types as high-level, mid-level, or low-level.

It's time to look at what type metrics are and the type of metrics that can be gathered.

Type metrics

Type refers to class types, interface types, array types, value types, enumeration types, type parameters, generic type definitions, and open or closed constructed generic types.

Types and how they are coded and used are behind all the problems we experience as programmers and end users. Understanding how they are used in our programs is an effective way of identifying a variety of issues with our code. When problems are identified, they can be rectified.

Type code quality metrics include the following:

- **Type rank**: A computed value that is computed based on the application of a ranking algorithm, similar to Google's PageRank algorithm, on types dependencies graph.
- **Afferent coupling**: The count of types that depend upon the current type.
- **Efferent coupling**: The count of types that the current type directly depends on.
- **Lack of cohesion methods**: For the code to adhere to the **single responsibility principle** (**SRP**), it will have only one reason to change, and no more.
- **Cyclomatic complexity**: The count of pathways through a method.
- **IL cyclomatic complexity**: The count of pathways through IL code.
- **Size of instance**: The size, in bytes, of the instances of the specified type.
- **Interfaces implemented**: The count of interfaces implemented.
- **Association between classes**: The count of members from other types that are directly used in the body of the methods of the current type.
- **The number of children**: The count of subclasses for a class, or the count of types that implement an interface.
- **Depth of inheritance tree**: The count of base classes for a class or structure.

Now we will look at what method metrics are and the types of method metrics that can be gathered.

Method metrics

Normally, methods are behind most performance issues. It is the method within a class that executes instructions that can cause any number of issues for your customers. These problems can include runtime errors, data errors, and performance issues. Being able to see and understand how a method interacts with other methods can be a real big help in solving various issues including performance issues. The method metrics that are available for analyzing the code quality of methods include the following:

- **Method rank**: A computed value based on the application of a ranking algorithm, similar to Google's PageRank algorithm, on the method dependencies graph.
- **Afferent coupling**: The count of methods that directly depend upon the current method.
- **Efferent coupling**: The count of methods that the current method directly depends on.
- **IL nesting depth**: The maximum count of encapsulated scopes inside a method body computed from the IL code.
- **Parameters**: The number of parameters used in the method signature.
- **Variables**: The method body variable count.
- **Overloads**: The method overload count.
- **Percentage branch coverage**: The percentage of branches covered by tests generated from opcodes.

The final metrics that we will look at are field metrics.

Field metrics

The metrics available for measuring coupling at the field level is **afferent coupling**. This refers to the count of methods that directly uses a variable. The higher the count, the more unstable the software becomes. So, this metric can be useful for improving the stability of the software.

The size of instance metric measures the size, in bytes, of the instances of a specified type.

In the next section, we will look at how to improve the architecture and code quality by performing static code analysis.

Performing static code analysis

The purpose of static code analysis is to help you improve your overall architectural quality, code quality, and performance by doing the following:

- Visualizing software architecture and its software dependencies
- Enforcing the designated architectural rules regarding laying, subsystems, calling rules, and more
- Identifying code that has been cloned and modified using cut, copy, and paste
- Identifying dead code that can be removed
- Calculating various software metrics
- Performing code style checks and flagging violations

Many companies employ static code analysis as part of their **Continuous Integration** (**CI**) process. There are various stages at which problems can come to light. These stages are listed as follows:

- When compiling source code in the IDE
- When running unit tests and end-to-end system tests
- When pushing source code to version control and issuing a pull request
- When a pull request has been issued and the code is issued to the build pipeline

Performing static code analysis during the coding phase helps to prevent issues from being flagged further down the development and release processes.

In Visual Studio via the **Project Properties | Code Analysis** page, you can run analyzers on the build and live analyses. You can enable .NET analyzers and set the analysis level to **preview**, **latest**, **5.0**, and **none**. Additionally, you can enforce CodeStyle on build. *Figure 5.1* shows the **Code Analysis** page:

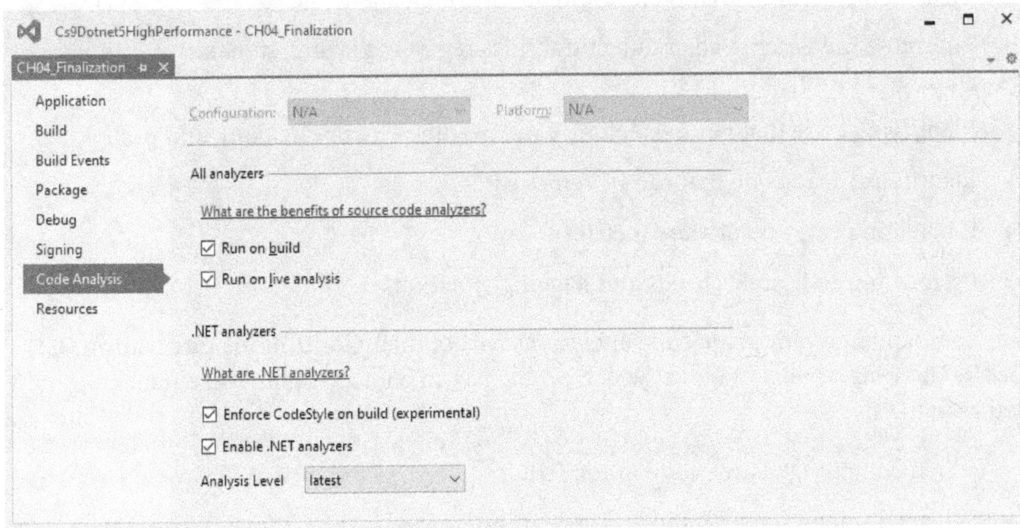

Figure 5.1 – The Visual Studio Code Analysis page on the Project Properties tab

The **Code Metrics Results** window is available from the **View** menu by selecting **View | Other Windows - Code Metrics Results**. The **Code Metrics Results** window is displayed in *Figure 5.2*:

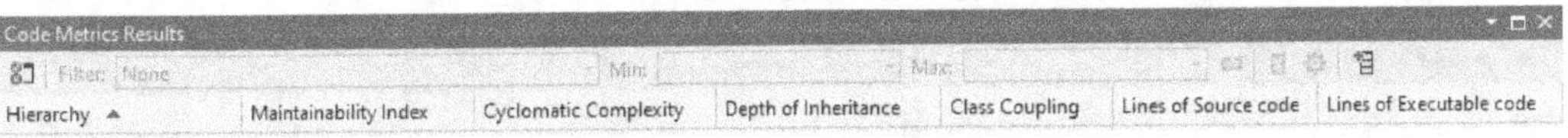

Figure 5.2 – The Code Metrics Results window

Right-click on the `CH04_Finalization` project and select **Analyze** and **Code Cleanup | Calculate Code Metrics** from the context pop-up menu. The **Code Metrics Results** window will be updated with the results of the analysis:

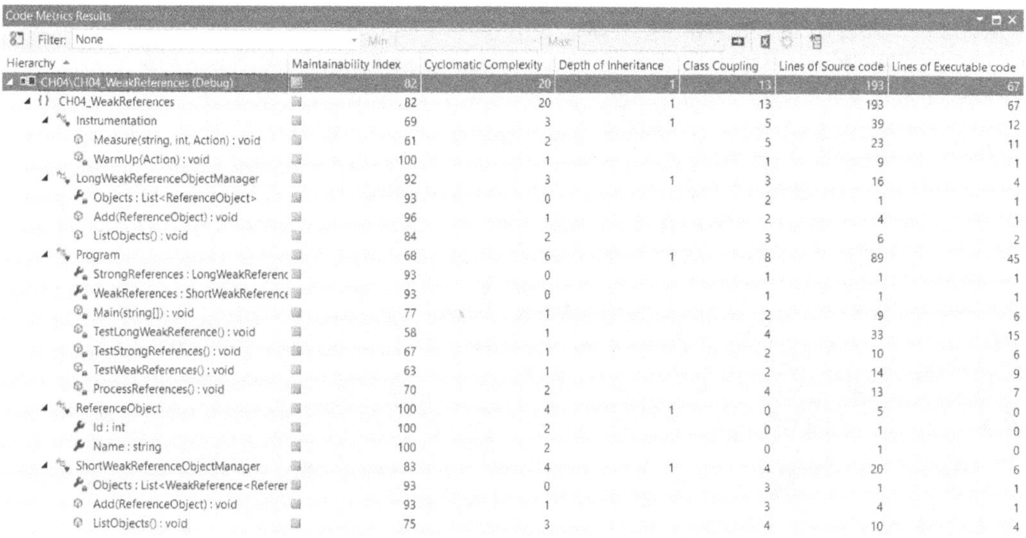

Code Metrics Results

Hierarchy	Maintainability Index	Cyclomatic Complexity	Depth of Inheritance	Class Coupling	Lines of Source code	Lines of Executable code
CH04\CH04_WeakReferences (Debug)	82	20	1	13	193	67
CH04_WeakReferences	82	20	1	13	193	67
Instrumentation	69	3	1	5	39	12
Measure(string, int, Action) : void	61	2		5	23	11
WarmUp(Action) : void	100	1		1	9	1
LongWeakReferenceObjectManager	92	3	1	3	16	4
Objects : List<ReferenceObject>	93	0		2	1	1
Add(ReferenceObject) : void	96	1		2	4	1
ListObjects() : void	84	2		3	6	2
Program	68	6	1	8	89	45
StrongReferences : LongWeakReferenc	93	0		1	1	1
WeakReferences : ShortWeakReference	93	0		1	1	1
Main(string[]) : void	77	1		2	9	6
TestLongWeakReference() : void	58	1		4	33	15
TestStrongReferences() : void	67	1		2	10	6
TestWeakReferences() : void	63	1		2	14	9
ProcessReferences() : void	70	2		4	13	7
ReferenceObject	100	4	1	0	5	0
Id : int	100	2		0	1	0
Name : string	100	2		0	1	0
ShortWeakReferenceObjectManager	83	4	1	4	20	6
Objects : List<WeakReference<Referer	93	0		3	1	1
Add(ReferenceObject) : void	93	1		3	4	1
ListObjects() : void	75	3		4	10	4

Figure 5.3 – Visual Studio 2022 Code Metrics Results for the CH04_WeakReference project

The **Code Metrics Results** window provides six code metrics that have been calculated for our project from `CH04_Finalization`.

> **Learn About the Metrics in Detail**
>
> If you want to learn more about the metrics (**Maintainability Index**, **Cyclomatic Complexity**, **Depth of Inheritance**, **Class Coupling**, **Lines of Source Code**, and **Lines of Executable code**), then you can find a dedicated chapter (*Chapter 12*) in my other book, *Clean Code in C#* (`https://www.packtpub.com/product/clean-code-in-c/9781838982973`), which is published by Packt.

From the traffic-light indicators of the **Maintainability Index** column, you can see that our project has green lights all the way. This means that our project is maintainable.

The cyclomatic complexity of our methods is between **1** and **2**, so our individual method code contains no risk. However, the overall cyclomatic complexity of our project is **31**, which is medium risk. This value is the summation of the overall cyclomatic complexity of each of the classes within our project. The cyclomatic complexity of each of our classes is the summation of the cyclomatic complexity of each of the methods. Since none of the classes have a cyclomatic complexity of more than **13**, our code is complex but only poses a low risk to our project. Because the overall complexity of the project is **31**, we should look to see whether the code can be refactored to lower the cyclomatic complexity. Sometimes, you will find that code is as simple as you can make it and that it is not possible to reduce cyclomatic complexity. That is okay. Just use your common sense and better judgment when you encounter such code.

The maximum depth of inheritance in our project is **2**. That is because our `FreeAllocateMemory` class inherits from our `DisposableBase` class, which inherits from the `System.Object` class. If we study what the `DisposableBase` class does, we can see that it will not cause us any issues.

The total number of lines of code in our project is about **200**. There are **50** lines of executable code. That's because we are making effective use of whitespace so that our code is easy to read. Easy-to-read code is easier to understand, extend, and maintain.

Open the **Error List** window by selecting **View | Error List**. Then, right-click on the project and select **Analyze** and **Code Cleanup | Run Code Analysis**. The **Error List** window will be updated with any errors, warnings, or informational messages for us to address. *Figure 5.4* shows the results of running code analysis on `CH06_Collections`:

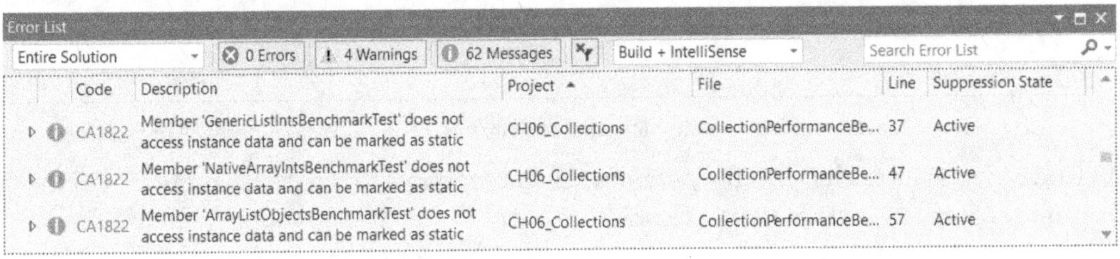

Error List

Entire Solution | 0 Errors | 4 Warnings | 62 Messages | Build + IntelliSense | Search Error List

Code	Description	Project	File	Line	Suppression State
CA1822	Member 'GenericListIntsBenchmarkTest' does not access instance data and can be marked as static	CH06_Collections	CollectionPerformanceBe...	37	Active
CA1822	Member 'NativeArrayIntsBenchmarkTest' does not access instance data and can be marked as static	CH06_Collections	CollectionPerformanceBe...	47	Active
CA1822	Member 'ArrayListObjectsBenchmarkTest' does not access instance data and can be marked as static	CH06_Collections	CollectionPerformanceBe...	57	Active

Figure 5.4 – The Visual Studio 2022 code analysis results for the CH04_Finalization project

In the preceding screenshot, we can see that we have 0 errors, 4 warnings, and 62 messages. The three informational messages inform us that three different methods do not access instance data and can be marked as static.

In the `CH04_Finalization.DisposableBase` class, we implement the `IDisposable` interface. In this class, code analysis raises two informational messages for code analysis rule CA1816. This code analysis rule informs us that the `Dispose` methods should call `SuppressFinalize`. Despite calling `GC.SuppressFinalize`, we are receiving this code analysis rule as an informational message. Therefore, to remove (suppress) the warning, we wrap the code in `#pragma` compiler directives. This can be done manually or by right-clicking on the message and selecting **Suppress | In Source**. Suppressing these messages updates the `DisposableBase` source file as follows:

```
#pragma warning disable CA1816
// Dispose methods should call SuppressFinalize
public void Dispose()
#pragma warning restore CA1816
// Dispose methods should call SuppressFinalize
{
```

```
    Dispose(true);
}

private void Dispose(bool disposing)
{
    if (disposing)
#pragma warning disable CA1816
// Dispose methods should call SuppressFinalize
        GC.SuppressFinalize(this);
#pragma warning restore CA1816
// Dispose methods should call SuppressFinalize
ReleaseManagedResources();
ReleaseUnmanagedResources();
}
```

Now that the `DisposableBase` class has been updated with these `#pragma` warning disable CA1816 statements, notice that the messages are no longer displayed in the error list.

Well, we have had a look at how to generate code metrics and run code analysis on our `CH04_Finalization` project using Visual Studio 2022. Now, let's move on to look at how to generate memory dumps and analyze them.

Generating and viewing memory dumps

When debugging in Visual Studio, if your program has stopped on a breakpoint or an exception, then the **Save Dump As** menu option becomes available in the **Debug** menu.

A minidump with a heap file provides a snapshot of an application's memory, shows the process that was running, and lists the modules that were loaded at a point in time. Dump files enable you to examine the stack, threads, and variables as they were within the application and memory at the point in time when the dump was saved.

You would save a minidump with heap files when testing software and a crash is encountered, and when a customer program crash cannot be replicated on your computer.

Let's go through the process of saving and loading a minidump with a heap file:

1. Using our `CH04_WeakReferences` project, put a breakpoint on the following line in the `program.cs` file:

   ```
   Console.WriteLine("Press any key to continue.");
   ```

2. Run the project to the breakpoint. Then, when the breakpoint is hit, select **Debug | Save Dump As**. Save the dump file to where you would like to save it. The filename will be called `CH04_WeakReference.dmp`. This file is a minidump with a heap file.
3. To read the file, select **File | Open | File**. Then, select the file you just saved. You should see the following window:

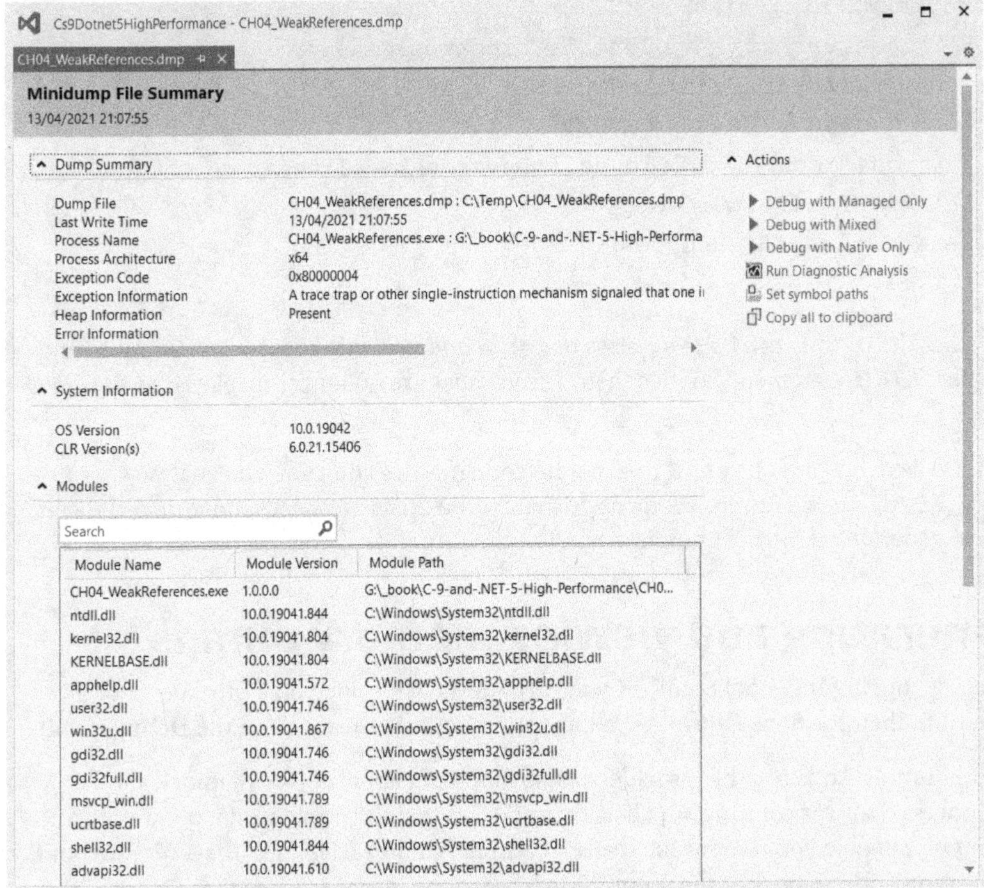

Figure 5.5 – A minidump with a heap file loaded in Visual Studio 2022

The preceding screenshot shows us that we can see the time at which the file was last updated, the process name, the computer architecture, the exception code and information, the heap information, and the error information. Then, we have the CLR and OS versions. Finally, there is a list of modules, including their names, versions, and paths.

You have just learned how to generate and read memory dumps in Visual Studio 2022. Now we will look at using the **Modules** window in Visual Studio 2022 to view what modules have been loaded by our projects.

Viewing loaded modules

To identify what might be causing performance issues such as excessive memory load, or that might be generating runtime errors, it can be useful to see what modules have been loaded into memory. In this section, you will learn how to view loaded modules and understand the items of information provided regarding those modules.

When you are debugging in Visual Studio 2022, the **Debug | Windows** menu contains the menus, as shown in *Figure 5.6*:

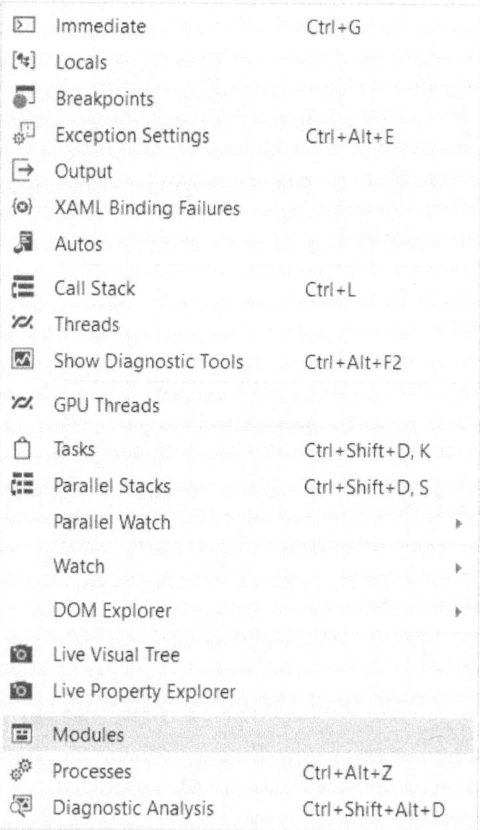

Figure 5.6 – The Windows menu during a debugging session

From the preceding menu, as shown in *Figure 5.6*, you can select **Modules** during a debugging session. This will load the **Modules** window, as shown in *Figure 5.7*:

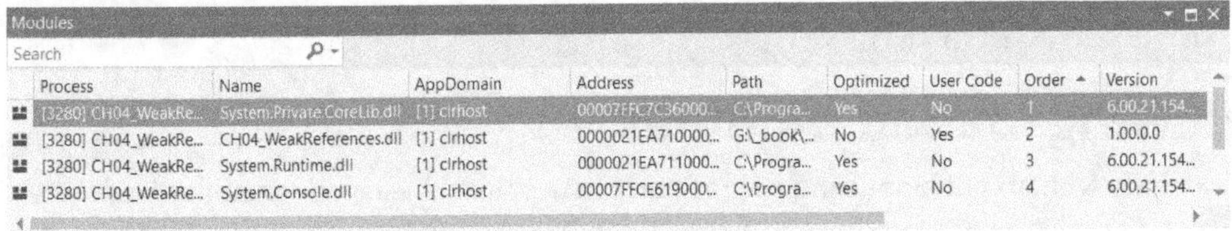

Figure 5.7 – The Modules window showing the loaded modules for the current process

As *Figure 5.7* shows, the `CH04_WeakReferences.exe` process runs in the **clrhost** AppDomain, and loads the following modules:

- `System.Private.CoreLib.dll`
- `CH04_WeakReference.dll`
- `System.Runtime.dll`
- `System.Console.dll`

The list of fields that are displayed in the **Modules** window is as follows:

- **Name**: The name of the loaded assembly (loaded module)
- **Path**: The path to the loaded module
- **Optimized**: Yes/no
- **User Code**: Yes/no
- **Symbol Status**: Skipped loading symbols/symbols loaded
- **Symbol File**: The path and filename of the loaded symbol file
- **Order**: The order of assembly loading
- **Version**: The assembly version
- **Address**: The memory address of the loaded module
- **Process**: The process identifier and executable name responsible for causing the modules to be loaded into memory
- **AppDomain**: The name of the application domain that the module is running under. This doesn't have any meaning in .NET Core and .NET 5 or higher. It is displayed because the debugger UI does not make the distinction between the .NET Framework and .NET Core.

You can use this information to see what modules are loaded, whereabouts they reside in memory, whether the symbols have been loaded, whether the code is system code or user code, and whether the code is optimized or not optimized. If you find user code that has not been optimized, then you can apply optimizations to improve performance.

In the next section, we will look at how to further debug your applications by briefly covering the tools available to you that you should already be familiar with.

Debugging your applications

It is assumed that you know how to debug your code by running through your code, stepping out and stepping over the code, running to the cursor, and setting breakpoints. However, there are other useful tools available when using the debugger. These include the following:

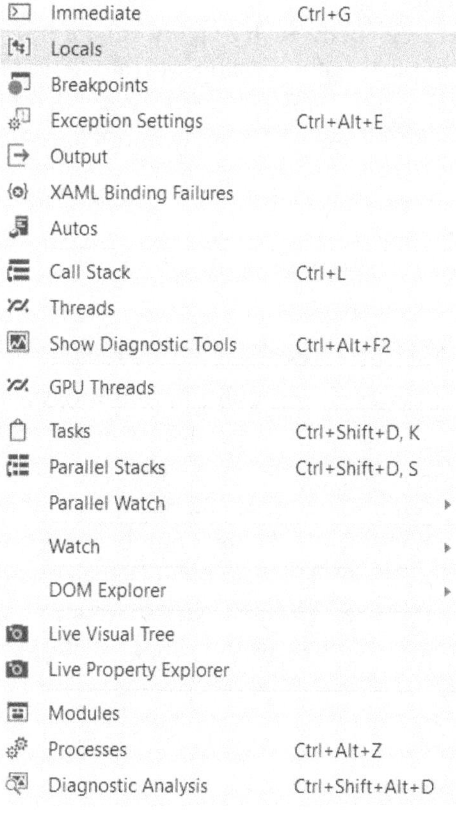

Figure 5.8 – The Debug | Windows menu

As you can see, there are a good number of different windows available to help debug your applications. The **Immediate** window is very good for executing commands when your program is paused. The **Locals** window is good for seeing the present state of your variables, and the call stack is useful for finding where an exception occurred, especially if it is in close code that is not yours! Take the time to run through your source code with these windows open. Different windows such as **XAML Binding Failures** are only used when working on the XAML-based code. But other windows, such as **Immediate**, **Locals**, **Output**, **Autos**, and **Call Stack**, can be used with all project types. The best way to get the most out of these tools is to use them for yourself and get to know them as you work through your code. Next, we will look at using tracing and diagnostics tools.

Using tracing and diagnostics tools

In this section, we will look at some profiling tools to help you trace and diagnose any issues with your code. By tracing and diagnosing your program, you can identify areas of performance concern and address them. Such concerns might be the number of memory allocations and the number of bytes they are using and identifying the number of objects surviving garbage collection. Such information can be useful in improving memory usage and performance and in preventing and removing memory leaks.

We will look at two offerings from JetBrains, called **dotMemory** and **dotTrace**, that are valuable tools in this respect. But first, we will start by looking at the built-in profiler that comes with Visual Studio 2022 called **Performance Profiler**.

Using the Visual Studio 2022 Performance Profiler

Now we are going to view the performance profile for our project. This will show us the number of objects over time and the way garbage collection is being utilized in our project, along with the number of objects that survive garbage collection. We can drill down on this profile to the assembly and method levels. This enables us to see the number of object allocations within a method and the total number of bytes those allocations use up. And because of this information, we can identify the areas of our program that generate the most memory usage. With such information, we can consider heavy allocation code for refactoring to improve memory performance.

To access the Visual Studio 2022 Performance Profile, select **Performance Profiler** from the Visual Studio 2022 **Debug** menu. This will bring up a tab, as shown in *Figure 5.9*:

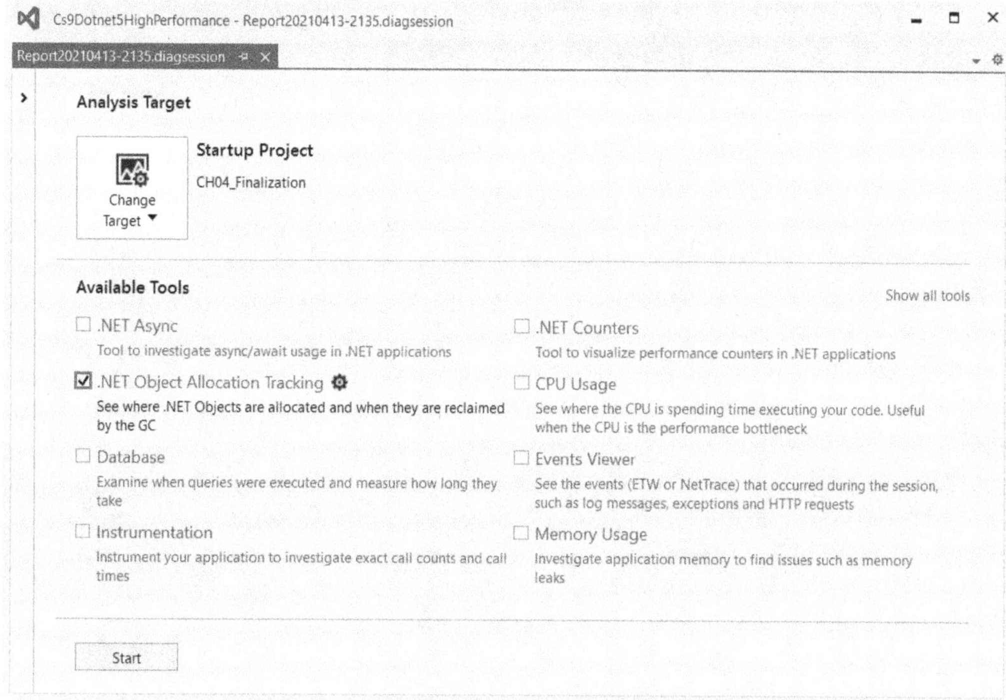

Figure 5.9 – The Visual Studio 2022 Performance Profiler

Now, we will run an analysis on the `CH04_Finalization` project:

1. Select your startup project.
2. Then, select the tool that you want to use. In our case, we have selected `CH04_Finalization`. And the tool we have selected is the tool for tracking .NET object allocations. This enables us to see where the .NET objects are allocated and when they are reclaimed.

3. Click on the **Start** button to start profiling the application. The profiler will run and then stop when the code stops. You will see a report similar to the one in *Figure 5.10*:

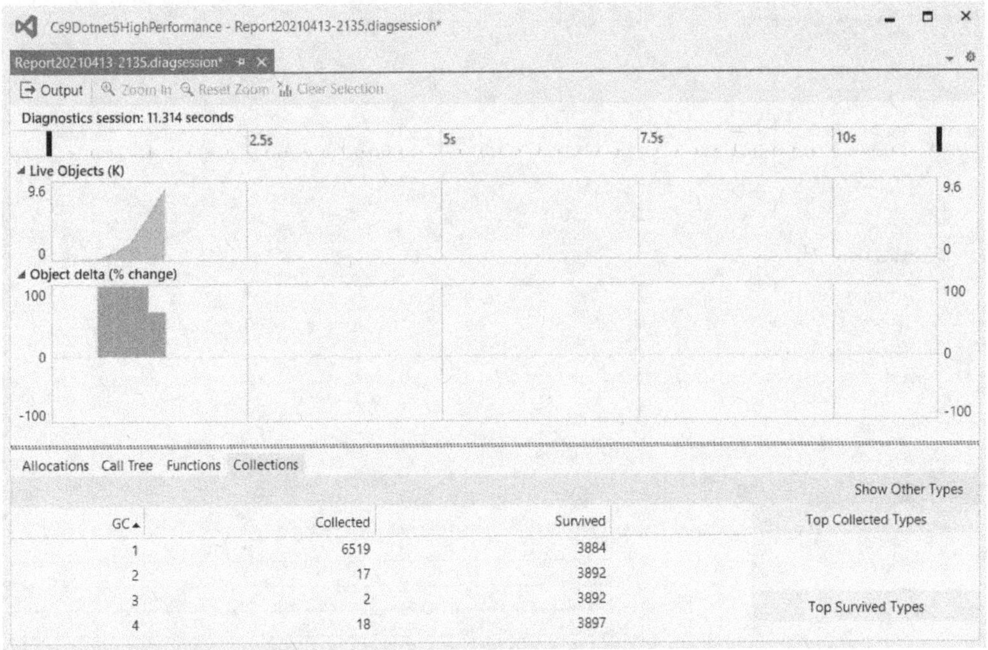

Figure 5.10 – The complete Visual Studio 2022 Performance Profiler report showing live objects over time

The main chart area shows the number of live objects over time. There are also four tabs that contain **Allocations, Call Tree, Functions**, and **Collections** data.

4. On the **Allocations** tab, you can see the types used and the number of their allocations. Clicking on a type brings up the **Backtrace** for that type. You can see the number of allocations for that type and the number of bytes allocated in your functions, as shown in *Figure 5.11*:

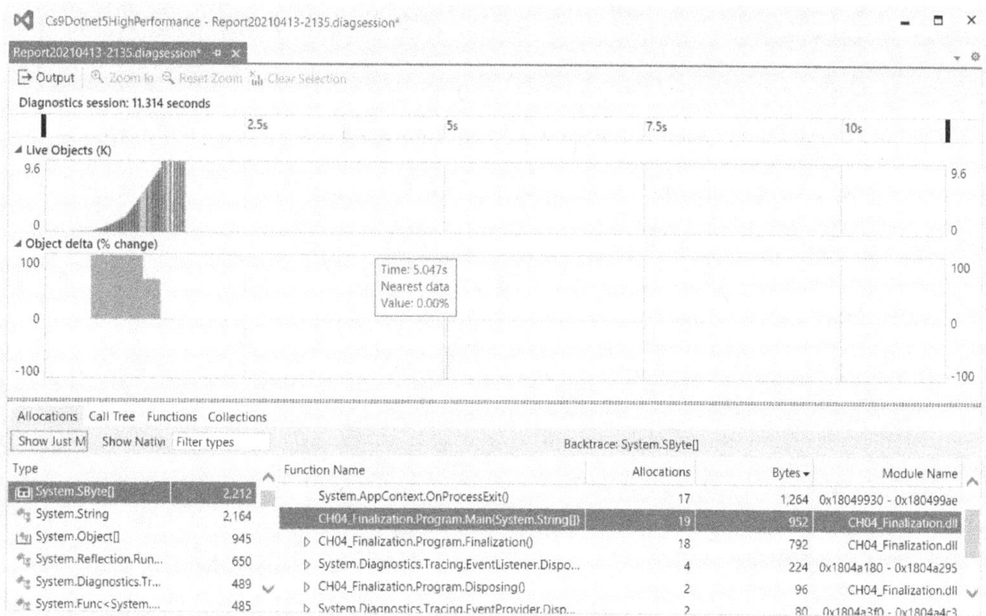

Figure 5.11 – The Visual Studio 2022 Performance Profiler allocations of System.Sbyte[]

In *Figure 5.11*, we can see that in our `Main` method, there are 19 allocations of the `System.Sbyte[]` type with an allocation size of **952** bytes.

5. Select the **Call Tree** tab. Showing just our code and the hot path with the hot path expanded, we can see that in the `DisplayGeneration(Product product)` method, there is one `System.Int32` allocation that is **24** bytes in size, as shown in *Figure 5.12*:

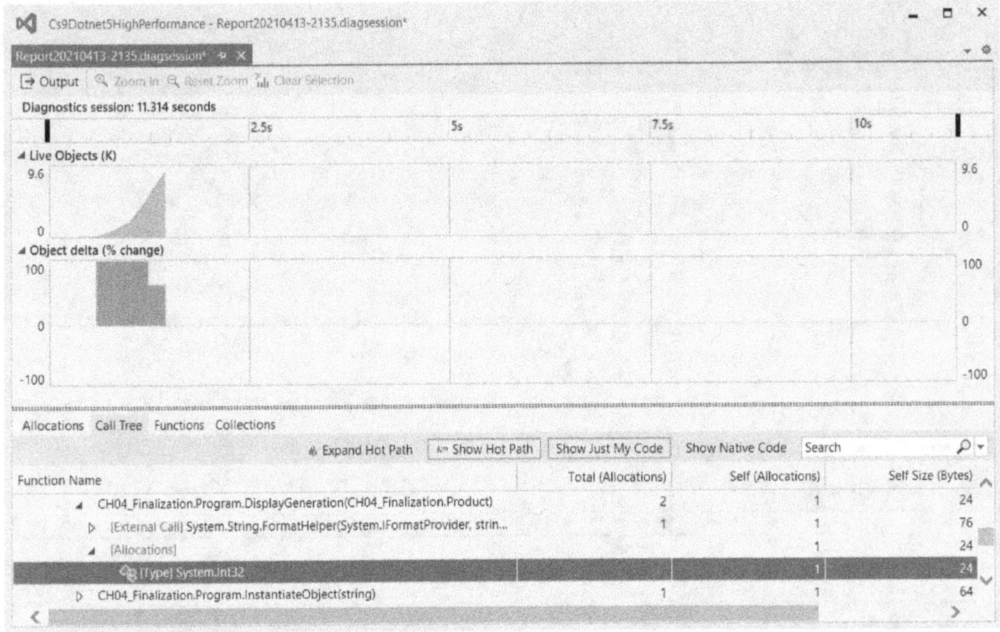

Figure 5.12 – The Visual Studio 2022 Performance Profiler Call Tree tab

6. Select the **Functions** tab. You will see that the `Main` method has a total of **347** allocations, **27** self-allocations, and is a total of **1,438** bytes in size, as shown in *Figure 5.13*:

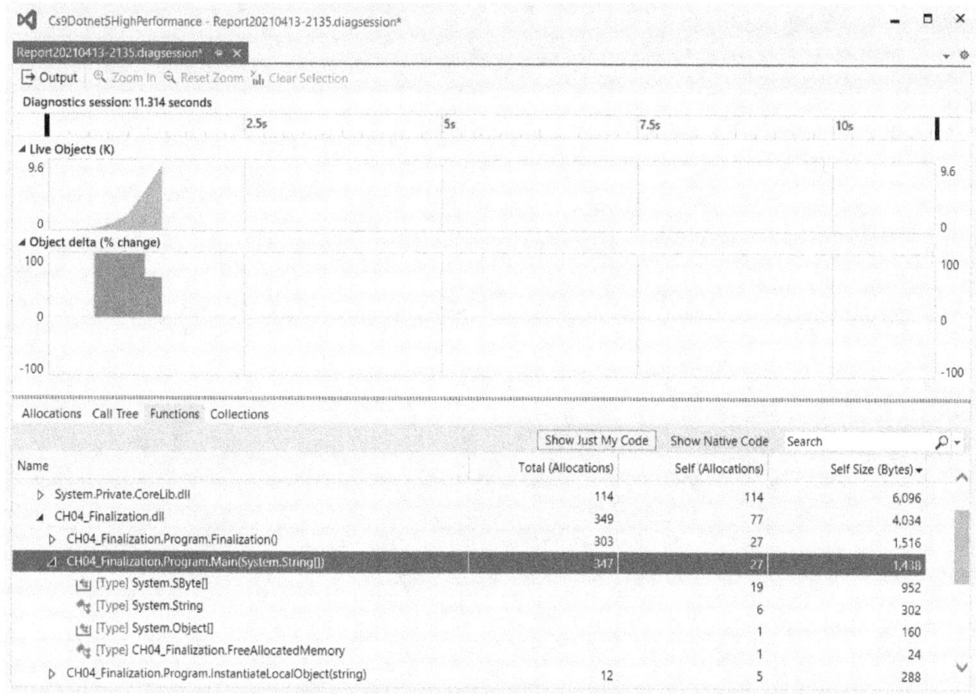

Figure 5.13 – Visual Studio 2022 Performance Profiler Functions tab showing allocations and sizes for various methods

7. Click on the **Collections** tab. Then, click on a row. You will see two pie charts for the top collected types and top survived types, as shown in *Figure 5.14*:

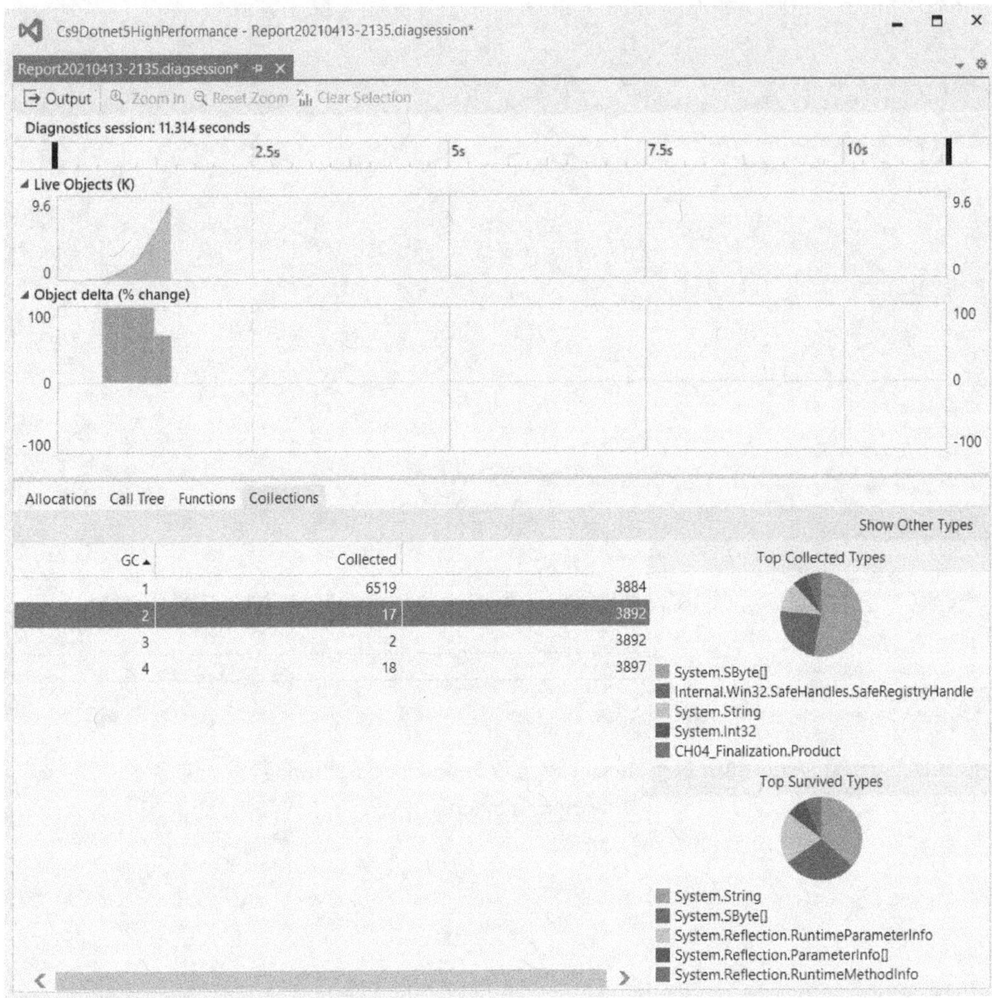

Figure 5.14 – Visual Studio 2022 Performance Profiler showing a breakdown of the garbage collection

In *Figure 5.14*, we can see the number of live objects over time along with the object delta (% change). Additionally, we can see the top collected types and top survived types in the two pie charts.

The Visual Studio 2022 Performance Profiler is a very useful tool that enables you to view allocations, byte sizes, and garbage collected and survived objects. You can also see the number of live objects over time. Now that you have been introduced to the profiler and know what it is capable of, let's move our attention to the JetBrains tool called **dotMemory**.

Using JetBrains dotMemory

We use dotMemory to profile and optimize memory and to help us identify memory leaks and other memory-related issues. In this section, we will be discussing the JetBrains dotMemory memory profiler.

The memory profiler will provide a chart with milliseconds on the *x* axis and megabytes on the *y* axis, which shows your application's memory usage over time. The following list of items is displayed on the chart:

- **Total used**: The total amount of memory used.
- **Unmanaged memory**: The total amount of memory placed on the stack.
- **Heap generation 0**: The amount of memory taken up by new objects. These objects will be less than 80,000 bytes in size.
- **Heap generation 1**: The objects that survive generation 0 garbage collection.
- **Heap generation 2**: Long-lived objects that survive level 1 garbage collection.
- **Large object heap (LOH)**: The amount of memory used by objects that are 80,000 bytes or larger in size.
- **Allocated in LOH since GC**: The amount of memory used on the LOH after garbage collection has taken place.

Let's see the dotMemory memory profiler in action. If you have not already done so, download and install dotMemory from JetBrains and the code for `chapter 4` from the GitHub page. Open dotMemory, and you will be presented with a screen similar to the one shown in *Figure 5.15*:

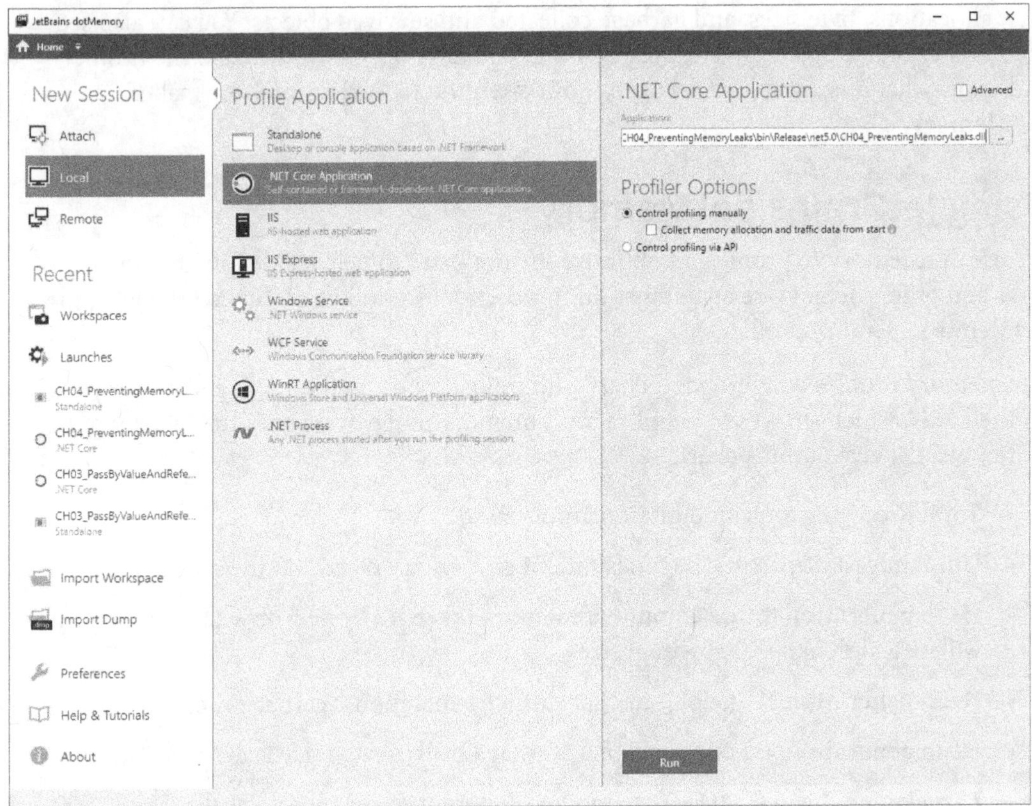

Figure 5.15 – The dotMemory Memory Profiler ready to profile .NET Core Application

In *Figure 5.15*, we have selected to profile **.NET Core Application**. The application selected for profiling is `CH04_PreventingMemoryLeaks.dll`. Click on the **Run** button. This will enable the profiler to start running and profiling your application. Once the application has been profiled, a report will be displayed showing the results in graphical form, as shown in *Figure 5.16*:

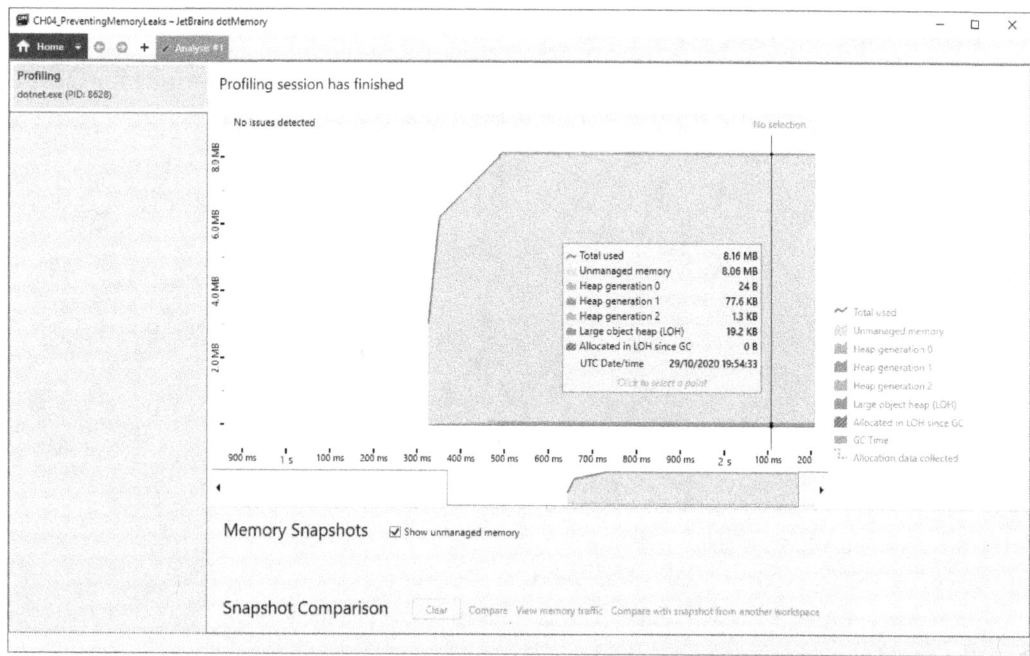

Figure 5.16 – The profile report for CH04_PreventingMemoryLeaks.dll

As you can see from the preceding screenshot, our application uses a total of **8.16 MB** of memory. This is not that much. Most of the memory is placed on the stack, as shown by the unmanaged memory usage at **8.06 MB**. The rest of the memory is on the heap. On the heap, **24 KB** has been allocated on generation **0**, **77.6 KB** has been allocated on generation **1**, and **1.3 KB** has been allocated on generation **2**. The most heap memory, **19.2 KB**, was placed on the LOH and did not remain after garbage collection.

Having seen the dotMemory tool in action, we can now turn our attention to what the JetBrains dotTrace tool has to offer us in terms of tracing and profiling.

Using JetBrains dotTrace

In this section, we will be looking at JetBrains dotTrace. You will learn how to use the JetBrains dotTrace tool to perform application tracing at runtime on your programs. This will help you to identify bottlenecks and memory issues in your executable programs.

The profiler options available in dotTrace include the following:

- **Sampling**: An accurate measurement of call time. This is optimal for most use cases.
- **Tracing**: An accurate measurement of call number. This is optimal for analyzing algorithm complexity.

- **Line-byline**: Advanced use cases only.
- **Timeline**: The measurement of temporal performance data. This is optimal for most use cases, including the analysis of multithreaded applications:

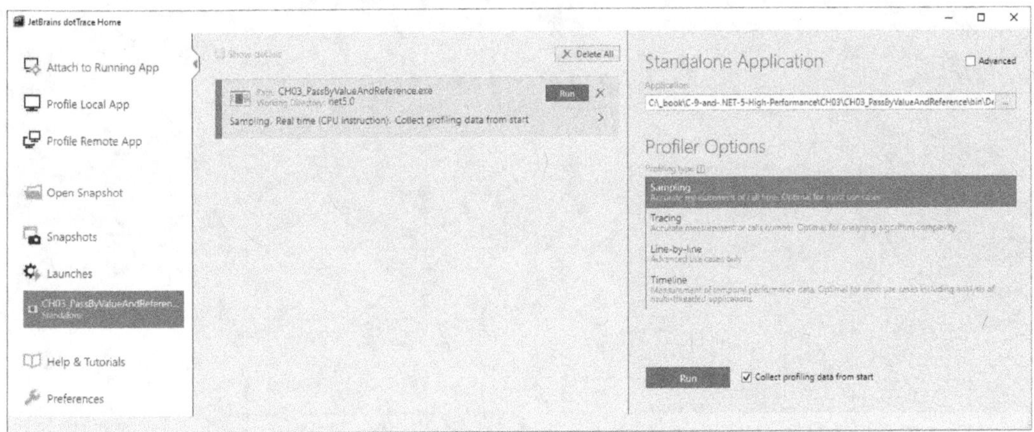

Figure 5.17 – JetBrains dotTrace ready to profile our application

Figure 5.17 shows the initial state of dotTrace. We have selected **CH03_PassByValueAndReference.exe** as our application to profile. And for our profiling option, we have selected to go with the default **Sampling** setting. Make sure that **Collect profiling data from start** is selected. Then, click on the **Run** button to start tracing.

When the tracing has been completed, the dotTrace Performance Viewer will automatically open, as shown in *Figure 5.18*:

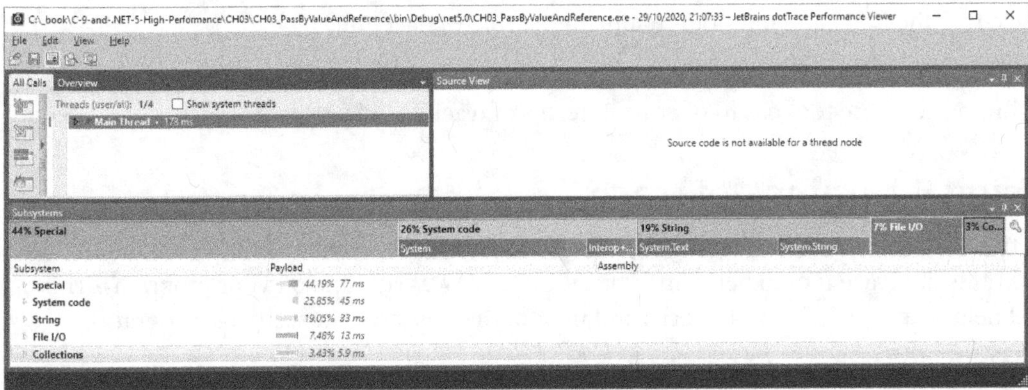

Figure 5.18 – JetBrains dotTrace Performance Viewer

The outcome of profiling the `CH03_PassByValueAndReference.exe` file is shown in the default view of *Figure 5.18*. If you click on the **Hot spots** icon and highlight the `Main` line, you will see the program code. The breakdown of the `Main` method shows that 19 ms (43.20%) of time was spent executing system code, 13 ms (29.56%) of time was spent performing File I/O, and 12 ms (27.24%) of time was executing the **String** subsystem, as shown in *Figure 5.19*:

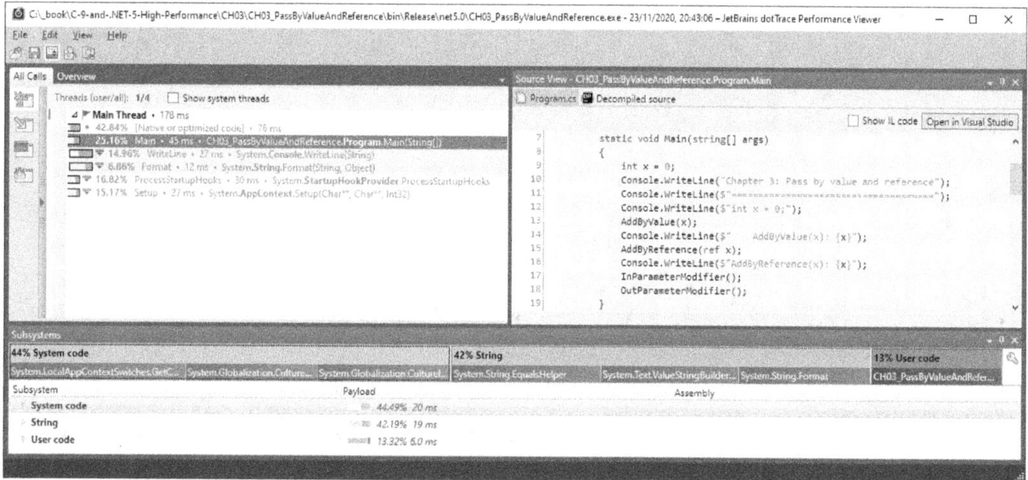

Figure 5.19 – Breakdown of the main method

Figure 5.19 shows the `Main` method source code and the fact that between `Main` and `InParameterModifier`, the `Main` method takes the most time to process. This information can be helpful to identify and work with bottlenecks.

We have seen two tools for memory profiling and tracing that can be used to measure performance and identify bottlenecks and problems. Now, let's move our attention to installing and using `dotnet-counters`.

Installing and using dotnet-counters

In this section, we will install and use `dotnet-counters`. These counters are very useful data-gathering tools that help us to monitor the health of our programs.

Open Developer Command Prompt for Visual Studio 2022. Then, type in the following command and press *Enter*:

```
dotnet tool install --global dotnet-counters --version 3.1.141901
```

This will download and install dotnet-tools. A successful installation will be presented, as shown in *Figure 5.20*:

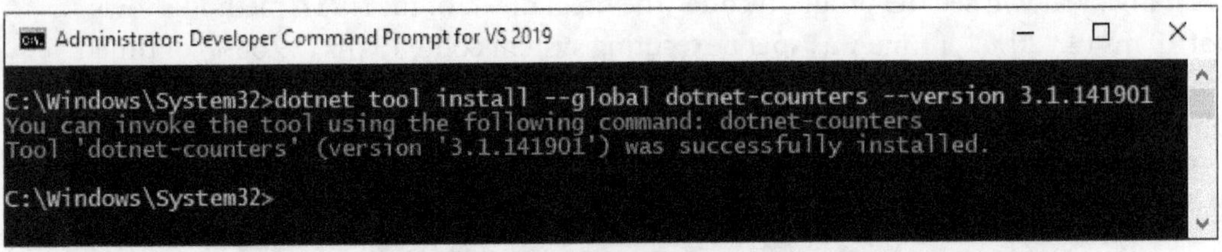

```
Administrator: Developer Command Prompt for VS 2019

C:\Windows\System32>dotnet tool install --global dotnet-counters --version 3.1.141901
You can invoke the tool using the following command: dotnet-counters
Tool 'dotnet-counters' (version '3.1.141901') was successfully installed.

C:\Windows\System32>
```

Figure 5.20 – The successful installation of dotnet-tools version 3.1.141901 using Developer Command Prompt

The purpose of using `dotnet-counters` is to perform health monitoring and a first-level performance investigation of your applications. If when using this program, potential performance problems are identified, then you can perform a more serious performance investigation using tools such as PerfView or dotnet-trace:

- To periodically collect selected counter values and export them to a file for post-processing, use the `dotnet-counters collect` command.
- The `dotnet-counters list` command displays a list of the counter names and descriptions that are grouped by the provider.
- And to display a list of .NET processes that can be monitored, you can use the `dotnet-counters ps` command.
- Using the `dotnet-counters monitor` command, you can display periodically refreshed values for selected counters.

To get a list of the available options for each command, append `-h` or `-help`. Let's put each of those commands to use. And before we do, add the following lines to the end of the `CH04_WeakRefereces` `Main` method in the `Program` class:

```
Console.WriteLine("Press any key to continue.");
Console.ReadKey();
```

Run the program. It will pause and wait for you to press a key before it continues.

Collecting data and saving it to a file for post-analysis

Now we will use `dotnet-counters` to save data to a file that we can analyze once our program has finished running:

1. Remove the breakpoint of `CH04_WeakReferences` in the `Program` class.
2. Update the `ProcessReferences()` method in the `Program` class as follows:

```
private static void ProcessReferences()
{
int x = 0;
while(x < 10000)
{
    StrongReferences.ListObjects();
    WeakReferences.ListObjects();
    Thread.Sleep(2000);
    GC.Collect();
    x++;
}
}
```

3. Add a breakpoint to the `while (x < 10000)` loop.
4. Then, run the program. Running the program will require some time – approximately 10,000 iterations x 2 seconds = 5.5h.
5. When the program stops on the breakpoint added in *step 3*, open Command Prompt as an admin and type in `dotnet-counters ps` followed by *Enter*. If you don't run as an admin, you will encounter counter access errors.
6. Obtain the process ID for the program.
7. Change the directory in Command Prompt to point to `C:\Temp`. Create the directory if it does not exist.
8. Enter the `dotnet-counters collect --process-id 1234` command (replace **1234** with the ID of your .NET process) followed by *Enter*.
9. The performance data will now be collected.

10. Remove the breakpoint added in *step 3* and continue the program. When you have let the program run a little while, press the *q* key. Your Command Prompt screen should look similar to *Figure 5.21*:

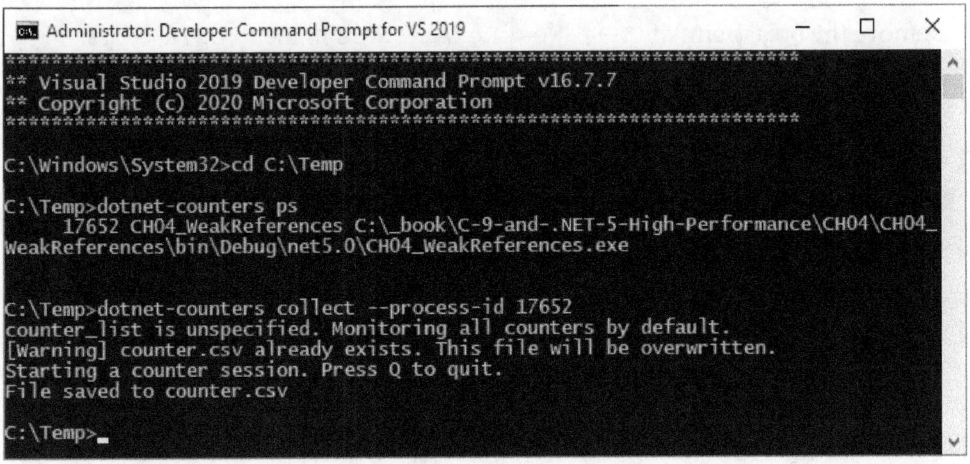

```
Administrator: Developer Command Prompt for VS 2019
**********************************************************************
** Visual Studio 2019 Developer Command Prompt v16.7.7
** Copyright (c) 2020 Microsoft Corporation
**********************************************************************

C:\Windows\System32>cd C:\Temp

C:\Temp>dotnet-counters ps
     17652 CH04_WeakReferences C:\_book\C-9-and-.NET-5-High-Performance\CH04\CH04_
WeakReferences\bin\Debug\net5.0\CH04_WeakReferences.exe


C:\Temp>dotnet-counters collect --process-id 17652
counter_list is unspecified. Monitoring all counters by default.
[Warning] counter.csv already exists. This file will be overwritten.
Starting a counter session. Press Q to quit.
File saved to counter.csv

C:\Temp>_
```

Figure 5.21 – The Developer Command Prompt having completed a collection

11. Open the file called `C:\Temp\counter.csv` in **Excel**. *Figure 5.22* shows an excerpt of the data contained within the spreadsheet:

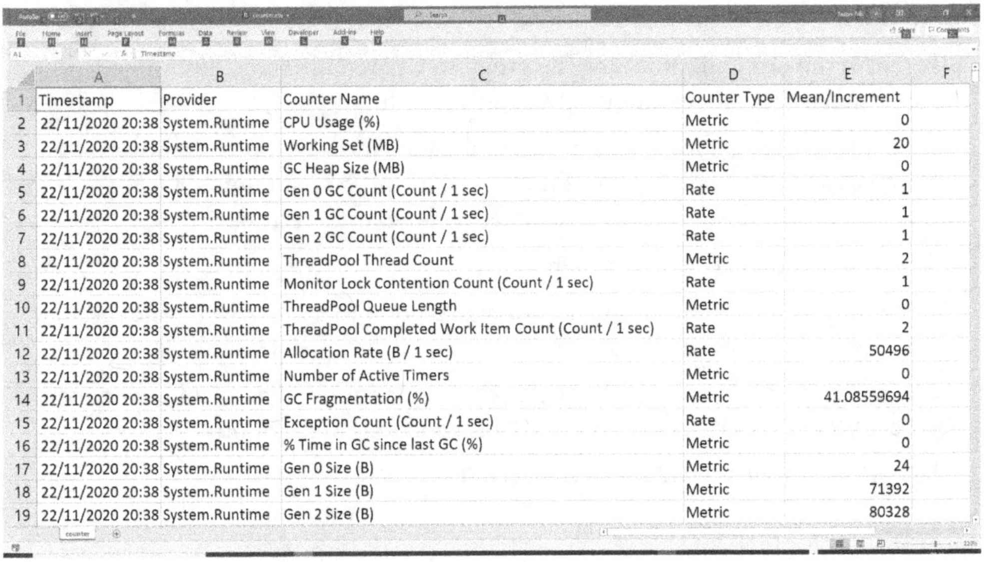

	A	B	C	D	E
1	Timestamp	Provider	Counter Name	Counter Type	Mean/Increment
2	22/11/2020 20:38	System.Runtime	CPU Usage (%)	Metric	0
3	22/11/2020 20:38	System.Runtime	Working Set (MB)	Metric	20
4	22/11/2020 20:38	System.Runtime	GC Heap Size (MB)	Metric	0
5	22/11/2020 20:38	System.Runtime	Gen 0 GC Count (Count / 1 sec)	Rate	1
6	22/11/2020 20:38	System.Runtime	Gen 1 GC Count (Count / 1 sec)	Rate	1
7	22/11/2020 20:38	System.Runtime	Gen 2 GC Count (Count / 1 sec)	Rate	1
8	22/11/2020 20:38	System.Runtime	ThreadPool Thread Count	Metric	2
9	22/11/2020 20:38	System.Runtime	Monitor Lock Contention Count (Count / 1 sec)	Rate	1
10	22/11/2020 20:38	System.Runtime	ThreadPool Queue Length	Metric	0
11	22/11/2020 20:38	System.Runtime	ThreadPool Completed Work Item Count (Count / 1 sec)	Rate	2
12	22/11/2020 20:38	System.Runtime	Allocation Rate (B / 1 sec)	Rate	50496
13	22/11/2020 20:38	System.Runtime	Number of Active Timers	Metric	0
14	22/11/2020 20:38	System.Runtime	GC Fragmentation (%)	Metric	41.08559694
15	22/11/2020 20:38	System.Runtime	Exception Count (Count / 1 sec)	Rate	0
16	22/11/2020 20:38	System.Runtime	% Time in GC since last GC (%)	Metric	0
17	22/11/2020 20:38	System.Runtime	Gen 0 Size (B)	Metric	24
18	22/11/2020 20:38	System.Runtime	Gen 1 Size (B)	Metric	71392
19	22/11/2020 20:38	System.Runtime	Gen 2 Size (B)	Metric	80328

Figure 5.22 – An excerpt from counter.csv

As you can see, there are various items that are recorded by the `dotnet-counters` collect process. These items include CPU usage, garbage collection data, heap information, exception information, the number of loaded assemblies, and JIT compilation information.

Listing .NET processes that can be monitored

To list .NET processes that can be monitored, open the Developer Command Prompt screen and type in the `dotnet-counters ps` command. You should see an output similar to the following:

```
Select Administrator: Command Prompt
Microsoft Windows [Version 10.0.19041.572]
(c) 2020 Microsoft Corporation. All rights reserved.

C:\WINDOWS\system32>dotnet-counters ps
      5364 CH04_WeakReferences C:\_book\C-9-and-.NET-5-High-Performance\
CH04\CH04_WeakReferences\bin\Release\net5.0\CH04_WeakReferences.exe


C:\WINDOWS\system32>
```

Figure 5.23 – The list of .NET processes that can be monitored

As *Figure 5.23* shows, the only process that can be monitored is process **5364**. Process **5364** is the program that we are currently debugging. If more .NET programs were running, then more would appear on this list.

Listing the available list of well-known .NET counters

To list the available .NET counters, run the following command:

```
dotnet-counters list
```

You will see a list of counters and their descriptions output to the console. For `Microsoft.AspNetCore.Hosting`, the available counters are listed as follows:

- **requests-per-second**: The request rate
- **total-requests**: The total number of requests
- **current-requests**: The current number of requests
- **failed-requests**: The failed number of requests

The available well-known counters for `System.Runtime` are listed as follows:

- **cpu-usage**: The amount of time the process has utilized the CPU in milliseconds
- **working-set**: The amount of working set used by the process in megabytes
- **gc-heap-size**: The total heap reported by the garbage collector in megabytes
- **gen-0-gc-count**: The number of generation 0 garbage collections per minute
- **gen-1-gc-count**: The number of generation 1 garbage collections per minute
- **gen-2-gc-count**: The number of generation 2 garbage collections per minute
- **loh size**: Large object heap size
- **alloc-rate**: The number of bytes allocated in the managed heap per second
- **assembly-count**: The number of assemblies loaded
- **exception-count**: The number of exceptions per second
- **threadpool-thread-count**: The number of thread pool threads
- **monitor-lock-contention-count**: The number of times there were contentions when trying to take the monitor lock per second
- **threadpool-queue-length**: The number of work items in the thread pool queue
- **threadpool-completed-items-count**: The number of completed work items in the thread pool
- **active-timer-count**: The number of timers that are currently active

Monitoring a .NET process

We are going to run our `CH04_WeakReferences` project. Once you have the project running, run the following command to get the process ID:

```
dotnet-counters ps
```

Then, once you have the process ID for your .NET program, run the following command:

```
dotnet-counters monitor -process-id 6719
```

For me, the process has an ID of **6719**. Replace **6719** with whatever your process ID is. The result should be that you see the .NET counters being displayed and updated in real time, as shown in *Figure 5.24*:

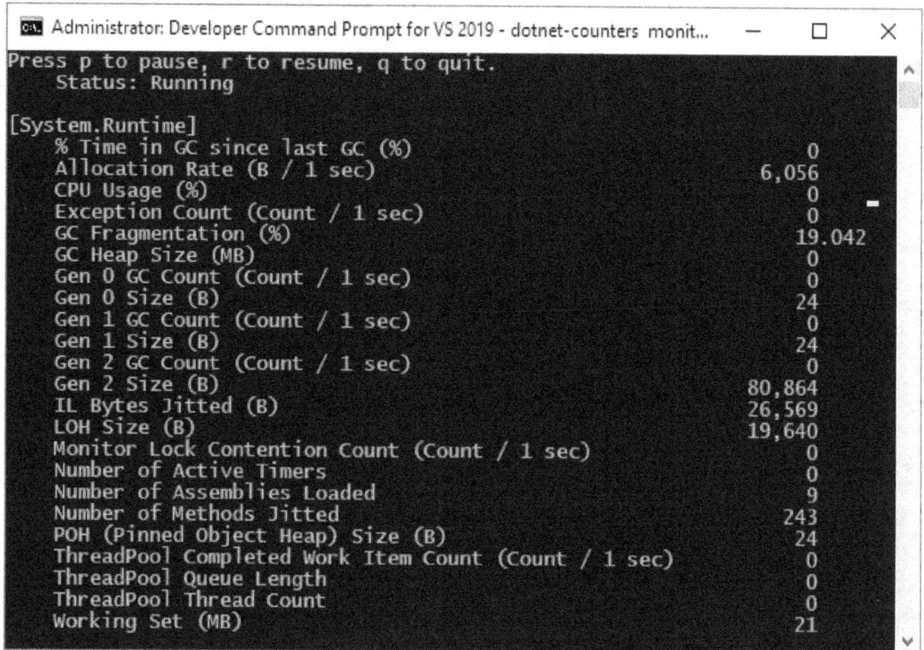

Figure 5.24 – The dotnet-counters being listed and updated in real time for our CH04_WeakReferences project

Press *q* to quit. As you can see, we have **19.042%** garbage collection fragmentation. There are **19,640** bytes on the LOH, and **80,864** bytes are assigned to generation **2**. We have **9** assemblies loaded and **24** bytes allocated to generation **0** and generation **1**. We have observed that memory fragmentation has occurred at **19.042%**, so this can be investigated further to see why we have fragmentation and to see whether we can avoid this.

In the next section, we are going to look at an example that tracks down a memory leak in a WPF application.

Tracking down and fixing a memory leak with dotMemory

In this section, we are going to run through an example of how to track down and fix memory leaks. A **memory leak** occurs when objects become inaccessible and remain in memory without being garbage collected. As the number of objects builds up, memory runs out and you end up with an `OutOfMemoryException` exception being thrown by the application.

Our example will be a WPF application called `CH05_GameOfLife`. To save time and space, download the source code for the WPF application. This will help you to focus on the task at hand, which is to track down the memory leak and fix it.

> **Note**
> When profiling and tracing, you are better off building your projects using **Release** mode. The reason for this is that **Debug** builds contain compiler instructions that might affect profiling results.

Perform the following steps:

1. Download and compile the `CH05_GameOfLife` project in **Release** mode.
2. Open **dotMemory**. The version used in this example is **2020.3.4**
3. Under **New Session**, select **Local**. Then, under **Profile Application**, select **.NET Core Application**. Select the **CH05_GameOfLife.exe** file under **.NET Core Application**, and for the **Profiler Options**, select **Collect memory allocation and traffic data from the start**. *Figure 5.25* shows dotMemory prepared to profile our application:

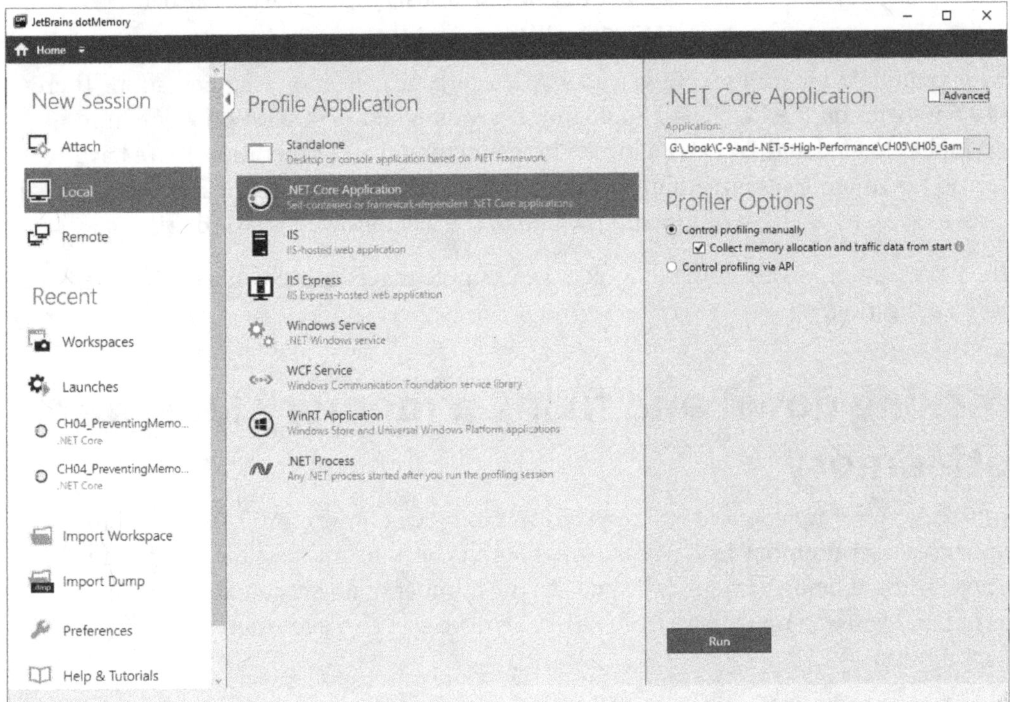

Figure 5.25 – dotMemory ready to profile our .NET 6.0 application CH05_GameOfLife.exe

4. Click **Run** to start profiling our application. You will see a new **Analysis** tab appear in dotMemory, as shown in *Figure 5.26*:

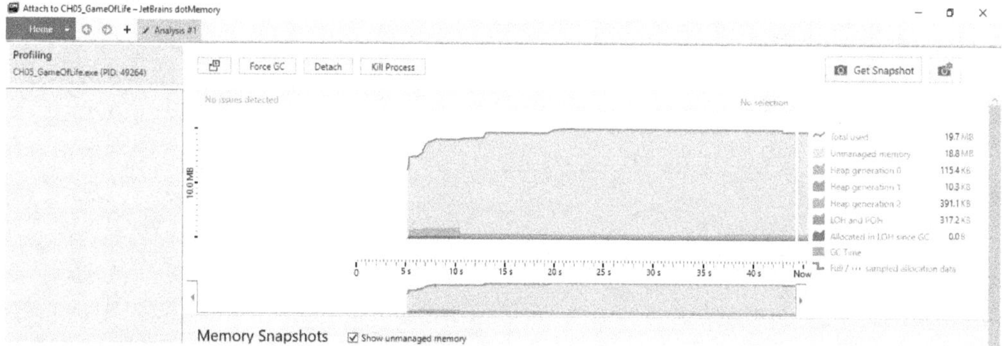

Figure 5.26 – dotMemory displaying the Analysis tab during the profiling of our app

5. When the profiler starts, it also starts our application. Click on the **Start** button of our application, as shown in *Figure 5.27*:

Figure 5.27 – Running CH05_GameOfLife

6. After *Game of Life* has been running for a while, click on the **Get Snapshot** button to take a memory snapshot. This will capture the application's managed heap at that moment in time.

7. Close the advert.
8. Take another snapshot so that we have two snapshots. Then, close the *Game of Life* application to stop the profiler. *Figure 5.28* shows the dotMemory **Analysis** tab with both snapshots taken:

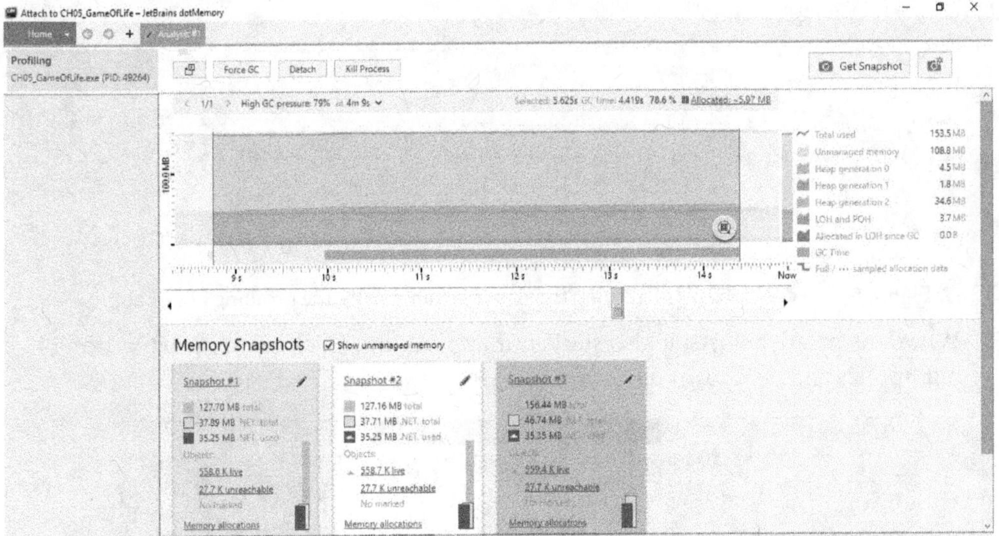

Figure 5.28 – The dotMemory Analysis tab displaying both memory snapshots

9. The next step is for us to compare the two different snapshots. *Figure 5.29* shows a close-up of the two snapshots side by side:

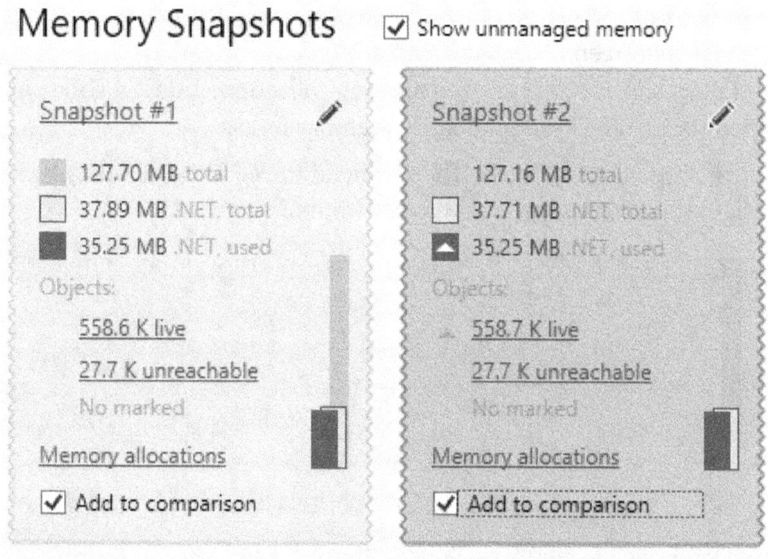

Figure 5.29 – dotMemory snapshots 1 and 2

10. Click on **Compare** to open the detailed side-by-side comparison of the two snapshots. You should see the comparison, as shown in *Figure 5.30*:

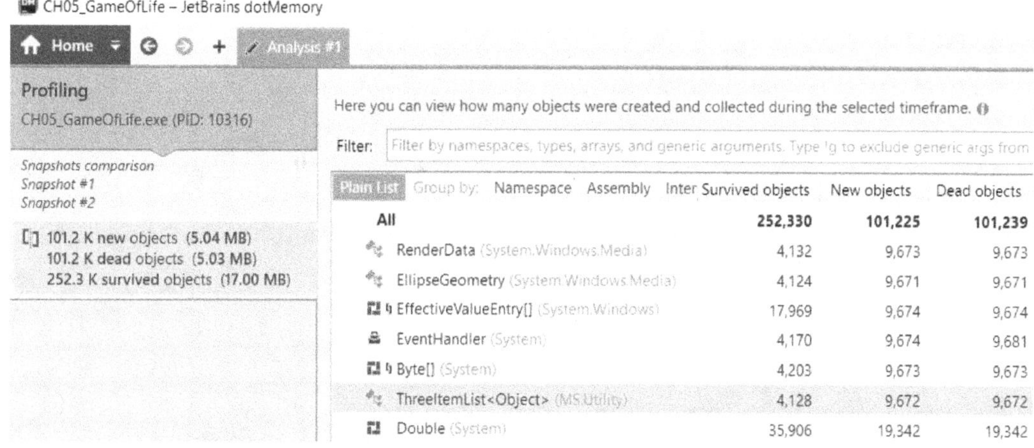

Figure 5.30 – The side-by-side snapshot comparison screen

As you can see, this view shows the number of new objects created, the number of objects that have been collected (dead objects) by the garbage collector, and the number of objects that have survived garbage collection. This is a good source of information that can be used to identify memory leaks.

11. Click on the **Namespace** column. Then, expand the **CH05_GameOfLife** namespace and highlight the **AdWindow** entry, as shown in *Figure 5.31*:

Plain List Group by: Namespace Assembly Interface	Survived objects	New objects	Dead objects	Objects delta	Survived bytes	New bytes	Dead bytes	Bytes delta
All	**249,621**	**103,951**	**103,902**	**49**	**17,650,871**	**5,460,726**	**5,424,282**	**36,444**
{} System	161,748	80,132	80,060	72	13,977,999	4,535,530	4,500,458	35,072
{} MS	73,930	10,024	10,047	-23	3,004,360	483,756	482,384	1,372
{} CH05_GameOfLife	13,801	13,795	13,795		663,904	441,440	441,440	
Cell (CH05_GameOfLife)	13,795	13,795	13,795		441,440	441,440	441,440	
App (CH05_GameOfLife)	1				208			
MainWindow (CH05_GameOfLife)	1				720			
Grid (CH05_GameOfLife)	1				56			
Cell[,] (CH05_GameOfLife)	2				220,800			
AdWindow (CH05_GameOfLife)	1				680			

Figure 5.31 – The analysis by Namespace with CH05_GameOfLife highlighted

12. In the **Survived objects** column, click on number **1** in the **AdWindow** row. This will bring up the dialog, as shown in *Figure 5.32*:

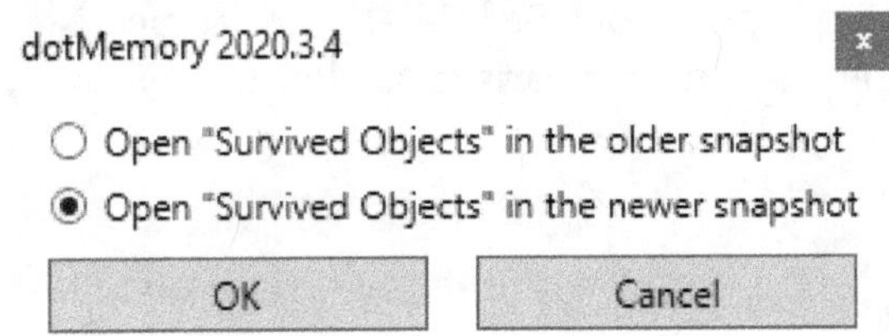

Figure 5.32 – dotMemory dialog prompting the opening of a snapshot

13. Select the newer snapshot option.
14. Then, click on the **Key Retention Paths** tab. The JetBrains dotMemory view will change to a view that is similar to *Figure 5.33*:

Figure 3.33 – The Key Retention Paths tab

You can see that `EventHandler` is keeping `AdWindow` alive, and `EventHandler` is referenced by the `DispatcherTimer` class. The `DispatcherTimer` class is referenced by the `Tick` event.

15. Click on the `DispatcherTimer` box. This will take you to the **Outgoing References** tab for the `DispatcherTimer` class, as shown in *Figure 3.34*:

Outgoing References | Key Retention Paths | Incoming References | Creation Stack Trace | Shortest Paths | Merged Shortest Paths

Here you can view what objects are referenced from the instance as well as values of instance's fields.

Type	Address	References count	Bytes	Retained bytes
DispatcherTimer (System.Windows.Threading)	0000017e2f807188	4	80	6,176
Fields				
Tick EventHandler (System)	0000017e2f8071f0	1	64	6,072
Fields				
_methodPtr Int64 140,715,014,550,256 = 0x7FFAC474F2F0				
_methodPtrAux Int64 0 = 0x00				
_invocationCount Int64 0 = 0x00				
_target AdWindow (CH05_GameOfLife)	0000017e2f7fb0d8	11	680	6,008
Fields				
_dispatcher Dispatcher (System.Windows.Threading)	0000017e2e8fe4c0	15	232	1,904
_dType DependencyObjectType (System.Windows)	0000017e2f7fb400	2	40	40
_effectiveValues EffectiveValueEntry[24] (System.Windows)	0000017e2fbf5af0	19	408	848

Figure 3.34 – The Outgoing References table displaying the details of DispatcherTimeruse

This tab certainly shows that `Tick EventHandler` is retaining bytes, which is leading to our `DispatcherTimer` object being kept alive in memory.

16. Click on the **Creation Stack Trace** tab. This will help us to identify the method responsible for our `EventHandler` creation. The method appears at the top, as shown in *Figure 3.35*:

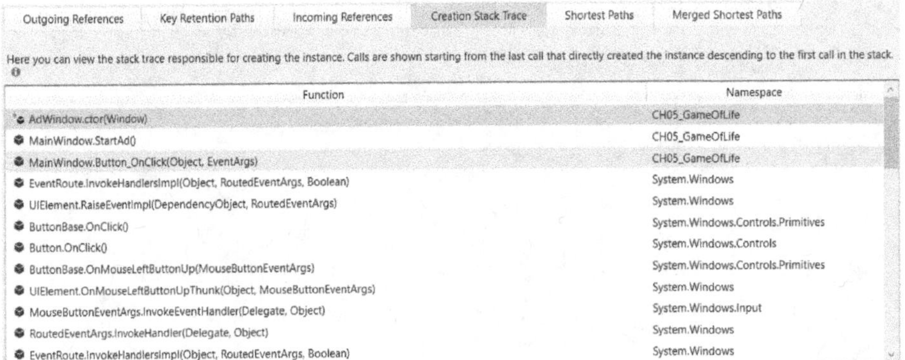

Outgoing References | Key Retention Paths | Incoming References | Creation Stack Trace | Shortest Paths | Merged Shortest Paths

Here you can view the stack trace responsible for creating the instance. Calls are shown starting from the last call that directly created the instance descending to the first call in the stack.

Function	Namespace
AdWindow.ctor(Window)	CH05_GameOfLife
MainWindow.StartAd()	CH05_GameOfLife
MainWindow.Button_OnClick(Object, EventArgs)	CH05_GameOfLife
EventRoute.InvokeHandlersImpl(Object, RoutedEventArgs, Boolean)	System.Windows
UIElement.RaiseEventImpl(DependencyObject, RoutedEventArgs)	System.Windows
ButtonBase.OnClick()	System.Windows.Controls.Primitives
Button.OnClick()	System.Windows.Controls
ButtonBase.OnMouseLeftButtonUp(MouseButtonEventArgs)	System.Windows.Controls.Primitives
UIElement.OnMouseLeftButtonUpThunk(Object, MouseButtonEventArgs)	System.Windows
MouseButtonEventArgs.InvokeEventHandler(Delegate, Object)	System.Windows.Input
RoutedEventArgs.InvokeHandler(Delegate, Object)	System.Windows
EventRoute.InvokeHandlersImpl(Object, RoutedEventArgs, Boolean)	System.Windows

Figure 3.35 – The Creation Stack Trace tab showing the AdWindow constructor that creates the timer

17. Locate the `AdWindow` constructor in the `AdWindow` class of the **CH05_GameOfLife** project:

```
public AdWindow(Window owner)
```

```
{
    ...
    _adTimer = new DispatcherTimer {
        Interval = TimeSpan.FromSeconds(3)
    };
    _adTimer.Tick += ChangeAds;
    _adTimer.Start();
}
```

As you can see from the preceding code snippet, we are subscribing to the `Tick` event, which is handled by the `ChangeAds` method. But the one thing we are not doing is unsubscribing from the event when we no longer require it. This is the reason for the memory leak.

18. To rectify our memory leak, all we have to do is unsubscribe from the event when we no longer need it. And to do this, we update the `OnClosed` method, as shown in the following code:

```
protected override void OnClosed(EventArgs e)
{
    _adTimer.Tick -= ChangeAds;
    base.OnClosed(e);
}
```

We have now rectified our memory leak by unsubscribing from the `Tick` event when we close the `AdWindow` constructor. Repeat the steps to profile this memory leak, and you will see that it has now been fixed, as shown in *Figure 5.36*:

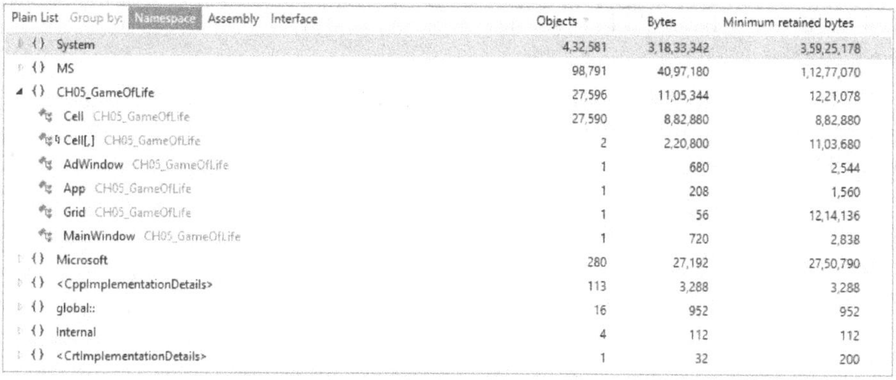

Plain List Group by: Namespace Assembly Interface	Objects	Bytes	Minimum retained bytes
{} System	4,32,581	3,18,33,342	3,59,25,178
{} MS	98,791	40,97,180	1,12,77,070
{} CH05_GameOfLife	27,596	11,05,344	12,21,078
Cell CH05_GameOfLife	27,590	8,82,880	8,82,880
Cell[,] CH05_GameOfLife	2	2,20,800	11,03,680
AdWindow CH05_GameOfLife	1	680	2,544
App CH05_GameOfLife	1	208	1,560
Grid CH05_GameOfLife	1	56	12,14,136
MainWindow CH05_GameOfLife	1	720	2,838
{} Microsoft	280	27,192	27,50,790
{} <CppImplementationDetails>	113	3,288	3,288
{} global::	16	952	952
{} Internal	4	112	112
{} <CrtImplementationDetails>	1	32	200

Figure 5.36 – dotMemory showing that the memory leak has been fixed

> **Note**
>
> We have effectively tracked down and fixed a memory leak with dotMemory. The memory leak was because we did not unsubscribe from an event we were subscribed to. This is a very common source of memory leaks in C#. To learn more about dotMemory and how to use it in various scenarios, please visit the official How-To documentation by JetBrains at `https://www.jetbrains.com/help/dotmemory/Examples.html`.

In the next section, we will look at how to track down and fix a UI freeze using dotTrace.

Finding the cause of a UI freeze with dotTrace

In this section, we will be using dotTrace to hunt down the reason for a UI freeze so that we can fix it. Again, to save time, we will use a project that has already been provided for you. Obtain the book's source code from the URL specified in the *Technical requirements* section. In the source code for `CH05`, you will find a project called `CH05_BatchFileProcessing`.

This project opens a number of text files specified by the user and then reverses each of the strings it finds. When the user clicks on the **Process Files** button, a separate `BackgroundWorker` thread is started that runs on a separate thread. In the left-hand corner, the progress of file processing is displayed. This changes to **All files were successfully processed when done**. However, a problem exists whereby the UI freezes while the files are being processed.

To find the source of this UI freeze and fix it, we are going to use timeline profiling, which is available using dotTrace:

1. Build the **CH05_BatchFileProcessing** project in **Release** mode.
2. Open dotTrace.
3. Select **Profile Local App | .NET Core Application | Timeline**, and select the executable you just compiled. Make sure to tick **Collect profiling data from start**. *Figure 5.37* shows dotTrace being configured before we start running it:

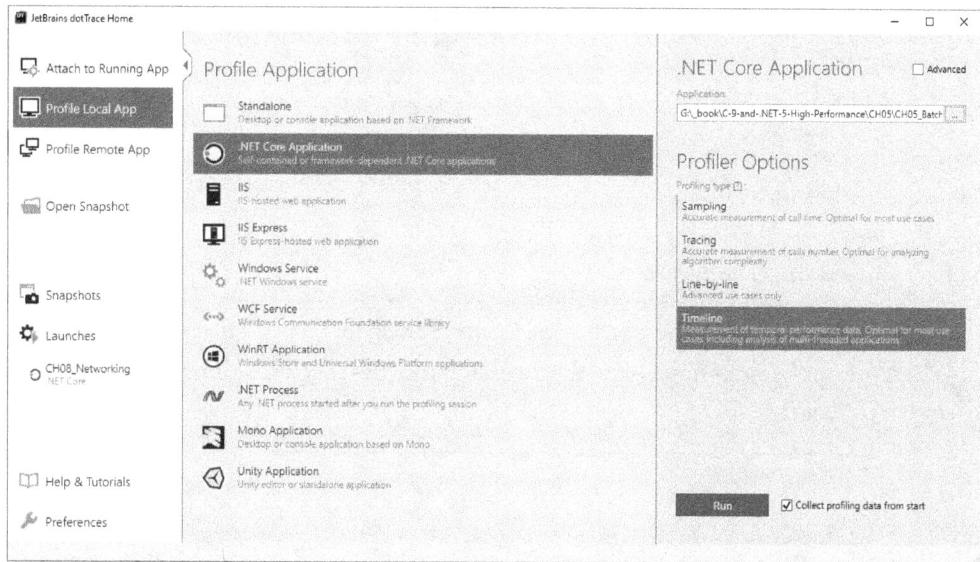

Figure 5.37 – dotTrace prior to us running the Timeline profiler

4. Click on the **Run** button to begin the timeline profiling. The profiler will be opened, as shown in *Figure 5.38*:

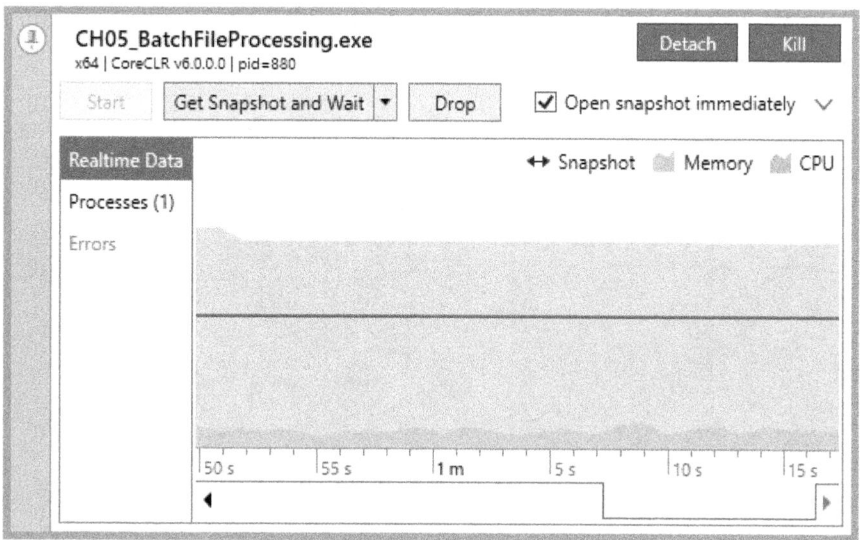

Figure 5.38 – The dotTrace Timeline profiler

The profiler will start the **CH05_BatchFileProcessor** program, as shown in *Figure 5.39*:

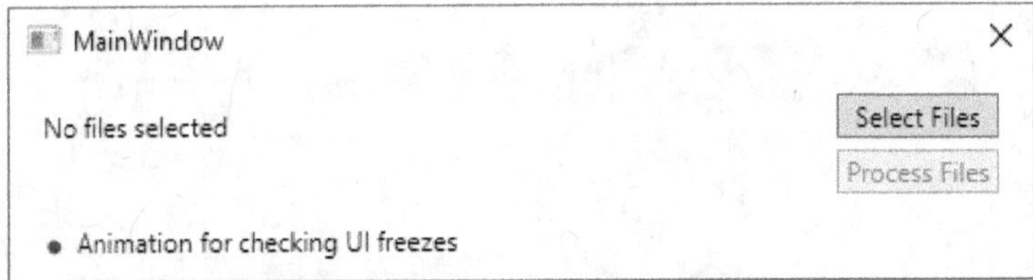

Figure 5.39 – The batch file processor

When the application has finished processing the files, the UI will be displayed, as shown in *Figure 5.40*:

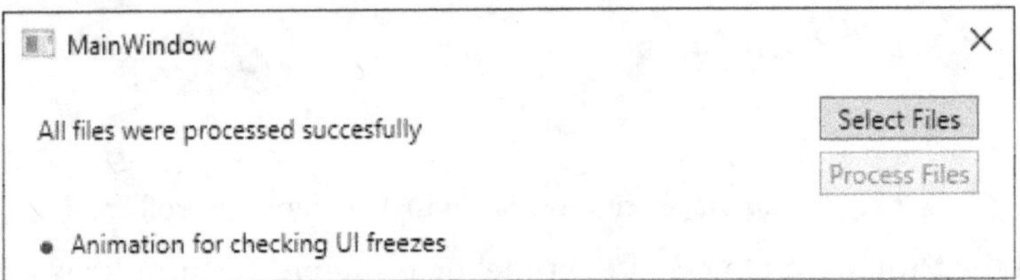

Figure 5.40 – CH05_BatchFileProcessor

5. Click on the **Get Snapshot** and **Wait** buttons on the timeline profiler. This will save the snapshot and open it in the dotTrace **Timeline Viewer** application, as shown in *Figure 5.41*:

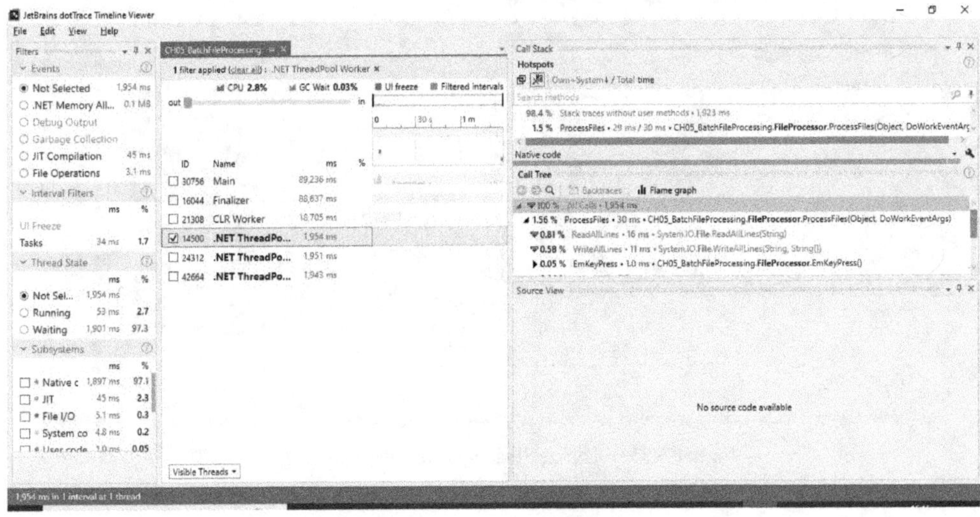

Figure 5.41 – The dotTrace Timeline Viewer application with a loaded timeline snapshot

6. You can close the **CH05_BatchFileProcessor** and dotTrace profiler applications down. But keep the dotTrace **Timeline Viewer** application open.
7. All filter values are calculated for all currently visible threads. We are only interested in threads that have activity on them. So, hide all threads that have no activity on them by selecting them, right-clicking, and selecting **Hide** selected threads.
8. Our **BackgroundWorker** thread is the **.NET ThreadPoolWorker** thread with an ID of **12764**, as shown in *Figure 5.42*:

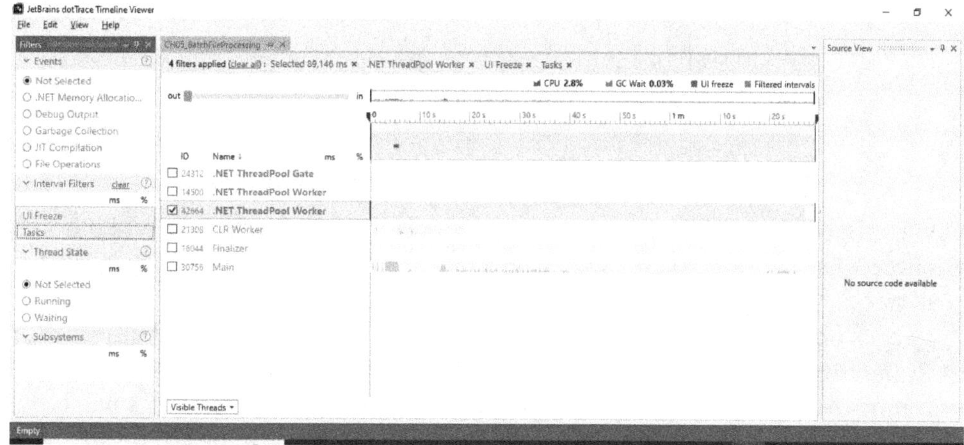

Figure 5.42 – The dotTrace Timeline Viewer application with our BackgroundWorker thread highlighted

9. Zoom into the timeline for the **.NET ThreadPool Worker**. You can see that the timeline consists of three states. These states are **Running**, **Waiting for CPU**, and **Waiting**. You can see our thread's timeline in *Figure 5.43*:

Figure 5.43 – Our thread's activity within the timeline trace

On the left-hand side of the screen, you will see the **Thread State** section within the **Filters** panel. Select each of the states in turn, and you will see the timeline highlighted accordingly. Have a play with all of the different filters available. Investigate what each option provides you. This is a good way to learn. The collapsed **Filters** panel is displayed in *Figure 5.44*:

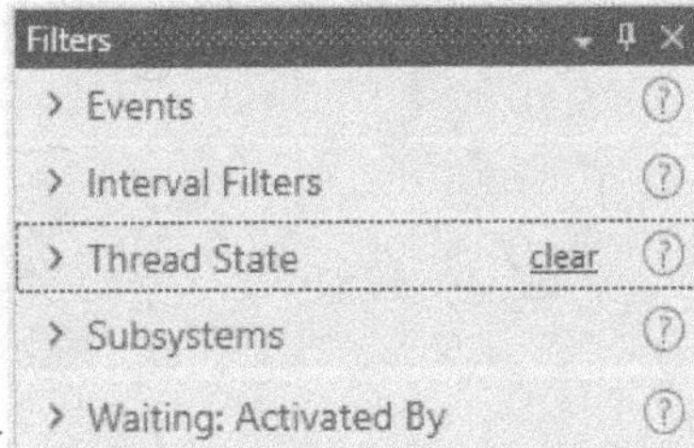

Figure 5.44 – The collapsed dotTrace Filters panel

10. On the right-hand side of the screen, you will see the **Call Stack** panel and the **Source View** panel. If you click anywhere on the thread's timeline, you will see the call stack at that point in time. The call tree will be displayed for that stack trace. If you click on an entry in the call stack, the code will be decompiled and displayed within the **Source View** tab. This functionality enables you to see what code is running at what point in time. Also, this view displays the full assembly's name, namespace, and class name for the code you are looking at. *Figure 5.45* displays the **Call Stack** panel:

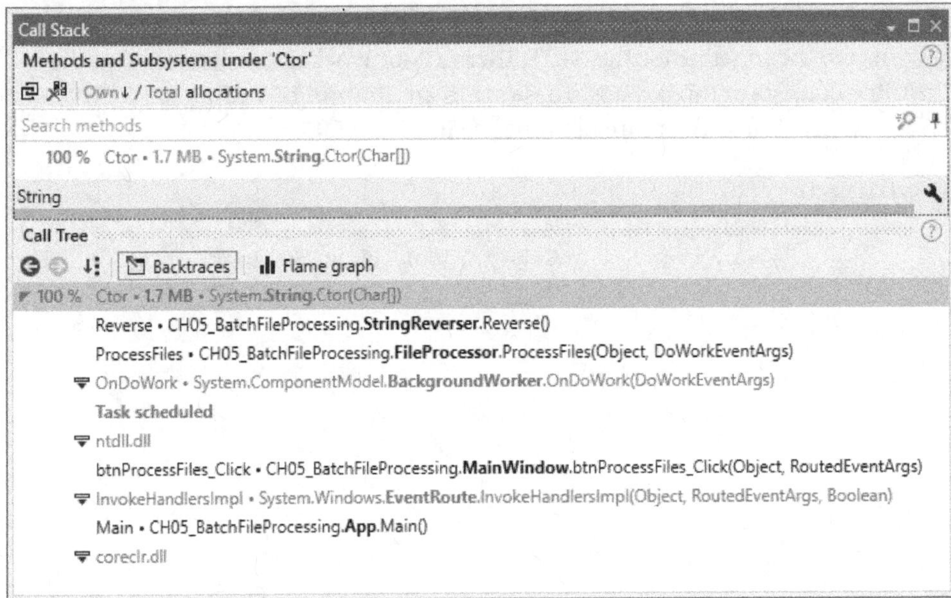

Figure 5.45 – The dotTrace Call Stack panel with the Backtraces tab displayed

Figure 5.46 displays the **Source View** panel:

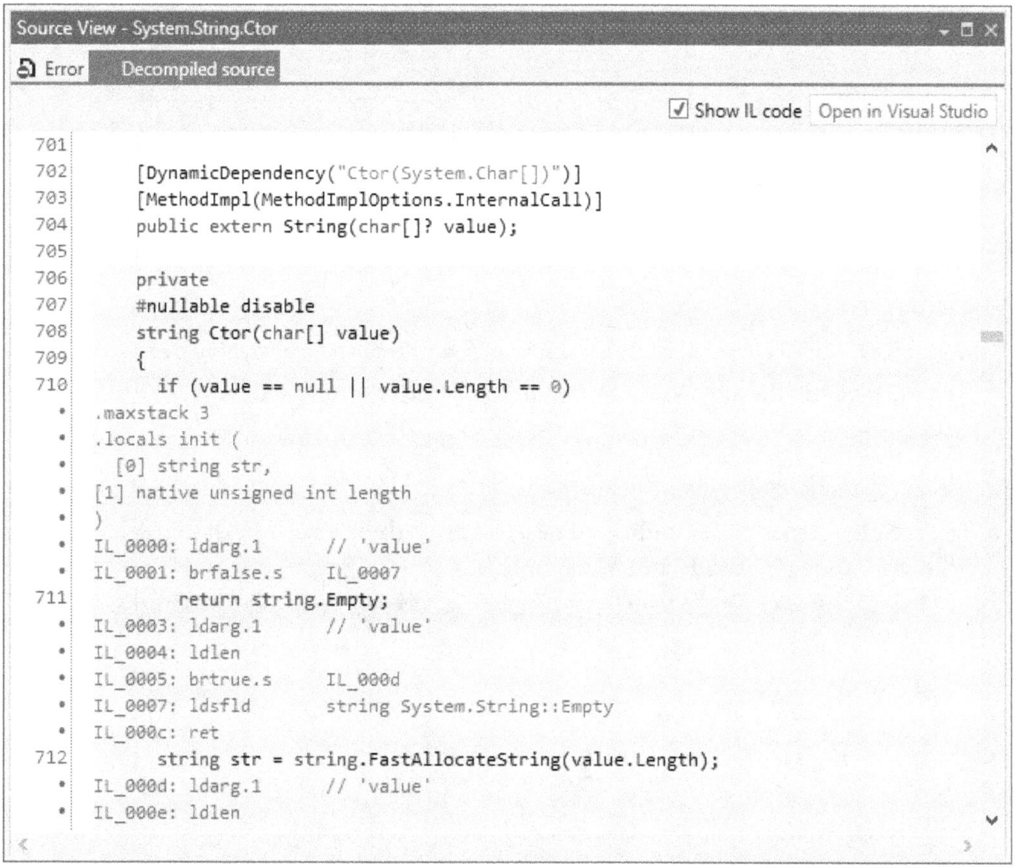

Source View - System.String.Ctor

Error | Decompiled source

Show IL code | Open in Visual Studio

```

    [DynamicDependency("Ctor(System.Char[])")]
    [MethodImpl(MethodImplOptions.InternalCall)]
    public extern String(char[]? value);

    private
    #nullable disable
    string Ctor(char[] value)
    {
      if (value == null || value.Length == 0)
  .maxstack 3
  .locals init (
    [0] string str,
  [1] native unsigned int length
  )
  IL_0000: ldarg.1      // 'value'
  IL_0001: brfalse.s    IL_0007
        return string.Empty;
  IL_0003: ldarg.1      // 'value'
  IL_0004: ldlen
  IL_0005: brtrue.s     IL_000d
  IL_0007: ldsfld       string System.String::Empty
  IL_000c: ret
      string str = string.FastAllocateString(value.Length);
  IL_000d: ldarg.1      // 'value'
  IL_000e: ldlen
```

Figure 5.46 – The dotTrace Source View screen showing decompiled C# and IL source code

> **Note**
>
> The colored bar that runs across the **Call Stack** panel, as shown in *Figure 5.45*, displays the different subsystems in use; in this case, `String`. Depending on what is happening at a particular point in time, this line might be multicolored if multiple subsystems are in use. This bar is also useful for showing thread locks, among other things.

11. Now we are ready to investigate why our UI is freezing. The purple lines in *Figure 5.47* represent moments in time when our UI is freezing:

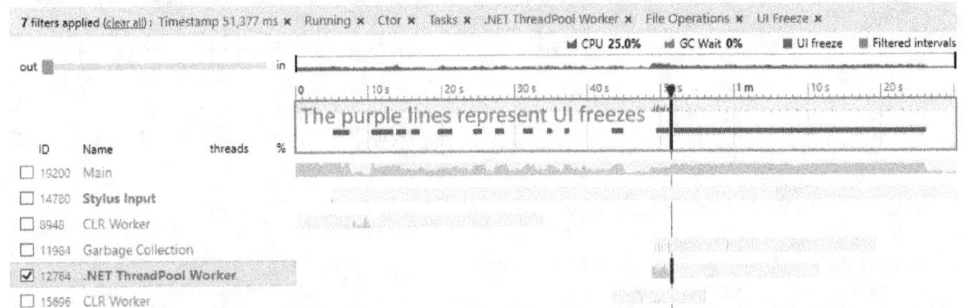

Figure 5.47 – The dotTrace filtered view displaying our thread and highlighting UI freezes

The purple line that we are interested in is the last very long one.

12. In the **Filters** section, select **Events | .NET Memory Allocation**.
13. Then, select **Thread State | Running**.
14. Select **Subsystems | User code**, and deselect everything else. You should see the following under **Methods and Subsystems**:

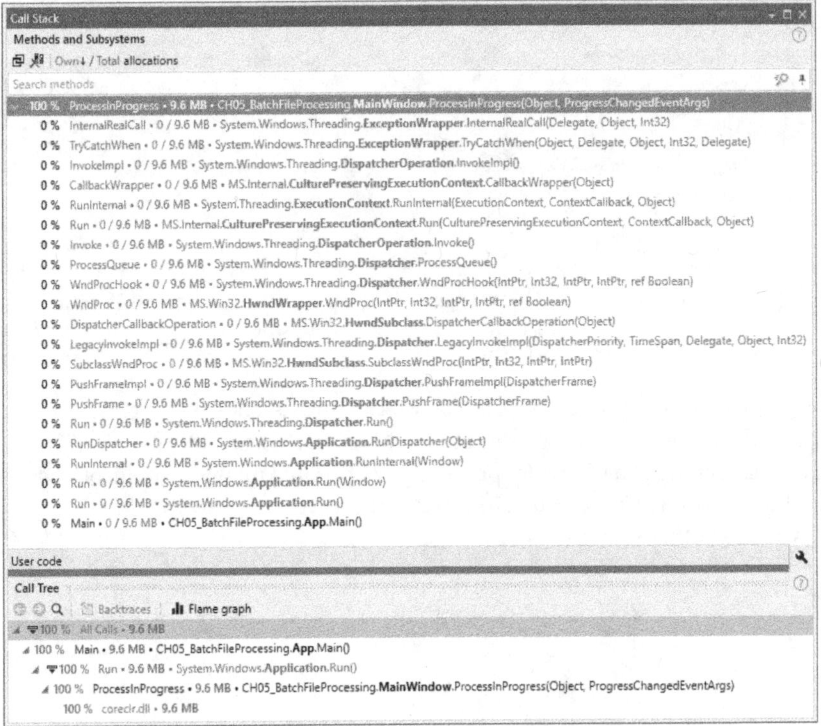

Figure 5.48 – The dotTrace Methods and Subsystems screen highlighting problematic user code

Looking at the preceding highlighted method called `ProcessInProgress`, we are calling it 100% of the time during the time period when the UI freeze occurs. Clicking on `ProcessInProgress` will display the contents of the `MainWindow.xaml.cs` file. Our offending code is as follows:

```
private void ProcessInProgress(
object sender,
ProgressChangedEventArgs e
)
{
var upd = (ProgressUpdater)e.UserState;
lblProgress.Content = $"File {upd.CurrentFileNmb} of {upd.
    TotalFiles}: {e.ProgressPercentage}%";
}
```

Our code is updating the progress label with the value passed into the method, which is of the `ProgressChangedEventArgs` type. So, what is calling this method? It is the `ProcessFiles` method in the `FileProcessor` class:

```
...
for (var i = 0; i < FilePaths.Count; i++)
{
    ...
for (var j = 0; j < _lines.Length; j++)
{
    var line = _lines[j];
    var stringReverser = new StringReverser(line);
    _lines[j] = stringReverser.Reverse();
    if (j % 5 == 0)
    {
        var p = (float)(j + 1) / _lines.Length * 100;
        Worker.ReportProgress((int)p, _updater);
    }
}
File.WriteAllLines(path, _lines);
}
```

This method iterates through the files that the user has selected. Each file is read along with each line, line by line. Each line has its text reversed. The problem is that we are calling this method far too often. So, the solution is to change `(j % 5 == 0)` to `(j% 1000 == 0)`.

15. Make the change to the code recompile and rerun the profiler. This time, there will be no lag. And you will see that the UI freeze has been fixed.

Now you have used dotTrace and the Timeline profile to track down and fix a UI freeze. In the final section, we will look at using dotTrace to optimize application performance and memory traffic.

Optimizing application performance and memory traffic with dotTrace

In this section, we are going to continue tracing our `CH05_BatchFileProcessing` project. We have fixed the UI freeze and will be running another trace to see whether we can identify any further issues. When analyzing the trace, we will see that a lot of memory traffic is being generated that is affecting the performance of our application. So, we will address this issue and fix it:

1. Open dotTrace. Your previous session should be saved. Select it, and click on the **Run** button to start tracing. The sample application will then be started.
2. Select the text files, and click on the **Process Files** button.
3. Once the files have been processed, kill the application. This will flush the data and load our trace in the trace viewer. Then, close dotTrace.
4. Once the trace snapshot has been loaded into **Timeline Viewer**, click on the button to **Show Snapshot**.
5. In the **Filters** view, select **Events | .NET Memory Allocations and Thread State | Running**.
6. Hide all threads except our **.NET ThreadPool Worker** thread.
7. In the **Call Stack** view under **Methods and Subsystems**, click on **Own** to view the percentage of memory allocations made by our code. You will see that our method for reversing a string allocates **28.5%** of the network traffic. The largest amount of memory traffic is generated by the **Concat** method within the `System.String` class. This will be the result of our `CH05_BatchFileProcessing.StringReverse.Reverse()` call. *Figure 5.49* shows the results of our trace in which we can see our methods and the percentages of memory traffic they generate:

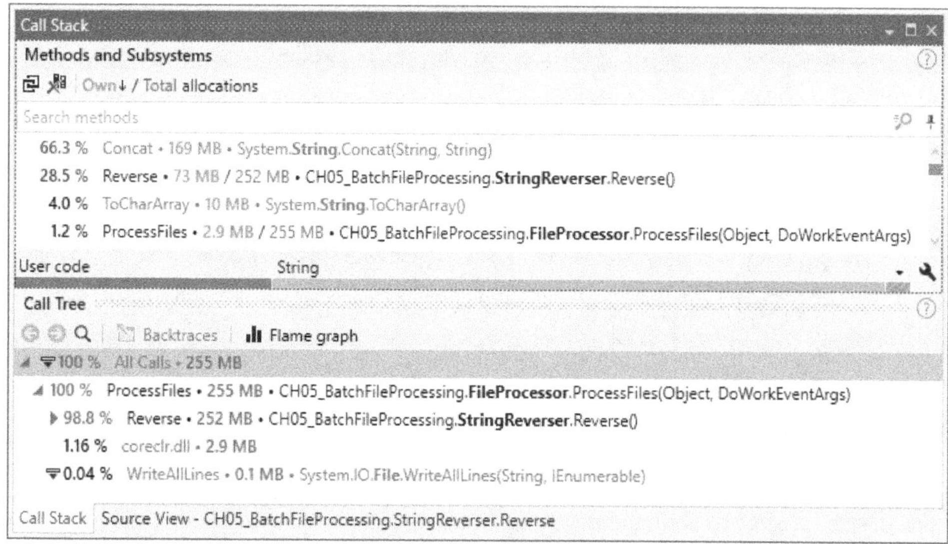

Figure 5.49 – The dotTrace Timeline Viewer Call Stack screen showing our methods and memory traffic percentage

The two different MB sizes are our own memory allocation in this method excluding memory allocations in the child method calls from this method/the amount of memory allocated by this method or any child methods called from this method. As you can see, the memory allocation is **73 MB/252 MB** for the `Reverse()` method and **2.9 MB/255 MB** for the `ProcessFiles()` method.

8. Open this class in Visual Studio. The code for the `Reverse()` method is as follows:

```
public string Reverse()
{
char[] charArray = _original.ToCharArray();
string stringResult = null;
for (int i = charArray.Length; i > 0; i--)
{
    stringResult += charArray[i - 1];
}
return stringResult;
}
```

As you can see, this method reverses a string by assigning it to an array. The array is then iterated backward, with each character assigned to a string using string concatenation. And herein lies the problem with our application's performance.

It is well documented that the most performant way to build up a string is to use the `StringBuilder` class. And we could do that here. However, there is another way to improve the performance of this method. Replace the existing `Reverse()` string method with the following version:

```
public string Reverse()
{
     char[] charArray = _original.ToCharArray();
     Array.Reverse(charArray);
return new string(charArray);
}
```

In our revised code, we reverse the array and return a new string from the reverse array.

9. Build your project in **Release** mode and then run a new trace. *Figure 5.50* shows the results of the new trace:

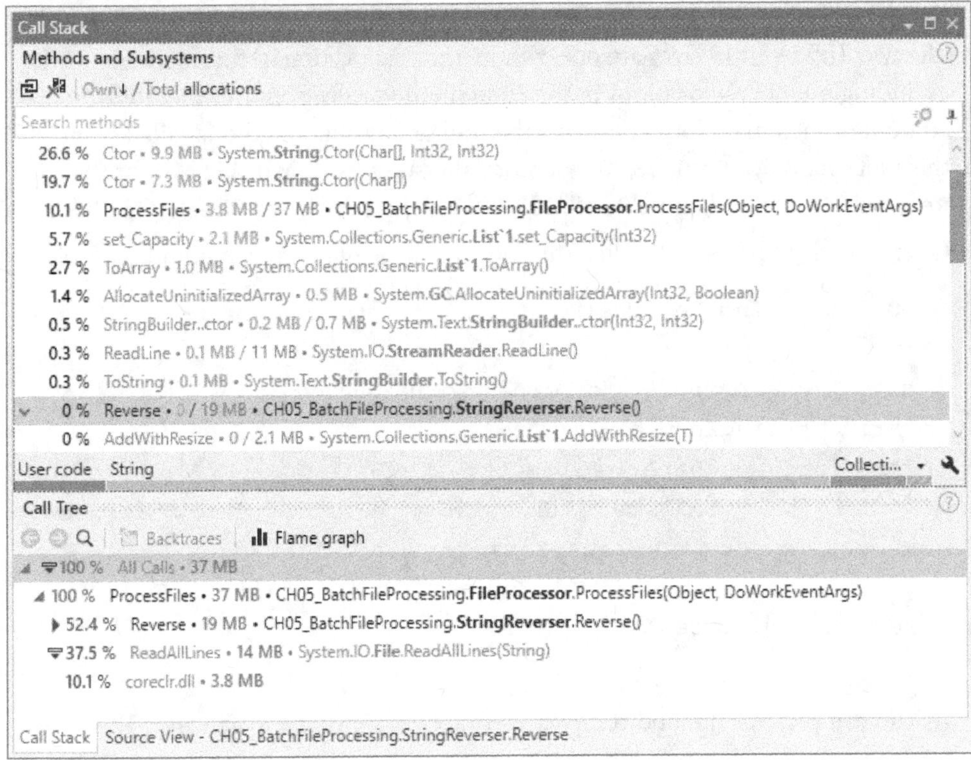

Figure 5.50 – The new trace showing our improved performance

We can see from our trace that the memory allocation for the `ProcessFiles` method went from **2.9 MB/255 MB**, generating **1.2%** of the memory traffic, to **3.8 MB/37 MB** of memory allocation, generating **10.1%** of the memory traffic.

Plus, our `Reverse()` method went from allocating **73 MB/252 MB**, and generating **28.5%** of the memory traffic, to allocating **0 MB/19 MB** of memory, generating **0%** of the memory traffic.

That is a good performance improvement!

In this chapter, we have covered various methods of measuring and analyzing code. With the data we obtained, we have managed to fix a memory leak caused by not unsubscribing to event handlers, fix a UI freeze caused by too frequent UI updates, and improve the application performance and memory traffic caused by the way we were batch processing string reversal. Now, it is time to summarize what we have learned.

Summary

We started with application profiling and tracing by looking at the various code metrics that are available to us. Various tools have different metrics available. These metrics cover the application, assemblies, namespaces, types, methods, and fields.

Then, we moved on to look at how we can perform static code analysis. We demonstrated static code analysis using Visual Studio 2022's built-in code analysis tool. We saw how to generate the following metrics: the maintainability index, cyclomatic complexity, the depth of inheritance, class coupling, lines of source code, and lines of executable code.

The next thing we looked at was the generation of memory dumps and how to view them from within Visual Studio 2022. We can view the dump time, the dump's location, the name of the process, the processor architecture, any exception information, the OS version, and the CLR version. Additionally, we can view loaded module names and their versions and physical paths.

Next, we looked at how to open the **Modules** window during a debugging session. The **Modules** window shows us the name and path of the module, whether the module is optimized, whether it is user code or system code, its symbol status, order, version, process, and AppDomain. We also saw the other options available in the **Debug | Windows** menu that add to our debugging capabilities.

Then, we looked at the tracing and diagnostics tools called Visual Studio 2022, JetBrains dotMemory, and JetBrains dotTrace. These tools provide an overall excellent debugging experience that provides all the information we need to track down any type of bug, including those that cause memory leakages and other memory-related issues.

Next, we looked at `dotnet-counters` and how to use this. We learned how to list the .NET processes that can be monitored. Then, we saw how to list the available well-known .NET counters. And our concluding section saw us collecting data and saving data to a file for post-analysis.

Finally, we worked through three examples of using JetBrains dotMemory and JetBrains dotTrace to fix a memory leak and UI freeze, improve performance, and reduce memory traffic.

In the next chapter, we will be taking a detailed look at the **Collections** framework. However, before then, take the time to further your reading and answer the following questions to reinforce what you have learned.

Questions

1. What aspects of our computer programs are covered by code metrics?
2. What metrics does the Visual Studio 2022 static code analysis produce?
3. What kinds of things can we view from the Visual Studio-generated minidumps with heap?
4. What columns are available in the **Modules** window?
5. What are the names of the four debugging, profiling, and tracing tools for performing the various diagnostic operations that we mentioned earlier?
6. What operations did we carry out using .NET counters?

Further reading

- Debugging Visual Studio 2019: `https://docs.microsoft.com/en-us/visualstudio/get-started/csharp/tutorial-debugger?view=vs-2019`.
- Dump files in the Visual Studio debugger: `https://docs.microsoft.com/visualstudio/debugger/using-dump-files?view=vs-2019`.
- `dotnet-counters`: `https://docs.microsoft.com/en-us/dotnet/core/diagnostics/dotnet-counters`.

- *.NET Core Counters internals: how to integrate counters in your monitoring pipeline*: `https://medium.com/criteo-engineering/net-core-counters-internals-how-to-integrate-counters-in-your-monitoring-pipeline-5354cd61b42e#:~:text=dotnet-counters%3A%20collect%20the%20metrics%20corresponding%20to%20some%20performance,how%20to%20fetch%20them%20via%20the%20EventPipe%20infrastructure`.
- *JetBrains dotTrace*: `https://www.jetbrains.com/profiler/`.
- *JetBrains dotMemory*: `https://www.jetbrains.com/dotmemory/`.
- *ndepend*: `https://www.ndepend.com/`.
- *Overview of .NET source code analysis*: `https://docs.microsoft.com/dotnet/fundamentals/code-analysis/overview`.

Part 2: Writing High-Performance Code

Part 2 covers putting the framework to work by programming high-performance code. We start by looking at collections. Then we move on to look at LINQ performance, followed by files and streams. Next, we look at networking followed by working with data. After that, we learn how to keep user interfaces active during long operations. Then we finish up by looking at distributed systems that scale.

This part contains the following chapters:

- *Chapter 6, The .NET Collections*
- *Chapter 7, LINQ Performance*
- *Chapter 8, File and Stream I/O*
- *Chapter 9, Enhancing the Performance of Networked Applications*
- *Chapter 10, Setting Up Our Database Project*
- *Chapter 11, Benchmarking Relational Data Access Frameworks*
- *Chapter 12, Responsive User Interfaces*
- *Chapter 13, Distributed Systems*

6
The .NET Collections

Collections are an integral part of .NET. There are different ways to use these collections. Microsoft .NET makes heavy use of arrays and collections when dealing with things such as datasets, arrays, lists, dictionaries, stacks, and queues. You will be hard-pressed to write a C# program without having to use the Collections Framework. The different ways of using the collections and arrays differ in terms of their performance degradation and performance improvement. Therefore, understanding when to use arrays and when to use collections will form an important aspect of your C# and .NET programming skills.

In this chapter, you will learn how to improve the performance of your collection operations. By using `BenchmarkDotNet` with different versions of the code, you will be able to see the differences in performance and be in a position to choose the best method that suits your needs.

We will be covering the following topics in this chapter:

- **Understanding the different collection offerings**: This section is purely informational and provides an overview of the `System.Collections`, `System.Collections.Generic`, `System.Collections.Concurrent`, and `System.Collections.Specialized` namespaces.
- **Setting up our sample database**: We will be using a SQL database that highlights the difference between `IEnumerable` and `IQueryable`. This section will show you how to develop our sample database with sample data that will be used later in this chapter.

- **Deciding between interfaces and concrete classes**: In this section, you will benchmark the performance between using classes and interfaces. Then, you will be able to decide on the method that best suits your needs.
- **Deciding between using arrays or collections**: There are strengths and weaknesses between using arrays and collections. In this section, you will benchmark the performance of arrays and collections and decide which to use based on your performance requirements.
- **Accessing objects using indexers**: In this section, we will discuss accessing objects in the same way we would access items in an array by using indexers.
- **Comparing IEnumerable and IEnumerator**: In this section, we will benchmark iterations using both IEnumerable and IEnumerator. You will see that there is a definite performance difference between these ways of enumerating.
- **Database query performance**: In this section, we will query a database using five different methods, benchmarking their performance to see which method produces the fastest performance.
- **Exploring the yield keyword**: In this section, you will learn about the `yield` keyword and how it relates to the performance of your applications, especially when it comes to iterating through collections and arrays.
- **Learning the difference between concurrency and parallelism**: In this section, you will understand the difference between concurrency and parallelism, and learn when to use one over the other.
- **Learning the difference between Equals() and ==**: In this section, you will understand the differences between the different equality operators, and learn when to use one over the other.
- **Studying LINQ performance**: LINQ is a C# query language that is heavily utilized when it comes to processing collections, but it can be slow or fast, depending on the way you code your queries. In this section, you will learn how to benchmark different ways of performing the same types of queries. In doing so, you will see the difference in performance between the different ways of writing the same queries.

By the end of this chapter, you will be able to do the following:

- Describe the different collections available and their uses
- Choose between using interfaces and collections
- Understand the trade-offs between arrays and collections
- Write indexers

- Choose the best form of iteration for your particular needs
- Use the `yield` keyword
- Know which equality operator to use for different types of equality checking
- Improve LINQ query performance

Technical requirements

To follow along with this chapter, you will need access to the following tools:

- Visual Studio 2022
- SQL Server (any version) Express or higher
- SQL Server Management Studio
- This book's source code: `https://github.com/PacktPublishing/High-Performance-Programming-in-CSharp-and-.NET/tree/master/CH06`

Understanding the different collection offerings

A collection is a group of records that can be treated as one logical unit. Examples of logical record groups include people, countries, products, ingredients, books, authors, and more.

There are four main types of collections, as follows:

- **Index-based** collections, such as an array or list. Index-based collections contain an internal index. The index can be either numeric or string-based. An index-based collection is more commonly accessed using a numerical index. Numerical indexes are zero-based. This means that a collection's index will start at zero for the first record and increase in value by the order of one for each subsequent record. Collections that can be accessed using numerical indexes include arrays and lists.
- **Key/value pair** collections, such as a hash table or sorted list. Key/value pair collections such as `Hashtable` and `SortedList` use a key to look up the value stored in a collection. So, for example, if you have a collection of products, you can access the product you need by using the product code that was assigned as the key when the product was added to the key/value pair collection.

- **Prioritized** collections, such as a stack or queue. Prioritized collections allow you to store and extract records in a particular sequence. A queue uses the **First In First First Out** (**FIFO**) sequence, while a stack uses the **Last In First Out** (**LIFO**) sequence.
- **Specialized** collections, such as string collections and hybrid dictionaries. Specialized collections are out-of-the-box collections for specific purposes. For example, there is the `CollectionsUtil` class, which creates collections that ignore the case in strings, and the `ListDictionary` class, which is recommended for collections that contain less than 10 items. It implements `IDictionary` using a singly linked list.

The .NET Collections Framework consists of the legacy `System.Collections` namespaces, as well as the newer `System.Collections.Generic`, `System.Collections.Concurrent`, and `System.Collections.Specialized` namespaces. Before we delve into the performance of collections, it is a good idea to reacquaint ourselves with the different collections that are available in each of the aforementioned namespaces.

The System.Collections namespace

The `System.Collections` namespace contains various classes, structures, and interfaces. In this section, we will briefly cover what is available. The collections in this namespace are not thread-safe. If you require thread-safe collections, it would be better to use the collections in the `System.Collections.Concurrent` namespace instead, as advised by Microsoft!

The `ICollection` interface defines the size, enumerators, and synchronization methods for all non-generic collections. To compare two objects, you can implement the `IComparer` interface. You can represent non-generic key/value pair collections using `Idictionary`. To enumerate a non-generic dictionary, you can use the `IDictionaryEnumerator` interface. Simple iteration over non-generic collections is provided by the `IEnumerator` interface, while equality between objects is implemented via the `IEqualityComparer` interface. The `IList` interface is used to implement non-generic collections of objects that can be individually accessed using the index. Structural comparison of objects and structural equality comparison of objects is implemented using the `IStructuralComparable` and `IStructuralEquatable` interfaces, respectively.

- The `ArrayList` class implements the `IList` interface using a dynamic array that can grow and shrink in size as required.

- On (0) and off (1), which are represented by the Boolean values `false` and `true`, respectively, are managed by the `BitArray` class.
- To compare two objects while ignoring string casing, you can use the `CaseInsensitiveComparer` class. Use `CaseInsensitiveHashCodeProvider` to generate hash codes using algorithms that ignore string casing.
- When you're building a strongly typed collection, inherit from the `CollectionBase` class.
- The `Comparer` class is used to compare two objects for equivalence with case-sensitive string comparison.
- Use `DictionaryBase` as the abstract class when developing strongly typed collections of key/value pairs.
- A collection of key/value pairs organized by key-based hash codes is represented by the `Hashtable` class.
- The `Queue` class provides a collection with FIFO access.
- The `ReadOnlyCollectionBase` abstract class is used as the base class for strongly typed non-generic, read-only collections.
- Use the `SortedList` class to hold a collection of key/value pairs that are sorted by the keys and are accessible by key or index.
- Use the `Stack` class if you need LIFO access for your collection.
- To compare two collection objects structurally, you can use the `StructuralComparisons` class.
- The `DictionaryEntry` structure defines a dictionary key/value pair that can be set or retrieved.

> **Note**
>
> `IHashCodeProvider` has now been marked obsolete and is no longer recommended by Microsoft for new development. Microsoft recommends that you use the `IEqualityComparer` and `IEqualityComparer<T>` interfaces instead.

We now know what is available in the `System.Collections` namespace. Now, let's look at what's available in the `System.Collections.Generic` namespace.

The System.Collections.Generic namespace

The classes and interfaces that are available in the System.Collections.Generic namespace provide collections that are strongly typed and that perform better than the classes within the System.Collections namespace. This namespace contains many classes, structs, and interfaces.

The CollectionExtensions class provides extension methods for generic collections. To compare two objects, you can use the Comparer<T> class, which implements the IComparer<T> interface. The IComparer<T> interface defines the method types to implement to compare two objects.

The IDictionary<TKey, TValue> interface provides methods for implementing generic dictionaries. For a dictionary to be read-only, it must implement the IReadOnlyDictionary<TKey, TValue> interface. A collection of keys and values is represented by the Dictionary<TKey, TValue> class. Dictionary<TKey, TValue>.KeyCollection cannot be inherited and represents the collection of keys within a Dictionary<TKey, TValue> collection. Finally, Dictionary<TKey, TValue>.ValueCollection cannot be inherited and represents the collection of values within a Dictionary<TKey, TValue> collection.

The IEqualityComparer<T> interface defines methods that you can use to compare objects for equality. A base class for implementations of the IEqualityComparer<T> interface is provided called EqualityComparer<T>.

HashSet<T> represents a set of values. When a key that's been used to access a collection cannot be found within the collection that's being searched, then a KeyNotFoundException is raised. A key/value pair instance is generated using the KeyValuePair class. For a doubly linked list, use the LinkedList<T> class. The non-inheritable LinkedListNode<T> class represents a node in a collection of the LinkedList<T> type.

IList<T> represents a collection of objects for implementing lists that can be accessed by index. Read-only lists implement the IReadOnlyList<T> interface. When you need a collection that is strongly typed that enables searching, sorting, and manipulating lists, then use the List<T> class. For FIFO collections, use the Queue<T> class.

ReferenceEqualityComparere is an IEqualityComparer<T> that uses reference equality by calling ReferenceEquals(Object, Object) instead of using value equality by calling Equals(Object) when comparing two object instances.

A key/value pair collection that's sorted on the key is represented by the SortedDictionary<TKey, TValue> class. This type of collection is represented by SortedDictionary<TKey, TValue>.KeyCollection, which cannot be inherited. The values that have been collected are represented by SortedDictionary<TKey, TValue>.ValueCollection, which cannot be inherited.

The SortedList<TKey, TValue> class represents a collection of key/value pairs that are sorted by key based on the associated IComparer<T> implementation. A collection of objects that has been maintained in sorted order is represented by the SortedSet<T> class. The Stack<T> class provides LIFO manipulation for instances of the same type.

There are several structures available for the various generic collection classes that allow you to enumerate the elements in the collection. These structures are called enumerators.

Asynchronously enumerating over values of a specific type can be done by implementing the IAsyncEnumerable<T> interface. IAsyncEnumerator<T> provides the necessary support to iterate over a generic collection. ICollection<T> defines the methods needed to manipulate generic collections. Strongly typed collections that are read-only implement the IReadOnlyCollection<T> interface. Sets implement the ISet<T> interface, while read-only sets implement the IReadOnlySet<T> interface.

Now that we've looked at what the System.Collections.Generic namespace has to offer, let's turn our attention to the System.Collections.Concurrent namespace.

The System.Collections.Concurrent namespace

The collections in the System.Collections.Concurrent namespace are thread-safe. Whenever multiple threads are concurrently accessing a collection, use the collections in this namespace over the collections in the System.Collections and System.Collections.Generic namespaces.

> **Note**
>
> Extension methods and explicit interface implementations of these collections are not guaranteed to be thread-safe. To ensure thread safety, synchronization may be required in these instances.

IProducerConsumerCollection<T> defines methods that form the basis of thread-safe collection manipulation in producer/consumer usage (also known as publisher/subscriber usage). Higher-level abstractions such as the BlockingCollection<T> class can use this collection as their underlying storage mechanism.

The `BlockingCollection<T>` class provides blocking and bounding capabilities to thread-safe collections that implement the `IProducerConsumerCollection<T>` interface.

Options to control partitioner buffering behavior are specified by the `EnumerablePartitionerOptions` enum.

Arrays, lists, and enumerable partitioning strategies are provided by the `Partitioner` class. The `Partitioner<Tsource>` class provides a particular manner of splitting a data source into multiple partitions, while `OrderablePartioner<Tsource>` splits an orderable data source into multiple partitions.

The `Concurrent<T>` class contains a thread-safe unordered list of objects. Thread-safe FIFO collections use the `ConcurrentQueue<T>` class, while thread-safe LIFO collections use the `ConcurrentStack<T>` class. To concurrently access key/value pairs in a thread-safe manner, use the `ConcurrentDictionary<Tkey, Tvalue>` class.

With that, we've covered the `System.Collections.Concurrent` namespace. Now, let's look at the `System.Collections.Specialized` namespace.

The System.Collections.Specialized namespace

The `System.Collections.Specialized` namespace contains specialized and strongly typed collections. Let's see what it has to offer.

The `CollectionChangedEventManager` class provides a `WeakEventManager` implementation. By using the `WeakEventListener` pattern, you can attach listeners for the collection-changed event.

To build a collection of strings that ignores the string casing, you can use the `CollectionUtils` class.

The `HybrdDictionary` class changes its behavior when the collection is small, and when the collection grows in size. It does this by implementing `IDictionary` using a `ListDictionary` when the collection is small; it uses a `Hashtable` when the collection grows in size and becomes large.

For fewer than 10 items, you can use `ListDictionary`, which implements `IDictionary` by using a singly linked list.

To hold a collection of the string keys of a collection, use `NameObjectCollectionBase.KeysCollection`.

When you need to provide data for the `CollectionChanged` event, use the `NotifyCollectionChangedEventArgs` class.

When you have an ordered collection of key/value pairs that you need to be accessible via either the key or the index, use `OrderedDictionary`.

You can use the `StringCollection` class to hold a collection of strings, and you can use the `StringEnumerator` class to perform a simple iteration of the `StringCollection` class.

To get a hash table of keys and strongly typed string values, use the `StringDictionary` class.

To store a Boolean value or small integer in 32 bits of memory, you can use the `BitVector32` structure. You can use `BitVector32.Section` of the vector to store an integer number.

Indexed collections of key/value pairs are represented by the `IOrderedDictionary` interface. The `INotifyCollectionChanged` interface is used to notify listeners of dynamic changes to a collection, such as when items are added, modified, or removed. The `NotifyCollectionChangedAction` enum describes the action that resulted in the `CollectionChanged` event being fired.

Now, let's look at custom collections and write one.

Creating custom collections

To create custom collections, you must inherit from `CollectionBase`. The `CollectionBase` class has a read-only `ArrayList` property called `InnerList`, and it implements the `IList`, `ICollection`, and `IEnumerable` interfaces. Then, you can add your own `Add`, `Remove`, `Clear`, and `Count` methods. We'll do this in our project. We will create a very simple custom collection that inherits from `CollectionBase` so that you can see how easy it is to create custom collections. Follow these steps:

1. Add a new class under the `CustomCollections` folder called `CustomCollections` that inherits from `CollectionBase`.
2. Add the `Add(object item)` method to the class:

```
public void Add(object item)
{
        InnerList.Add(item);
}
```

   This method adds an item to `InnerList`, which we have inherited from the `CollectionBase` class.

3. Add the `Remove(object item)` method to the class:

```
public void Remove(object item)
{
        InnerList.Remove(item);
}
```

   This method removes an item from the inherited `InnerList`.

4. Add the `Clear()` method:

```
public new void Clear()
{
InnerList.Clear();
}
```

   This method clears all the items from `InnerList`.

5. Add the `Count()` method:

```
public new int Count()
{
        return InnerList.Count;
}
```

   This method returns the count of the number of items in `InnerList`.

As you can see, creating custom collections does not have to be hard. Our implementation is very simple and basic. However, such a class can be made to hold specific types instead of the generic object type. You could also make your class generic so that it accepts classes that implement a specific interface.

The following is a detailed article by Microsoft on implementing custom collections by implementing `ICollection`: `https://docs.microsoft.com/troubleshoot/dotnet/csharp/implement-custom-collection`.

As you read through this chapter, you will see different aspects of collections. You will also measure their performance. This way, as you create custom collections, you can choose the most performant way of doing things for the tasks at hand.

Now that we've briefly covered the different collection offerings in the .NET Collections Framework, let's look at what Big O notation is.

Understanding Big O notation

Big O notation is used to determine algorithmic efficiency. It determines how time scales concerning input. Constant time equates to a Big O notation value of O(1). Data operations that scale linearly over time, depending on the size of the operation, have a Big O notation value of (N), where N equals the amount of data being processed.

For example, if you were iterating over several elements in an array or collection, you would use O(N), which is a linear time, where N is the size of the array or collection. If an iteration contains pairs such as x and y, where you iterate over x in the iteration and then y in the iteration, then your Big O notation would be O(N^2). Another scenario would be identifying the amount of time it takes to harvest a square plot of land. This could be written as O(a), where a is the area of land. Alternatively, you could write the Big O notation as O(s^2), where s is the length of one size.

There are some rules to consider when using Big O notation:

- Different steps in your algorithm are added together. So, if step 1 takes O(a) time, and step 2 takes O(b) time, then your Big O notation for the algorithm will be O($a+b$).
- Drop constants. For example, if you have two operations that are both constants in your algorithm, you do not write O($2N$). The notation remains O(N).
- If you have different inputs that are different variables, such as collection a and collection b, then your Big O notation would be O($a*b$).
- Drop non-dominant terms. So, O(n^2) is equivalent to O($n + n^2$), which is equivalent to (n^2+n^2).

Now that we understand what Big O notation is and the various collections available to us, let's look at choosing the right collections for our work items.

Choosing the right collection

The key to performance when working with multiple items of data in memory is to choose the correct storage mechanism that offers the fastest processing time for your requirements. Here's the list of the different types of collections and their strengths to help you choose the right collections for the right tasks:

- A `Dictionary` is an unordered collection with contiguous storage that is directly accessible via a key. A dictionary's lookup efficiency using a key is O(1) and its manipulation efficiency is also O(1). Dictionaries are best used for high-performance lookups.

- A `HashSet` is unordered, has contiguous storage, and is directly accessible via a key. It has a lookup efficiency using a key of O(1), and a manipulation efficiency of O(1). `HashSet` is a unique unordered collection, called `Dictionary`, except the key and the value are the same object.
- A `LinkedList` lets the user have complete control over how it is ordered, does not have contiguous storage, and is not directly accessible. It has a lookup efficiency value of O(*n*), and a manipulation efficiency of O(1). It's best to use lists when you need to insert or remove items and no direct access is required.
- A `List` lets the user have complete control over how it is ordered, has contiguous storage, and is directly accessible via an index. It has a lookup efficiency using an index of O(1), and a lookup efficiency using a value of O(*n*). Its manipulation efficiency is O(*n*). It is best to use this list when direct access is required, the list is small, and there is no sorting.
- A `Queue` is ordered according to FIFO, has contiguous storage, and only has direct access from the front of the queue. It has a lookup efficiency at the front of the queue of O(1), and a manipulation index of O(1). It is essentially the same as `List<T>`, except it is only processed using FIFO.
- A `SortedDictionary` is ordered, does not have contiguous storage, and can be directly accessed using a key. It has a lookup efficiency using the key of O(*log n*) with a manipulation efficiency of O(*log n*). This collection makes a trade-off between speed and ordering and uses a binary search tree.
- A `SortedList` is ordered, has contiguous storage, and is directly accessible via a key. It has a lookup efficiency using the key of O(*log n*) and a manipulation efficiency of (O(*n*). The tree is implemented as an array, making lookups faster on preloaded data, but slower on loads.
- A `SortedSet` is ordered, does not have contiguous storage, and is directly accessible via a key. It has a lookup efficiency using a key of O(*log n*), and a manipulation efficiency of O(*log n*). It's a unique sorted collection, similar to a `SortedDictionary`, except the key and value are the same object.
- A `Stack` is ordered according to LIFO, has contiguous storage, and can only be directly accessed from the top of the stack. It has a lookup efficiency of the top item of O(1) and a manipulation efficiency of O(1)*. It is essentially the same as `List<T>`, except it is only processed using LIFO.

> **Note**
>
> For mission-critical code, it is advised that you avoid using classes in the `System.Collection` namespace. Instead, you should be using the classes from the `System.Collections.Generic` namespace. Although this may sound like tried and tested advice, you are advised to run benchmark tests to see which method is best for your particular scenario.

Now that you have been introduced to arrays and collections, we will set up our sample database before we continue looking at collections from a performance perspective.

Setting up our sample database

In this chapter, we will be demonstrating the difference between how different collection interfaces handle data. For our demonstrations, we require access to database data. To do so, we will create a database, add a table to it, and populate it with data. We will use SQL Server for our database engine and SQL Server Management Studio to develop our sample database.

To add our database, follow these steps:

1. Open **SQL Server Management Studio** and connect to your database engine.
2. Right-click on the **Databases** folder in **Object Explorer**, as shown in the following screenshot:

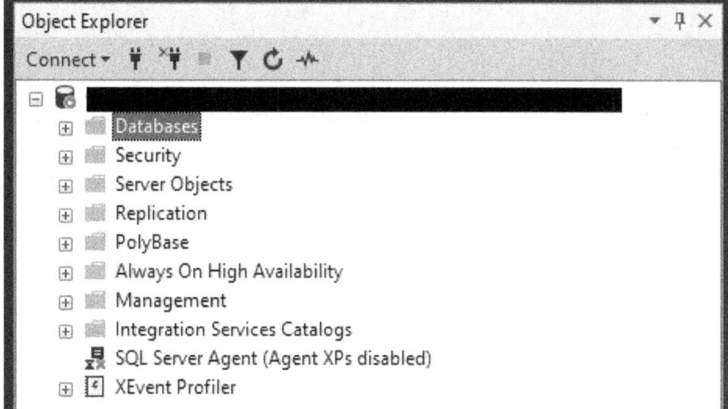

Figure 6.1 – SQL Server Management Studio – Object Explorer

3. Select **New Database** from the context menu. This will display the **New Database** dialog, as shown in the following screenshot:

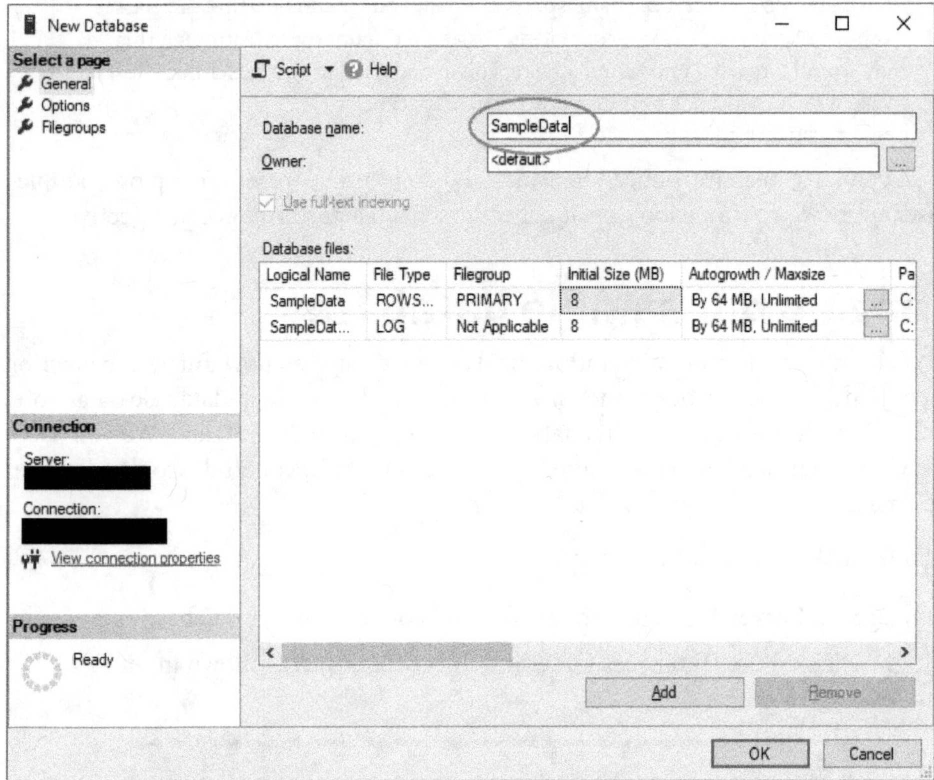

Figure 6.2 – SQL Server Management Studio – the New Database dialog

4. Once you have entered `SampleData` under **Database name**, click on the **OK** button to create the database.

5. Locate the database by expanding the **Databases** folder, and then expand the database. Right-click on the **Tables** folder and select **New | Table**. Add a new table called `Products`, as shown here:

Name	Type	Null	Primary Key	Auto Increment
Id	int	No	Yes	Yes
Name	nvarchar(50)	No	No	No
Description	nvarchar(255)	No	No	No
UnitPrice	money	No	No	No

Table 6.1 – The Products table's design

6. **Save** the table, and then expand the **Tables** folder. Right-click on the **Product** table and select **Edit Top n records**, where *n* will be the number of configured records to edit. This is 200 by default.
7. Add the data shown in the following table to the **Product** table:

Id	Name	Description	UnitPrice
1	Roasted Peanuts	500g bag of dry roasted peanuts.	0.69
2	Cashew Nuts	75g bag of cashew nuts.	0.75
3	Milk (Whole)	2 liters of whole milk.	1.25
4	Bread (50/50)	50% white and 50% wholemeal bread.	1
5	Butter (Salted)	100g salted butter.	2.5
6	Roast Chicken	5kg frozen roast chicken.	4.99
7	Potatoes	5kg Maris variety potatoes.	1.75
8	Roasting Vegetables	1kg bag of frozen roasting vegetables.	1.5
9	Coffee	1kg of Arabic coffee.	2.99
10	Demera Sugar	1kg bag of Demera sugar.	1
11	Chicken Gravy	1 tub of chicken gravy granules.	0.89
12	Yorkshire Puddings	1 bag of 12 frozen Yorkshire puddings.	1.35
13	Sage and Onion Stuffing	1 box of sage and onion stuffing.	0.59

Table 6.2 – The Product table's row data

We now have a database with a single table filled with data that we will later use in this chapter. Now, let's understand collections from a performance perspective. Let's start by looking at how we decide between using arrays or collections.

Deciding between interfaces and concrete classes

In this section, we will show that declaring a collection using an interface declaration rather than a concrete class declaration provides better time-based performance. We will accomplish this by benchmarking the generation of collections using an `IList` interface, as well as by using a `List` concrete class, so that you can see the difference in the performance of the different approaches. Follow these steps:

1. In the `CH06_Collections` project, add a new folder called `ConcreteVsInterface`.

2. In the `ConcreteVsInterface` folder, add the `ITax` interface:

```
internal interface ITax
{
        int Id { get; set; }
        TaxType TaxType { get; set; }
        TaxRate TaxRate { get; set; }
        decimal LowerLimit { get; set; }
        decimal UpperLimit { get; set; }
        decimal Percentage { get; set; }
        decimal Calculate(decimal amount);
}
```

   This interface defines a contract that various concrete tax classes will have to adhere to. It enforces impact analysis since a change in this interface will be felt by all the classes that implement it.

3. Next, add the `BaseTax` class:

```
internal abstract class BaseTax : ITax
{
     public int Id { get; set; }
     public TaxType TaxType { get; set; }
     public TaxRate TaxRate { get; set; }
     public decimal LowerLimit { get; set; }
     public decimal UpperLimit { get; set; }
     public decimal Percentage { get; set; }
     public abstract decimal Calculate(decimal amount);
}
```

   This abstract class implements the `ITax` interface but marks `Calculate(decimal amount)` as abstract so that its implementation is left up to the subclasses.

4. Now, add the `TaxRate` enum:

```
using System;
[Flags]
internal enum TaxRate
{
     TaxFreePersonalAllowance,
```

```
    StarterRate,
    BasicRate,
    IntermediateRate,
    HigherRate,
    AdditionalRate
}
```

The `TaxRate` enum provides the different types of tax rates for UK income tax.

5. Add the `TaxtType` enum:

```
[Flags]
internal enum TaxType
{
    CorporationTax,
    ValueAddedTax,
    IncomeTax,
    NationInsuranceContributions,
    ExciseDuties,
    RoadTax,
    StampDuty
}
```

The `TaxType` interface provides the different kinds of UK taxes. Add the `BaseRate` class. This class will inherit from the `BaseTax` class.

6. Then, add the following constructor:

```
public BasicRate()
{
    this.LowerLimit = 14550M;
    this.UpperLimit = 24944M;
    this.TaxType = TaxType.IncomeTax;
    this.TaxRate = TaxRate.BasicRate;
    this.Percentage = 0.2M;
}
```

This constructor sets the properties contained within `BaseClass` to the values applicable to basic rate income tax.

7. Now, implement the `Calculate(decimal amount)` method:

```
public override decimal Calculate(decimal amount)
{
        if (Percentage > 1)
                throw new Exception("Invalid percentage.
                    Percentage must be between 0 and 1.");
if (amount < LowerLimit & amount > UpperLimit)
    return 0;
return Percentage * amount;
}
```

This method checks if the percentage is less than one and throws an exception if it is not. The lower and upper amounts a person earns that are taxed are checked. If the amount is outside of this range, then zero is returned. The amount of tax on earnings is then returned and the method exits.

8. Add a new class called `TaxMan`:

```
using BenchmarkDotNet.Attributes;
using BenchmarkDotNet.Engines;
using BenchmarkDotNet.Order;
using CH06_Collections.Linq;
using System.Collections.Generic;
using System.Threading;
[MemoryDiagnoser]
[Orderer(SummaryOrderPolicy.FastestToSlowest)]
[RankColumn]
public class TaxMan { }
```

Our class is now configured to perform benchmarking using `BenchmarkDotNet`.

9. Add the following method:

```
[Benchmark]
public void BasicRateInterface()
{
        IList<BasicRate> basicRate = new
             List<BasicRate>();
}
```

The `BasicRateInterface()` method declares a list of `BasicRate` objects using the `IList` interface.

10. Add the `BasicRateConcrete()` method:

```
[Benchmark]
public void BasicRateConcrete()
{
        List<BasicRate> basicRate = new
            List<BasicRate>();
}
```

The `BasicRateConcrete()` method declares a list of `BasicRate` objects using the concrete `List` class.

11. In the `Program` class, comment out the code in the `Main` method and add the following line of code:

```
BenchmarkRunner.Run<TaxMan>();
```

This line of code will run our benchmarks. Do a release build, and then run the executable from the command line. You should see the following output or similar:

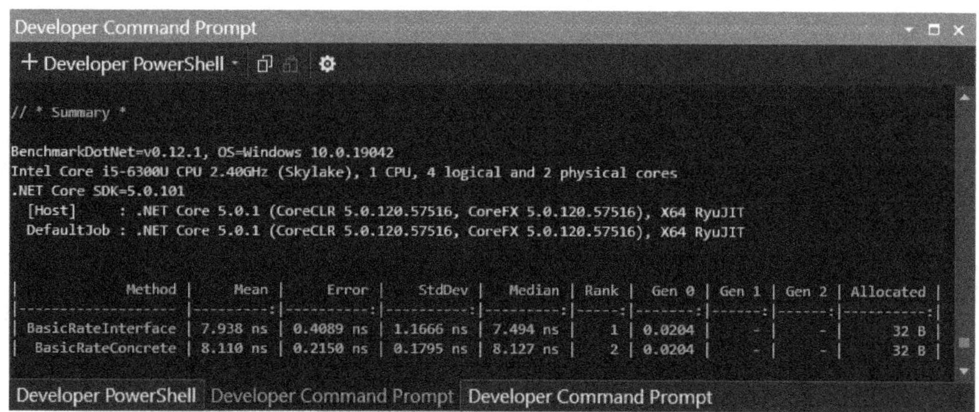

Figure 6.3 – The BenchmarkDotNet summary report showing the time difference between assigning IList<T> and List<T>

As we can see from the report, memory utilization is the same for both the interface and the concrete class implementations. But the faster instantiation time is obtained by assigning `IList<T>` instead of `List<T>`. Although the value will not be noticeable to the naked eye, it will become more noticeable over some time if there are a large number of assignments, such as when a large data iteration is taking place.

Now, let's look at array and collection performance.

Deciding between using arrays or collections

In this section, we'll discuss the pros and cons of using arrays and collections. We will also perform various benchmarks that measure array and collection performance. Armed with benchmark information, you can then make informed decisions as to whether arrays or collections are best suited to your specific needs. We will start by looking at arrays.

The downsides to using arrays are as follows:

- Arrays are fixed in size, meaning that once the size of the array has been changed, its size cannot be changed.
- Since arrays are fixed in size, they are not recommended for efficient memory usage.
- Arrays can only hold heterogeneous data types, and data types can be primitive and object types.
- Data elements of the `object` type can hold different types of data elements.
- Arrays lack many useful methods.

The benefits of using arrays are as follows:

- Arrays have a small memory footprint and have undergone some serious performance improvements in C# 9.0 and .NET 5.
- However, as arrays are fast and have undergone speed improvements, they are recommended when performance matters.

The downside to using collections is as follows:

- When it comes to performance, they are not recommended over arrays.

The benefits of using arrays are as follows:

- Collections effectively wrap arrays; `generic List<T>` is a good example.
- They are growable, which means that we can shrink and grow our collections as required. Because of this, collections are recommended over arrays when it comes to efficient memory utilization.
- Data elements (item data) in a collection can be homogeneous and heterogeneous.
- Collection classes have ready-made method support for most operations and can easily be extended. By this, we mean that arrays lack some useful methods that we get for free when we use collections.

> **Note**
>
> It is recommended that you do not use the collections in the `System.Collections` namespace. Instead, you are encouraged to use the collections in the `System.Collections.Generic` namespaces.

The standard collection that most programmers will be familiar with is the generic `List<T>` class. In this section, we will create a new project. Then, we will build up a `uint` array and a `List<uint>` collection and iterate through them. This process will be benchmarked using `BenchmarkDotNet`.

We will be benchmarking adding items, iterating through, and retrieving items from arrays and collections. So, let's begin:

1. Add a new class under the project root called `ArraysVsCollections` with the following `using` statements:

```
using BenchmarkDotNet.Attributes;
using BenchmarkDotNet.Order;
using System;
using System.Collections;
using System.Collections.Generic;
using System.Linq;
```

These `using` statements give us what we need to work with arrays and collections and benchmark them.

2. Add the following member variables:

```
private int[] array;
private List<int> collection;
```

The array of `int` and the list of `int` will be used to benchmark adding, getting, and iterating arrays and collections.

3. Next, add the `GlobalSetup()` method:

```
[GlobalSetup]
public void GlobalSetup()
{
array = new int[1000];
collection = new List<int>(1000);
for (int i = 0; i < 1000; i++)
{
```

```
        array[i] = i;
        collection.Add(i);
    }
}
```

The `GlobalSetup()` method is attributed to the `[GlobalSetup]` attribute. This informs `BenchmarkDotNet` to run this method before all other benchmark methods. It initializes the array and collection with a size of `1000` and adds a value of `i` in the current iteration to both the array and collection.

4. Although we will not be utilizing the `GlobalCleanup()` method, we will add it for completeness so that you know how to perform cleanup operations when benchmarking:

```
[GlobalCleanup]
public void GlobalCleanup()
{
// Disposing logic
}
```

The `GlobalCleanup()` method is where you would provide your cleanup logic if it were needed.

5. Now, add the `ArrayAdd1000Logic()` method:

```
[Benchmark]
public void ArrayAdd1000Logic1()
{
int[] list = new int[1000];
for (int i = 0; i < 1000; i++)
{
    list[i] = i;
}
}
```

The `ArrayAdd1000Logic()` method declares an array of 1000 `int` values and later proceeds to add integer values to each element in the array.

6. Add the `CollectionAdd1000Logic()` method:

```
[Benchmark]
public void CollectionAdd1000Logic()
{
```

```
Ilist<int> list = new new List<int>();
for (int i = 0; i < 1000; i++)
        list.Add(i)
}
```

The `CollectionAdd1000Logic ()` method declares a list of `int` elements. Then, it loops 1,000 times using a `for` loop and adds the current value to the collection.

7. Add the `ArrayIterationLogic()` method:

```
[Benchmark]
public int ArrayIterationLogic()
{
int res = 0;
for (int i = 0; i < 1000; i++)
     res += array[i];
return res;
}
```

The `ArrayIterationLogic()` method declares an `int` variable and assigns it a value of `0`. A `for` loop is used to iterate 1,000 times and add the value of the array at the index position to the `res` value. Once the iteration is over, the `res` variable is returned.

8. Now, add the `CollectionIterationLogic()` method:

```
[Benchmark]
public int CollectionIterationLogic()
{
int res = 0;
for (int i = 0; i < 1000; i++)
     res += collection[i];
return res;
}
```

`CollectionIterationLogic()` declares an `int` variable and assigns it a value of `0`. A `for` loop is used to iterate 1,000 times and add the value of the array at the index position to the `res` value. Once the iteration is over, the `res` variable is returned.

9. Add the `ArrayGetElement500Logic()` method:

```
[Benchmark]
public int ArrayGetElement500Logic()
{
return array[500];
}
```

The `ArrayGetElement500Logic()` method returns the value of the array at position `500`.

10. Now, add the `CollectionGetElement500Logic()` method:

```
[Benchmark]
public int CollectionGetElement500Logic()
{
return collection[500];
}
```

The `CollectionGetElement500Logic()` method returns the value of the collection at position `500`.

11. Replace the code in the `Main` method with the following line of code:

```
BenchmarkRunner.Run<ArraysVsCollections>();
```

This call will run our benchmarks. Release build your code and run it from the console. You should see a report with similar timings to those shown in the following screenshot:

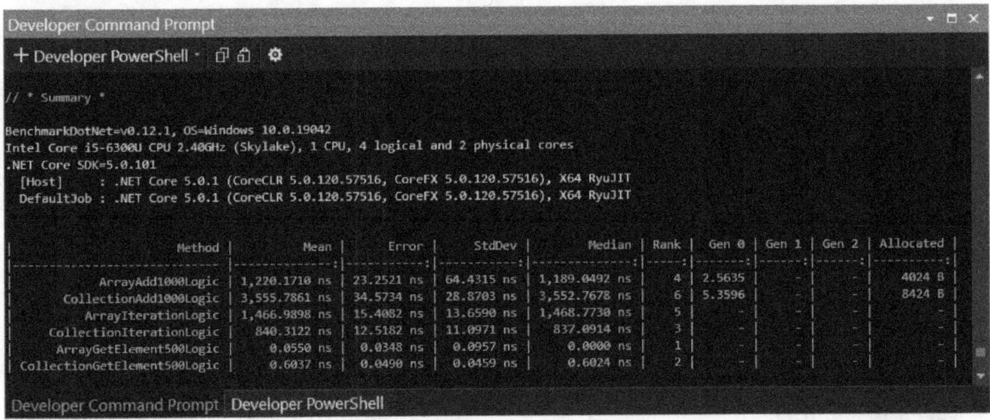

Method	Mean	Error	StdDev	Median	Rank	Gen 0	Gen 1	Gen 2	Allocated
ArrayAdd1000Logic	1,220.1710 ns	23.2521 ns	64.4315 ns	1,189.0492 ns	4	2.5635	-	-	4024 B
CollectionAdd1000Logic	3,555.7861 ns	34.5734 ns	28.8703 ns	3,552.7678 ns	6	5.3596	-	-	8424 B
ArrayIterationLogic	1,466.9898 ns	15.4082 ns	13.6590 ns	1,468.7730 ns	5	-	-	-	-
CollectionIterationLogic	840.3122 ns	12.5182 ns	11.0971 ns	837.0914 ns	3	-	-	-	-
ArrayGetElement500Logic	0.0550 ns	0.0348 ns	0.0957 ns	0.0000 ns	1	-	-	-	-
CollectionGetElement500Logic	0.6037 ns	0.0490 ns	0.0459 ns	0.6024 ns	2	-	-	-	-

Figure 6.4 – The BenchmarkDotNet summary report for array and collection operations

Looking at the performance in terms of time, adding items to an array is faster than adding items to a collection. Iterating a collection is faster than iterating over an array and getting an item from an array using its index is faster than getting a collection from a collection by its index. Based on these findings, you need to decide what your requirements are, and then choose the best type based on these requirements.

Now, let's look at indexers.

Accessing objects using indexers

Indexes enable objects in classes to be accessed in the same way you access items in an array. An indexer will have a modifier, a return type, the `this` keyword to indicate the object of the current class, and an argument list. You will always use the `this` keyword when creating an indexer. Indexer is the term given to a parameterized property. The index is created using the `get` and `set` accessors. You are not allowed to use the `ref` or `out` keywords to modify indexer parameters. A minimum of one parameter should be specified. An indexer cannot be static since it is an instance member. However, the indexer properties can be static. You would implement an indexer if you need to operate on a group of elements. The main difference between a property and an indexer is that you identify and access a property by its name. On the other hand, with an indexer, it is identified by its signature and accessed using indexes. Moreover, you can overload indexers.

Now, let's write a simple indexer example. In this example, we will have a class that has a constructor that takes a size. This size will set the size of an internal array of strings. We will be able to get the index of a string in the array by name and get an item from the array by index using indexers. Follow these steps:

1. Add a new class called `Indexers` and add a `using` statement to `System namespace`. Then, add the following array and constructor at the top of the class:

```
private string[] _items;
public Indexers(int size)
{
        _items = new string[size];
}
```

The `_items` array will contain several strings. The size of the array is set by the value that's passed into the constructor that initializes the array.

2. Add the indexer to get a string by index:

```
public string this[int index]
{
    get
    {
        if (IsValidIndex(index))
            return _items[index];
        else
            return string.Empty;
    }
    set
    {
        if (IsValidIndex(index))
            _items[index] = value;
    }
}
```

   This indexer uses an `int` value to get an item from the array and set the value of the array at the given index. Items are only set and retrieved if the index is valid.

3. We can check the index by passing it into the `IsValidIndex(int index)` method, which returns a `bool`. Let's add the `IsValidIndex(int index)` method:

```
private bool IsValidIndex(int index)
{
    return index > -1 && index < _items.Length;
}
```

   This method returns `true` if the index is greater than -1 and less than the length of the array. Otherwise, it returns `false`.

4. Now, add the index that takes a `string` and returns the string's index:

```
public int this[string item]
{
    get
```

```
{
    return Array.IndexOf(_items, item);
}
}
```

This indexer takes a `string`. Then, it looks up the index for the string and returns the index. There is no setter for this index.

5. In the `Program` class, add the `IndexerExample()` method:

```
public static void IndexerExample()
{
     Indexers indexers = new Indexers(1000);
     for (int i = 0; i < 1000; i++)
          indexers[i] = $"Item {i}";
Console.WriteLine($"The item at position 500 is
     \"{indexers[500]}\".");
Console.WriteLine($"The index of \"Item 500\" is
     {indexers["Item 500"]}.");
}
```

This method creates a new `Indexer` object with an internal array size of `1000`. Then, it loops 1,000 times and sets the value of each item in the array. After that, it prints out the value of the array at position 500 and prints out the value of `Item 500`.

6. Comment out the code in the `Main` method, and then add the following line:

```
IndexerExample();
```

This statement calls the method that executes our `Indexer` method. You should see the following output:

```
The item at position 500 is "Item 500".
The index of "Item 500" is 500.
```

That concludes our look at indexers. As you can see, they are pretty simple. You can use any data item that you like for an indexer. However, it will be up to you to see how well such indexers perform. Now, let's look at the difference between the `IEnumerable` and `IEnumerator` interfaces.

Comparing IEnumerable and IEnumerator

The `IEnumerable` and `IEnumerator` interfaces can both be used for iteration but in different ways. Let's understand each in brief.

An object of the `IEnumerable` type will know how to traverse the collection that it holds, regardless of what its internal structure is like. There is one method that makes up an enumerable: `GetEnumerator()`. It returns as an instance of a class that implements the `IEnumerable` interface. Iteration is normally carried out using a `foreach` loop. Iterations of an enumerable are carried out using a `foreach` loop. However, an enumerable does not remember its location when iterating.

Objects of the `Ienumerator` type declare two methods: `MoveNext()` and `Reset()`. There is one property called `Current` that gets the current item in the list that's being enumerated. The `MoveNext()` method moves to the next record in a collection and returns a Boolean value indicating the end of the collection. `Reset()` will reset the position to the first item in the collection. The `Current` property is called through an object that implements the `IEnumerable` interface, which returns the current element in the collection. An enumerator remembers its current location and uses a `while` loop when iterating.

Let's see which method of enumeration is fastest. Will it be looping using an enumerable, or will it be looping using an iterator?

1. Add a new class called `IEnumerableVsIEnumerable` with the following `using` statements:

   ```
   using BenchmarkDotNet.Attributes;
   using BenchmarkDotNet.Running;
   using System;
   using System.Collections;
   using System.Collections.Generic;
   using System.Diagnostics;
   ```

   These `using` statements provide the elements we will need to build and test the performance between `IEnumerable` and `IEnumerator`.

2. Add the following code to the class:

   ```
   private List<int> _years;
   public IEnumerableVsIEnumerator()
   {
     _years = new List<int> { 1970, 1971, 1972, 1973, 1974,
   ```

```
        1975, 1976, 1977, 1978, 1979 };
}
```

Here, we are declaring a list of `int` values that will hold several year values. Our constructor then initializes the array with the years `1970` to `1979`.

3. Add the `IterateEnumerator1970to1975()` method:

```
public void IterateEnumerator1970To1975()
{
    var years = _years.GetEnumerator();
    while (years.MoveNext())
{
        Debug.WriteLine(years.Current);
        if (years.Current > 1975)
            IterateEnumberator1976To1979(years);
}
}
```

This method iterates over the values `1970` to `1975` and prints the values out to the debug window.

4. If the current year is greater than `1975`, then the enumerator is passed into the `IterateEnumerator1976To1979(IEnumerator<int> years)` method, which we will add now:

```
public void IterateEnumberator1976To1979
        (IEnumerator<int> years)
{
while (years.MoveNext())
{
        Debug.WriteLine(years.Current);
}
}
```

This method takes in an enumerator and iterates through it. On each iteration, it prints the current year to the debug window.

5. Add the following line to the end of the `Main` method in the `Program` class:

```
IEnumerableVsIEnumeratorExample();
```

This line of code calls a method that will run our example and show how an enumerator remembers where it is in the iteration.

6. Add the `IEnumerableVsIEnumeratorExample()` method to the `Program` class:

```
private static void IEnumerableVsIEnumeratorExample()
{
    IEnumerableVsIEnumerator eve = new
      IEnumerableVsIEnumerator();
    eve.IterateEnumerator1970To1975();
}
```

This method runs our code. If you do a debug build and run the code, then you should see the years *1970* to *1979* printed to the output window.

Now that you have seen an enumerator in action, we will add two methods to the `IEnumerableVsIEnumerator` class.

7. Add the `BenchmarkIEnumerabled()` method:

```
[Benchmark]
public void BenchmarkIEnumerable()
{
    IEnumerable<int> enumerable = IEnumerable<int>) _years;
    foreach (int i in enumerable)
        Debug.WriteLine(i);
}
```

This method uses an enumerable and a `foreach` loop to iterate through the years and write them to the debug window.

8. Add the `BenchmarkIEnumerator()` method:

```
[Benchmark]
public void BenchmarkIEnumerator()
{
    IEnumerator<int> enumerator = _years.GetEnumerator();
    while (enumerator.MoveNext())
        Debug.WriteLine(enumerator.Current);
}
```

This method uses an enumerator and a `while` loop to iterate through the years and write them to the debug window.

9. Comment out the code in the `Main` method in the `Program` class, and then add the following line:

```
BenchmarkRunner.Run<IEnumerableVsIEnumerator>();
```

This line of code detects our benchmarks and runs them to produce a summary report on performance. Do a release build and run the program from the command prompt. You should see the following output:

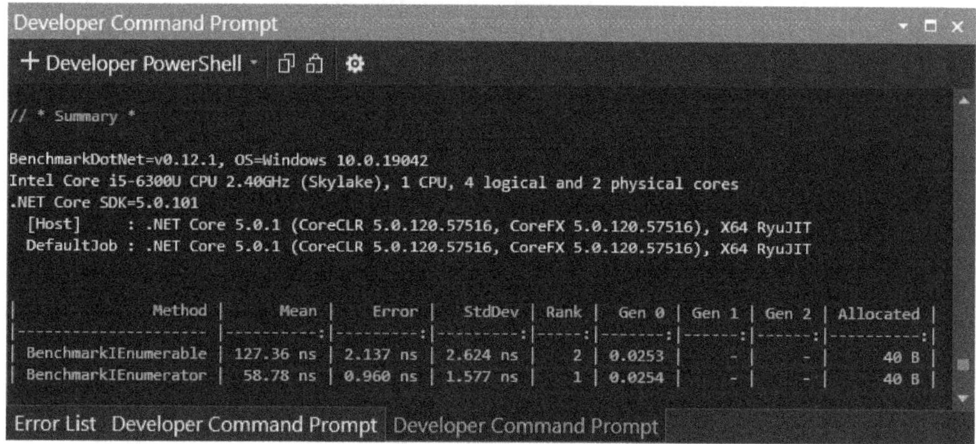

Figure 6.5 – The BenchmarkDotNet summary report showing that IEnumerator is faster than IEnumerable

As we can see, even though `IEnumerable` and `IEnumerator` both perform iterations on the same collection, they do so in different ways. And by viewing the benchmarking summary report, we can see that the clear winner in terms of performance is the `IEnumerator` interface. Now, let's look at the difference between `IEnumerable`, `IEnumerator`, and `IQueryable`, and the effects these differences have on performance when performing LINQ queries on a database.

Database query performance

In the previous section, we saw how `IEnumerator` is different from and performs faster than `IEnumerable` when iterating through an in-memory collection. Now, let's query a database and iterate through the resulting collection using various benchmarked techniques. To do so, we'll follow these steps:

1. Add a new class called `IEnumeratorVsIQueryable`.
2. We will be connecting to a SQL Server database, and we will have information we need to keep secret. Our `secret.json` files do not get checked into version control. So, right-click on the project and select **Manage User Secrets** from the context menu.
3. A dialog box will pop up, informing you that additional packages are required. Click on **Yes**:

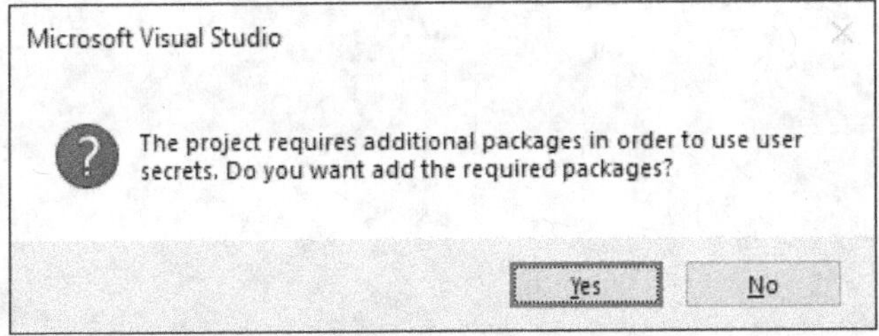

Figure 6.6 – A dialog box, informing you that additional packages are required to manage user secrets

4. Visual Studio will then open the `secrets.json` file in a new tab. This is where you will add your user secrets.
5. Open the Package Manager Console and add the following packages:
   - `Microsoft.EntityFrameworkCore`
   - `Microsoft.EntityFrameworkCore.SqlServer`
   - `Microsoft.EntityFrameworkCore.Tools`
   - `Microsoft.Extensions.Configuration`
   - `Microsoft.Extensions.Configuration.EnvironmentVariables`
   - `Microsoft.Extensions.Configuration.UserSecrets`
   - `Microsoft.Extensions.OptionsConfigurationExtensions`

   These packages allow you to connect to and extract data from our SQL Server database.

6. Update your `secrets.json` file with the connection string to the database that we created at the start of this chapter:

```
{
  "DatabaseSettings": {
    "ConnectionString": "YOUR _ CONNECTION _ STRING"
  }
}
```

   This connection string will be used to connect to our database, perform a query that returns some data, and allow us to iterate through that data and perform some operations on it.

7. Add a folder called `Configuration`. In that folder, add a class called `SecretsManager` with an empty static constructor and the following `using` statements:

```
using Microsoft.Extensions.Configuration;
using System;
using System.IO;
```

   We need these `using` statements for our file I/O and system configuration, such as obtaining secrets from a `secrets.json` file.

8. Add the following line at the top of the `SecretsManager` class:

```
public static IConfigurationRoot Configuration { get;
    set; }
```

   This line declares our static configuration property, which is used to obtain the configuration data within our application.

9. Now, add the following code:

```
public static T GetSecrets<T>(string sectionName)
    where T : class
{
var devEnvironmentVariable = Environment
    .GetEnvironmentVariable("NETCORE _ ENVIRONMENT");
var isDevelopment = string.IsNullOrEmpty
    (devEnvironmentVariable) || devEnvironmentVariable
        .ToLower() == "development";
var builder = new ConfigurationBuilder()
```

```
    .SetBasePath(Directory.GetCurrentDirectory())
.AddJsonFile("appsettings.json", optional: true,
    reloadOnChange: true)
.AddEnvironmentVariables();
if (isDevelopment) //only add secrets in development
{
    builder.AddUserSecrets<T>();
}
Configuration = builder.Build();
return Configuration.GetSection(sectionName).Get<T>();
}
```

This code gets the environment variables for the .NET Core environment. Then, it gets the code to see if it is running in a software development environment. The configuration is built for the environment it will be running in. If we are in development, then we must add our `secrets` class as defined by the `T` variable. Switch to the `Product` class in the `Models` folder.

10. Add a `using` statement for `System.ComponentModel.DataAnnotations`. Change the struct to a class, and add the `[Key]` attribute to the `Id` property. We need these changes since we are using Entity Framework to connect to a database and extract data.
11. Add the `DatabaseSettings` class to the `Configuration` folder:

```
public class DatabaseSettings
{
    public string ConnectionString { get; set; }
}
```

This class has a single property called `ConnectionString` that will hold our connection string to our `SampleData` database. Notice that the name of the class and property match the name of the JSON section and property!

12. Now, add `appsettings.json` to the root of your project with the following contents:

```
{
  "DatabaseSettings": {
    "ConnectionString": "Set in Azure. For
        development, set in User Secrets"
  }
}
```

This file contains the same layout as the `secrets.json` file and the `DatabaseSettings` class. This file is used to store our connection string. In development, it is set in our `secrets` file, while in production, it is set in Azure. Now that we have our database configuration in place, we can add our benchmarking code.

13. Add a new class to the root of the project called `DatabaseQueryAndIteration` that implements `IDisposable` with the following code:

```
using BenchmarkDotNet.Attributes;
using BenchmarkDotNet.Order;
using CH06_Collections.Configuration;
using CH06_Collections.Data;
using CH06_Collections.Models;
using Microsoft.Extensions.Options;
using System;
using System.Collections.Generic;
using System.Diagnostics;
using System.Linq;
[MemoryDiagnoser]
[Orderer(SummaryOrderPolicy.Declared)]
[RankColumn]
public class DatabaseQueryAndIteration : IDisposable
{
}
```

This code declares our class and defines the fact that it implements `IDisposable`. It is also configured to be benchmarked.

14. Implement the `IDisposable` interface in our class:

```
private bool disposedValue;
protected virtual void Dispose(bool disposing)
{
    if (!disposedValue) {
        if (disposing)
            _context.Dispose();
        disposedValue = true;
    }
}
```

```
public void Dispose(){
    Dispose(disposing: true);
    GC.SuppressFinalize(this);
}
```

This code disposes of our managed resources and suppresses the call to the class finalizer method.

15. We have everything in place to benchmark the methods in this class, access database resources, and clean up after ourselves. Add the following code to the class:

```
private DatabaseContext _context;
[GlobalSetup]
public void GlobalSetup()
{
    var connectionString = SecretsManager.
        GetSecrets<DatabaseSettings>(nameof
          (DatabaseSettings)).ConnectionString;
_context = new DatabaseContext(connectionString);
}
[GlobalCleanup]
public void GlobalCleanup()
{
        Dispose(true);
}
```

The `_context` variable provides us with our database access. The `GlobalSetup()` method gets our connection string from our secrets file and creates a new `DatabaseContext` using the safely stored connection string. The `GlobalSetup()` method will run before our benchmarks. The `GlobalCleanup()` method calls the `Dispose(disposing)` method to clean up our managed resources after our benchmarks have finished running.

16. Next, add the `QueryDb()` method:

```
[Benchmark]
public void QueryDb()
{
    var products = (from p in _context.Products
                      where p.Id > 1 select p);
```

```
foreach (var product in products)
    Debug.WriteLine(product.Name);
}
```

The `QueryDb()` method performs a simple LINQ query on the database by selecting products with an ID that's greater than `1`. Then, it iterates each product in the `lQueryable<Product>` list and writes the product name out to the debug window.

17. Now, add the `QueryDbAsList()` method:

```
[Benchmark]
public void QueryDbAsList()
{
List<Product> products = (from p in _context.Products
where p.Id > 1
select p).ToList<Product>();
foreach (var product in products)
Debug.WriteLine(product.Name);
}
```

`QueryDbAsList()` performs the same query as `QueryDb()`, except the processed type is of the `List<Product>` type.

18. Add the `QueryDbAsIEnumerable()` method:

```
[Benchmark]
public void QueryDbAsIEnumerable()
{
var products = (from p in _context.Products
                where p.Id > 1
                select p).AsEnumerable<Product>();
foreach (var product in products)
    Debug.WriteLine(product.Name);
}
```

The `QueryDbAsIEnumerable()` method performs the same query as `QueryDbAsList`, but the processed type is of the `Ienumerable<Product>` type instead.

19. Add the `QueryDbAsIEnumerator()` method:

```
[Benchmark]
public void QueryDbAsIEnumerator()
{
    var products = (from p in _context.Products
                    where p.Id > 1
                    select p).GetEnumerator();
while (products.MoveNext())
    Debug.WriteLine(products.Current.Name);
}
```

`QueryDbAsIEnumerator()` does the same as the previous methods but operates on the `IEnumerator<Product>` type and iterates using a `while` loop instead of a `foreach` loop.

20. Our final method in this class is the `QueryDbAsIQueryable()` method:

```
[Benchmark]
public void QueryDbAsIQueryable()
{
var products = (from p in _context.Products
                where p.Id > 1
                select p).AsQueryable<Product>();
foreach (var product in products)
    Debug.WriteLine(product.Name);
}
```

This method is the same as `QueryDb` but explicitly operates on the `IQueryable<Product>` type.

21. Replace the code in the `Main` method within the `Program` class with the following code:

```
BenchmarkRunner.Run<DatabaseQueryAndIteration>();
```

This code runs our benchmarks. Do a release build of the code and run the executable from the command line. You should see a summary report similar to the following:

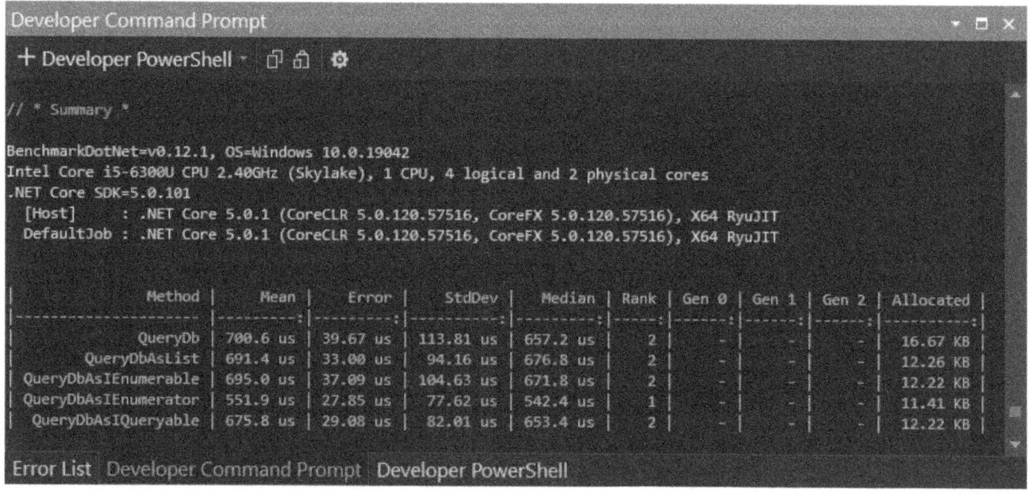
Developer Command Prompt
+ Developer PowerShell

// * Summary *

BenchmarkDotNet=v0.12.1, OS=Windows 10.0.19042
Intel Core i5-6300U CPU 2.40GHz (Skylake), 1 CPU, 4 logical and 2 physical cores
.NET Core SDK=5.0.101
  [Host]   : .NET Core 5.0.1 (CoreCLR 5.0.120.57516, CoreFX 5.0.120.57516), X64 RyuJIT
  DefaultJob : .NET Core 5.0.1 (CoreCLR 5.0.120.57516, CoreFX 5.0.120.57516), X64 RyuJIT

Method	Mean	Error	StdDev	Median	Rank	Gen 0	Gen 1	Gen 2	Allocated
QueryDb	700.6 us	39.67 us	113.81 us	657.2 us	2	-	-	-	16.67 KB
QueryDbAsList	691.4 us	33.00 us	94.16 us	676.8 us	2	-	-	-	12.26 KB
QueryDbAsIEnumerable	695.0 us	37.09 us	104.63 us	671.8 us	2	-	-	-	12.22 KB
QueryDbAsIEnumerator	551.9 us	27.85 us	77.62 us	542.4 us	1	-	-	-	11.41 KB
QueryDbAsIQueryable	675.8 us	29.08 us	82.01 us	653.4 us	2	-	-	-	12.22 KB

Error List Developer Command Prompt Developer PowerShell

Figure 6.7 – The different times and memory allocations of various database query types using LINQ

In terms of memory usage, the worst performer is the `QueryDb()` method, followed by the `QueryDbAsList()` method. `QueryDbAsIEnumerable()` and `QueryDbAsIQueryable()` are both slightly better than the previous two. However, the best performing method in terms of memory allocation out of all five methods is the `QueryDbAsIEnumerator()` method.

Speedwise, the `QueryDb()` method was the worst again, followed by `QueryDbAsIEnumerable()`, then `QueryDbAsList()`, and then `QueryDbAsIQueryable()`. And again, the best performer in terms of speed is the `QueryDbAsIEnumerator()` method.

Here, we can see that the best performing method for querying and iterating a database in terms of both speed and memory usage is the `QueryDbAsIEnumerator()` method. Now, let's look at the `yield` keyword.

Exploring the yield keyword

The **yield** keyword is contextual and is used with iterators. The following are the two ways to use the `yield` keyword:

- `yield return <expression>;`: This returns the value of the expression.
- `yield break;`: This will exit from the iteration

When using the `yield` keyword, there are some restrictions to be aware of. These are as follows:

- You cannot use the `yield` keyword in `unsafe` blocks of code.
- You cannot use the `ref` or `out` parameters for methods, operators, or accessors.
- You cannot return using the `yield` keyword in a `try-catch` block.
- You cannot use the `yield` keyword in anonymous methods.
- You can use `yield` in a `try` block if the `try` block is followed by the `finally` block.
- You can use `yield break` in a `try-catch` block but not the `finally` block.

In this section, we are going to add a class that shows the `yield` keyword in action. Then, we will benchmark two ways to return an `IEnumerable<long>` consisting of 1 million items, and show the vast difference in performance between them. Let's begin:

1. Add a new class called `Yield` to the root of the project:

```
using BenchmarkDotNet.Attributes;
using BenchmarkDotNet.Order;
using System;
using System.Collections.Generic;
[MemoryDiagnoser]
[Orderer(SummaryOrderPolicy.Declared)]
[RankColumn]
public class Yield { }
```

This class will benchmark the use of the `yield` keyword.

2. Now, add the `YieldSample()` method:

```
public void YieldSample()
{
DoCountdown();
PrintMonthsOfYear();
DoBreakIteration();
}
```

The `YieldSample()` method will be called from our `Program` class. It will run all three methods.

3. Add the `Countdown()` method:

```
private IEnumerable<int> Countdown()
{
    for (int x = 10; x >= 0; x--)
    yield return x;
}
```

This method loops from `10` to `0`. Each iteration is returned using the `yield` keyword.

4. Add the `DoCountdown()` method:

```
private void DoCountdown()
{
foreach (int x in Countdown())
    Console.WriteLine(x);
}
```

The `DoCountdown()` method prints the countdown from `10` to `0` to the console window.

5. Add a class called `Month`:

```
internal class Month
{
    public string Name { get; set; }
    public int MonthOfYear { get; set; }
}
```

This class holds the name of a month of the year and its number.

6. Now, add the `Months` class:

```
internal class Months
{
   public IEnumerable<Month> MonthsOfYear
   {
    get
    {
    yield return new Month { Name = "January",
        MonthOfYear = 1 };
    yield return new Month { Name = "February",
```

```
            MonthOfYear = 2 };
        yield return new Month { Name = "March",
            MonthOfYear = 3 };
        yield return new Month { Name = "April",
            MonthOfYear = 4 };
        yield return new Month { Name = "May",
            MonthOfYear = 5 };
        yield return new Month { Name = "June",
            MonthOfYear = 6 };
        yield return new Month { Name = "July",
            MonthOfYear = 7 };
        yield return new Month { Name = "August",
            MonthOfYear = 8 };
        yield return new Month { Name = "September",
            MonthOfYear = 9 };
        yield return new Month { Name = "October",
            MonthOfYear = 10 };
        yield return new Month { Name = "November",
            MonthOfYear = 11 };
        yield return new Month { Name = "December",
            MonthOfYear = 12 };
        }
    }
}
```

This class returns a collection of `Month` objects using the `yield` keyword. Switch back to the `Yield` class.

7. Add the `PrintMonthsOfYear()` method:

```
private void PrintMonthsOfYear()
{
foreach (Month month in new Months().MonthsOfYear)
    Console.WriteLine($"{month.Name} is month
        {month.MonthOfYear} of the year.");
}
```

This method iterates through the months of the year and prints them out to the console window.

8. Add the `BreakIteration()` method:

```
private IEnumerable<int> BreakIteration()
{
int x = 0;
while (x < 20)
{
    if (x < 15)
        yield return x;
    else
        yield break;
    x++;
}
}
```

This method iterates `20` times. A check is made upon each iteration. If the value is less than `15`, the result is yielded and the variable is incremented. Otherwise, the iteration is exited.

9. Add the `DoBreakIteration()` method:

```
private void DoBreakIteration()
{
        foreach (int x in BreakIteration())
            Console.WriteLine($"Line {x}:");
}
```

The `DoBeakIteration()` method iterates through `BreakIteraton()` and writes the value to the console window.

10. In the `Program` class, add a method called `Yield()`, and call it from your `Main` method:

```
private static void Yield()
{
        var yieldToMe = new Yield();
        yieldToMe.YieldSample();
}
```

This method runs our `yield` keyword examples. Do a debug build and step through the code so that you can see how it behaves. You will see that each time the `yield` keyword is encountered, it returns to the calling method. Then, it continues the iteration from where it left off.

11. Now, let's add our benchmarking to test the performance of the `yield` keyword. Add the `GetValues()` method:

```
public IEnumerable<long> GetValues()
{
        List<long> list = new List<long>();
        for (long i = 0; i < 1000000; i++)
              list.Add(i);
return list;
}
```

This method creates a collection of `long` values using a generic `List`. It iterates 1 million items and adds them to the collection. Once complete, the collection is returned to the caller as an `IEnumerable<long>` collection.

12. Add the `GetValuesYield()` method:

```
public IEnumerable<long> GetValuesYield()
{
        for (long i = 0; i < 1000000; i++)
            yield return i;
}
```

This method iterates through 1 million items and returns a collection of `IEnumerable<long>`. The iteration uses the `yield` keyword, so each iteration is returned to the caller.

13. Add the `GetValuesBenchmark()` method:

```
[Benchmark]
public void GetValuesBenchmark()
{
        var data = GetValues();
}
```

This method benchmarks the `GetValues()` method.

14. Add the `GetValuesYieldBenchmark()` method:

```
[Benchmark]
public void GetValuesYieldBenchmark()
{
    var data = GetValuesYield();
}
```

This method benchmarks the `GetValuesYield()` method.

15. Replace the code in the `Main` method in the `Program` class with the following line of code:

```
BenchmarkRunner.Run<Yield>();
```

This line of code runs our benchmarks. Do a release build and then run the executable from the command line. You should see the following summary report:

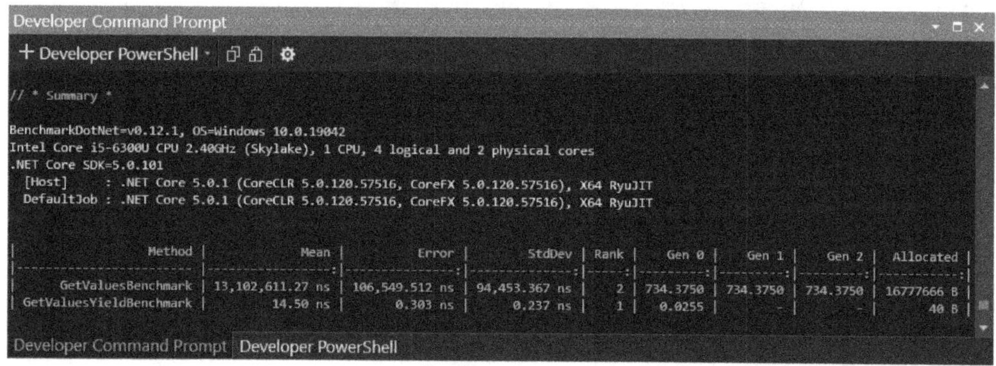

Figure 6.8 – The BenchmarkDotNet summary report showing the performance benefits of using the yield keyword

As you can see from the report, building a list of 1 million `long` values is much slower compared to using the `yield` keyword. The `yield` keyword significantly speeds up how collections are processed. That's a 13,102,611.27 ns / 14.50 ns = 903,628.26 times increase in performance! So, you can see that the use of the `yield` keyword is very beneficial to the performance of your computer programs.

In the next section, we will look at the difference between concurrency and parallelism and the effects they have on performance.

Learning the difference between concurrency and parallelism

Concurrency and parallelism are often mistaken for the same thing, but they are different. Concurrency does many tasks at the same time using multi-threading. Multi-threading allots time to various threads based on time/context switching. This presents the illusion that the computer is doing multiple things at the same time. But it is, in reality, only doing one thing. Parallelism, on the other hand, does many things all at the same time.

Concurrency is used to manage multiple computations simultaneously. It accomplishes this using interleaving operations. The benefit of concurrency is that it increases the amount of work that can be completed over time. It uses context switching to perform interleaving operations. Concurrency can work with a single processor. You are already aware of concurrency at work, as you will have had multiple applications running at the same time. All these programs are making use of concurrency.

The main usage of concurrency is to have usable applications that are non-blocking. For example, if you have an application that performs a long-running operation, this operation can be run on a background thread to allow the user to still use the application and get work done. So, concurrency is not necessarily about performance – it is more about not blocking your users from being able to do what they intend with your application.

Parallelism performs multiple computations at the same time in parallel to each other. To accomplish parallelism, multiple processors are required. The benefit of using parallelism is increased computational processing speed. Running document crawlers over a cluster and performing parallel queries and big data are examples of using parallelism.

The main goal of parallelism is performance. In other words, the intention of using parallelism is to complete an operation in the shortest amount of time. An example of parallelism in use would be data-intensive number crunching for report generation.

You should never mix concurrency with performance. If you do, your design will either be bad or over-engineered. So, if you want user interfaces to be non-blocking, use concurrency. However, if you want non-UI tasks to complete in the shortest possible time, use parallelism. Later in this book, we will devote whole chapters to concurrency, parallelism, and asynchronous processing. But for now, let's turn our attention to the difference between `Equals()` and `==`.

Learning the difference between Equals() and ==

The == operator compares object references, known as shallow comparison, while the `Equals()` method compares object content, known as deep comparison. Both the operator and the method can be overloaded.

> **Note**
>
> If you overload the == operator, then you should overload the `Equals()` method and vice versa.

The == operator returns `true` in the following situations:

- `Value Type Value == Value Type Value`
- `Reference Type Instance == Reference Type Instance`
- `String == String`

The `Equals()` method returns `true` in the following situations:

- `ReferenceType.Equals(ReferenceType)` both refer to the same object reference
- `ValueType.Equals(ValueType)` are both the same type and have the same value

Now, let's add a new class called `Equality` to the root of the *CH06_Collections* project to demonstrate the difference in performance between the == operator and the `Equals()` method. Let's get started:

1. Add the `Equality` class, as follows:

```
using BenchmarkDotNet.Attributes;
using BenchmarkDotNet.Order;
using System;
using System.Collections;
using System.Collections.Generic;
using System.Diagnostics;
using System.Linq;
[MemoryDiagnoser]
[Orderer(SummaryOrderPolicy.Declared)]
```

```
[RankColumn]
public class Equality { }
```

With that, our class has been configured to perform benchmarking.

2. Add the following code to the top of the class:

```
private List<string> _listOne;
private List<string> _listTwo;
private int _value1;
private int _value2;
private string _string1;
private string _string2;
```

Here, we have our value types, reference types, and string types in place that will have their equality tested.

3. Now, add the `GlobalSetup()` method:

```
[GlobalSetup]
public void GlobalSetup()
{
        _listOne = new List<string>
{
     "Alpha", "Beta", "Gamma", "Delta", "Eta", "Theta"
};
  _listTwo = _listOne;
  _value1 = 123;
  _value2 = _value1;
  _string1 = "Hello, world!";
  _string2 = _string1;
}
```

This method assigns our variables in preparation for our equality benchmarks.

4. Add the `ValueOperatorValue()` method:

```
[Benchmark]
public void ValueOperatorValue()
{
        bool value = _value1 == _value2;
}
```

The `ValueOperatorValue()` method benchmarks the equality checking of two values using the `equality` operator.

5. Add the `ValueEqualsValue()` method:

```
[Benchmark]
public void ValueEqualsValue()
{
        bool value = _value1.Equals(_value2);
}
```

The `ValueEqualsValue()` method benchmarks the equality checking of two values using the `Equals(value)` method.

6. Add the `ReferenceOperatorReference()` method:

```
[Benchmark]
public void ReferenceOperatorReference()
{
        bool value = _listOne == _listTwo;
}
```

The `ReferenceOperatorReference()` method benchmarks the equality checking of two reference values using the equality operator.

7. Add the `ReferenceEqualsReference()` method:

```
[Benchmark]
public void ReferenceEqualsReference()
{
        bool value = _listOne.Equals(_listTwo);
}
```

The `ReferenceEqualsReference()` method benchmarks the equality checking of two values using the `Equals(reference)` method.

8. Add the `StringOperatorString()` method:

```
[Benchmark]
public void StringOpertatorString()
{
        bool value = _string1 == _string2;
}
```

The `StringOperatorString()` method benchmarks the equality testing of two strings using the == operator.

9. Next, add the `StringEqualsString()` method:

```
[Benchmark]
public void StringEqualsString()
{
        bool value = _string1.Equals(_string2);
}
```

The `StringEqualsString()` method benchmarks the equality testing of two strings using the `Equals()` method.

10. Add `BenchmarkRunner.Run<Equality>();` to the `Main` method of the `Program` class, do a `Release` build, and then run your executable from the command line. You should end up with the following benchmark report:

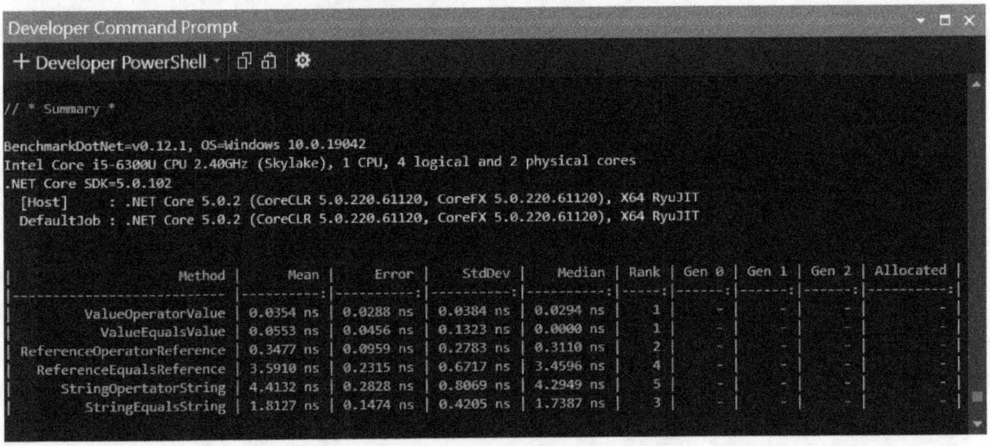

```
Developer Command Prompt
+ Developer PowerShell

// * Summary *

BenchmarkDotNet=v0.12.1, OS=Windows 10.0.19042
Intel Core i5-6300U CPU 2.40GHz (Skylake), 1 CPU, 4 logical and 2 physical cores
.NET Core SDK=5.0.102
  [Host]     : .NET Core 5.0.2 (CoreCLR 5.0.220.61120, CoreFX 5.0.220.61120), X64 RyuJIT
  DefaultJob : .NET Core 5.0.2 (CoreCLR 5.0.220.61120, CoreFX 5.0.220.61120), X64 RyuJIT


|                      Method |      Mean |     Error |    StdDev |    Median | Rank | Gen 0 | Gen 1 | Gen 2 | Allocated |
|---------------------------- |----------:|----------:|----------:|----------:|-----:|------:|------:|------:|----------:|
|          ValueOperatorValue | 0.0354 ns | 0.0288 ns | 0.0384 ns | 0.0294 ns |    1 |     - |     - |     - |         - |
|            ValueEqualsValue | 0.0553 ns | 0.0456 ns | 0.1323 ns | 0.0000 ns |    1 |     - |     - |     - |         - |
| ReferenceOperatorReference | 0.3477 ns | 0.0959 ns | 0.2783 ns | 0.3110 ns |    2 |     - |     - |     - |         - |
|   ReferenceEqualsReference | 3.5910 ns | 0.2315 ns | 0.6717 ns | 3.4596 ns |    4 |     - |     - |     - |         - |
|      StringOpertatorString | 4.4132 ns | 0.2828 ns | 0.8069 ns | 4.2949 ns |    5 |     - |     - |     - |         - |
|          StringEqualsString | 1.8127 ns | 0.1474 ns | 0.4205 ns | 1.7387 ns |    3 |     - |     - |     - |         - |
```

Figure 6.9 – The BenchmarkDotNet summary report for various equality checks

As we can see, it is quicker to test value type equality using the == operator, quicker to use the == operator to test reference type equality, and quicker to use `Equals(string)` when comparing strings.

With that, we have completed this chapter. But before we move on to *Chapter 7, LINQ Performance*, let's summarize what we have learned in this chapter.

Summary

In this chapter, we learned about the different types of collections and their usage. We saw that we should prefer using generic collections over non-generic collections. Then, we briefly touched on Big O Notation and how to use it to determine algorithmic efficiency. After that, we looked at choosing the right type of collection for what we needed.

After that, we set up a sample database to test the querying and iteration of data using further on in the chapter. Then, we looked at how to choose between using interfaces and concrete classes and choosing between arrays and collections. Next, we looked at indexers and then moved on to look at IEnumerable<T>, IEnumerator<T>, and IQueryable<T> and their performance.

The next topic we looked at was using the yield keyword. We touched on the differences between concurrency and parallelism and mentioned that these will be looked at in more depth in later chapters. Finally, we looked at the difference between the == operator and the Equals() method in terms of performance.

In the next chapter, we will be looking at LINQ performance. But for now, see if you can answer the following questions, and check out the *Further reading* section to solidify what you have learned in this chapter.

Questions

Answer the following questions to test your knowledge of this chapter:

1. List the different namespace collections.
2. What is Big O notation used for?
3. What does algorithmic efficiency measure?
4. Is it preferable to use IList<T> or List<T> in terms of instantiation speed?
5. Should we use collections or arrays?
6. What does an indexer do?
7. Which method of iteration is fastest on an in-memory collection between IEnumerable<T> and IEnumerator<T>?
8. In terms of memory and speed performance, what database query method performs best?
9. When building a collection using iteration, what is the quickest way to build the collection up and return the results?

Further reading

To learn more about the topics that were covered in this chapter, take a look at the following resources:

- *Indexers*: `https://docs.microsoft.com /dotnet/csharp/programming-guide/indexers/`.
- *ConsoleSecrets*: `https://github.com/jasonshave/ConsoleSecrets`.
- *Equality Operators*: `https://docs.microsoft.com/dotnet/standard/design-guidelines/equality-operators`.
- *Interesting Performance Implications of C# 9 Records Equality Check*: `https://gmanvel.medium.com/interesting-performance-implications-of-c-9-records-equality-check-f0d0a3612919`.
- *Improving Struct Equality Performance in C#:* `http://dontcodetired.com/blog/post/Improving-Struct-Equality-Performance-in-C`.
- *String Equality and Performance in C#:* `https://rhale78.wordpress.com/2011/05/16/string-equality-and-performance-in-c/`.
- *Performance Implications of Default Struct Equality in C#:* `https://devblogs.microsoft.com/premier-developer/performance-implications-of-default-struct-equality-in-c/`.
- *Performance Best Practices in C#:* `https://kevingosse.medium.com/performance-best-practices-in-c-b85a47bdd93a`.
- *8 Techniques to Avoid GC Pressure and Improve Performance in C# .NET*: `https://michaelscodingspot.com/avoid-gc-pressure/`.

7
LINQ Performance

LINQ has a reputation for being slow. But contrary to people's views, there are ways to use LINQ that ensure optimal performance.

In this chapter, you will learn how to perform LINQ queries with performance in mind. Depending on how you use LINQ, different methods that return the same result can behave and perform differently. And so, in this chapter, you will learn how best to perform queries on LINQ to improve the performance of your applications.

Here, you will benchmark different ways to determine the most performative ways to obtain the last element of a LINQ query. You will learn about the performance penalty of using the `let` keyword in LINQ statements, and why you should avoid using it. Benchmarking different `Group By` methods, you will gain insight into the most performant way to perform GroupBy queries using LINQ. When performing queries and data manipulation using LINQ, there may be times when you need to use closures. By writing parametrized and non-parameterized closures, you will see that parameterized closures perform much better than non-parameterized closures.

We will be covering the following topics in this chapter:

- Setting up our sample database
- Setting up our in-memory sample data
- Querying a database using LINQ
- Getting the last value of a collection

- Avoid using the let keyword in LINQ queries
- Increasing Group By performance in LINQ queries
- Filtering lists
- Understanding closures

By the end of this chapter, you will have the skills to securely store secrets and query databases and in-memory data using efficient LINQ. You will also be able to understand the performance impact of using the `let` keyword in your queries and performing efficient filtering and grouping of data using LINQ.

Technical requirements

In order to follow along with this chapter, you will need access to the following tools:

- Visual Studio 2022
- SQL Server 2019
- SQL Server Management Studio
- The book's source code: `https://github.com/PacktPublishing/High-Performance-Programming-in-CSharp-and-.NET/tree/master/CH07`

Setting up a sample database

In this chapter, we will be demonstrating the difference between how different collection interfaces handle data, and for the demonstrations, you require access to database data. To do so, you will create a database, add a table, and populate it with data. You will use SQL Server for your database engine, and use SQL Server Management Studio to develop your sample database.

> **Note**
>
> In the `CH07_LinqPerformance.Data` source code folder, you will find a database creation script called `SampleData.Product.sql` that creates the database and populates it with data. You can run this script in SQL Server Management Studio. This will save you from having to run through setting up the database in this section. But if you are new to SQL Server, you may want to run through this section.

To add your database, follow these steps:

1. Open SQL Server Management Studio and connect to your database engine.
2. Right-click on the **Databases** folder in **Object Explorer** as shown in *Figure 7.1*:

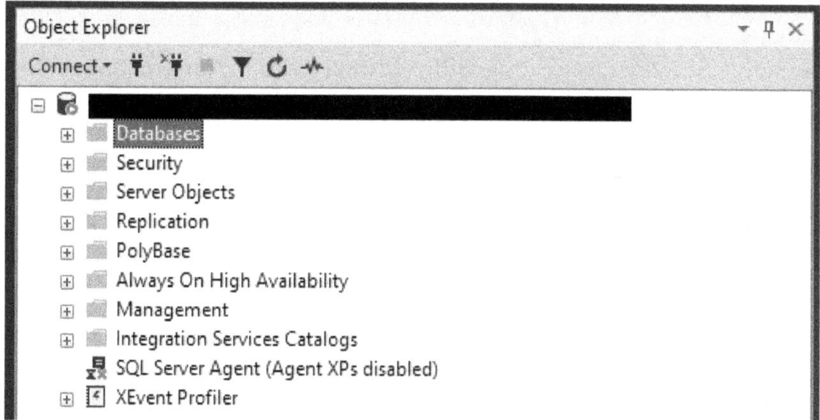

Figure 7.1: The SQL Server Management Studio Object Explorer tab

3. Select **New Database** from the context menu. This will display the **New Database** dialog as shown in *Figure 7.2*:

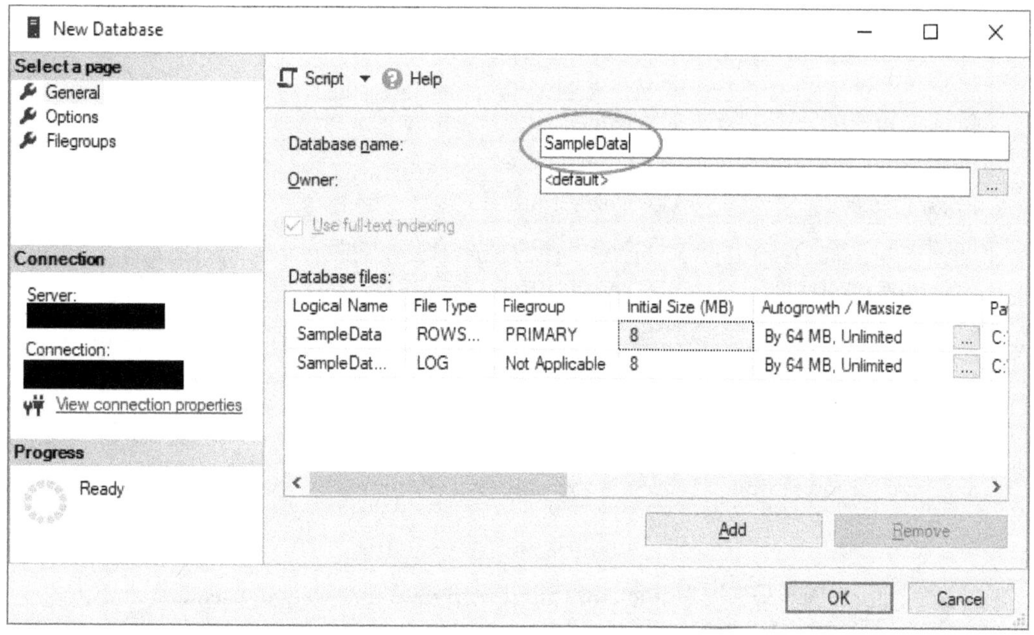

Figure 7.2: The SQL Server Management Studio New Database dialog

4. Once you have entered `SampleData` for the database name, click on the **OK** button to create the database.
5. Locate the database by expanding the **Databases** folder, and then expand the database. Right-click on the **Tables** folder and select **New | Table**. Add a new table called `Products` as shown in the following figure:

Name	Type	Null	Primary Key	Auto Increment
Id	int	No	Yes	Yes
Name	nvarchar(50)	No	No	No
Description	nvarchar(255)	No	No	No
UnitPrice	money	No	No	No

Table 7.1: The Products table design

6. Save the table and then expand the **Tables** folder. Right-click on the **Products** table and select **Edit Top n records** where *n* will be the number of configured records to edit, which is *200* by default.
7. Add the data shown in the following figure to the **Product** table:

Id	Name	Description	UnitPrice
1	Roasted Peanuts	500 g bag of dry roasted peanuts	0.69
2	Cashew Nuts	75 g bag of cashew nuts	0.75
3	Milk (Whole)	2 litres of whole milk	1.25
4	Bread (50/50)	50% white and 50% wholemeal bread	1
5	Butter (Salted)	100 g salted butter	2.5
6	Roast Chicken	5 kg frozen roast chicken	4.99
7	Potatoes	5 kg Maris Piper variety potatoes	1.75
8	Roasting Vegetables	1 kg bag of frozen roasting vegetables	1.5
9	Coffee	1 kg of Arabic coffee	2.99
10	Demerara Sugar	1 kg bag of Demerara sugar	1
11	Chicken Gravy	1 tub of chicken gravy granules	0.89
12	Yorkshire Puddings	1 bag of 12 frozen Yorkshire puddings	1.35
13	Sage and Onion Stuffing	1 box of sage and onion stuffing	0.59

Table 7.2: The Product table row data

We now have a database with a single table filled with data that we will use later in the chapter. In the next section, we will be adding our in-memory sample data.

Setting up our in-memory sample data

You will be studying LINQ performance, therefore, you are going to need a collection to work with. You will work with a collection of `Person` objects. Each person will be named from the Greek alphabet. A `Person` object will consist of a `FirstName`, `LastName`, and `FullName` property. The `FullName` property will be an interpolated string that combines the first and last name of the person.

Let us now begin coding our LINQ coding combined with **benchmarking**, so that we can measure the performance of our LINQ statements:

1. Create a new .NET 6.0 console application called `CH07_LinqPerformance`.
2. Install the NuGet package `BenchmarkDotNet`.
3. Add the following `Person` struct:

```
public struct Person
{
        public string FirstName { get; set; }
        public string LastName { get; set; }
        public string FullName { get { return
          $"{FirstName} {LastName}"; } }
public Person(string firstName, string lastName)
{
     FirstName = firstName;
     LastName = lastName;
}
}
```

   This structure defines the `Person` with their `FirstName`, `LastName`, and computed `FullName`.

4. Now, add a new class called `LinqPerformance` with the following `using` statements:

```
using BenchmarkDotNet.Attributes;
using BenchmarkDotNet.Order;
using System.Collections.Generic;
using System.Linq;
```

   These `using` statements provide you with access to benchmarking, generic collections, and LINQ classes.

5. Add the following code to the top of the class:

```
private List<Person> _people = new List<Person>();
private string[] _group1 = new string[] { "iota",
    "epsilon", "sigma", "upsilon" };
private string[] _group2 = new string[] { "alpha",
    "omega" };
```

   You have declared a list of people and two arrays. Both these arrays contain the surnames of people in lowercase that belong to those groups.

6. Now, add the global setup class that will prepare your collection for benchmarking various LINQ queries:

```
[GlobalSetup]
public void PrepareBenchmarks()
{
  _people.Add(new Person("Alpha", "Beta"));
  _people.Add(new Person("Chi", "Delta"));
  _people.Add(new Person("Epsilon", "Phi"));
  _people.Add(new Person("Gamma", "iota"));
  _people.Add(new Person("Kappa", "Lambda"));
  _people.Add(new Person("Mu", "Nu"));
  _people.Add(new Person("Omicron", "Pi"));
  _people.Add(new Person("Theta", "Rho"));
  _people.Add(new Person("Sigma", "Tau"));
  _people.Add(new Person("Upsilon", "Omega"));
  _people.Add(new Person("Xi", "Psi"));
  _people.Add(new Person("Zeta", "Iota"));
  _people.Add(new Person("Alpha", "Omega"));
      _people.Add(new Person("Omega", "Chi"));
      _people.Add(new Person("Sigma", "Tau"));
}
```

You now have your sample database and in-memory sample data in place for the topics we will be covering in this chapter. So, let us start by investigating various ways of querying a database and their effects on LINQ query performance.

Database query performance

We saw in *Chapter 6, The .NET Collection*, how `IEnumerator` is different from `IEnumerable`, and how `IEnumerator` performs faster than `IEnumerable` when iterating through an in-memory collection. Now, we will query a database and iterate through the resulting collection using various benchmarked techniques. To do so, we will follow these steps:

1. Add a new class called `IEnumeratorVsIQueryable`.
2. You will be connecting to a SQL Server database and will have the information you need to keep secret. Your `secret.json` files do not get checked into version control. So, right-click on the project and select **Manage User Secrets** from the context menu.
3. A dialog will pop up informing you that additional packages are required. Click on **Yes**.

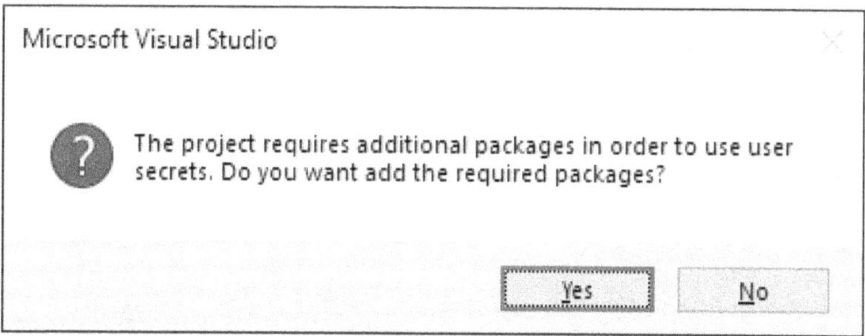

Figure 7.3: Dialog Informing you that additional packages are required to manage user secrets

4. Visual Studio will then open the `secrets.json` file in a new tab. This is where you will add your user secrets.
5. Open **Package Manager Console** and add the following packages:

```
Microsoft.EntityFrameworkCore
Microsoft.EntityFrameworkCore.SqlServer
Microsoft.EntityFrameworkCore.Tools
Microsoft.Extensions.Configuration
Microsoft.Extensions.Configuration.EnvironmentVariables
Microsoft.Extensions.Configuration.UserSecrets
Microsoft.Extensions.OptionsConfigurationExtensions
```

These packages enable you to connect to and extract data from the SQL Server database.

6. Update your `secrets.json` file with the connection string to the database you created at the start of the chapter:

```
{
   "DatabaseSettings": {
      "ConnectionString": "YOUR_CONNECTION_STRING"
   }
}
```

   This connection string will be used to connect to your database, perform a query that returns some data, and enable you to iterate through that data and perform operations on it.

7. Add a folder called `Configuration`, and in that folder, add a class called `SecretsManager` with an empty static constructor and the following `using` statements:

```
using Microsoft.Extensions.Configuration;
using System;
using System.IO;
```

   You need these `using` statements for your file I/O and system configuration such as obtaining secrets from a `secrets.json` file.

8. Add the following line at the top of the `SecretsManager` class:

```
public static IConfigurationRoot Configuration
     { get; set; }
```

   This line declares your static configuration property that is used to obtain your configuration data within your application.

9. Now add the following code:

```
public static T GetSecrets<T>(string sectionName)
     where T : class
{
var devEnvironmentVariable = Environment
      .GetEnvironmentVariable("NETCORE_ENVIRONMENT");
var isDevelopment = string.IsNullOrEmpty
        (devEnvironmentVariable) || devEnvironment
              Variable.ToLower() == "development";
var builder = new ConfigurationBuilder()
```

```
        .SetBasePath(Directory.GetCurrentDirectory())
    .AddJsonFile("appsettings.json", optional: true,
        reloadOnChange: true)
    .AddEnvironmentVariables();
    if (isDevelopment) //only add secrets in development
    {
        builder.AddUserSecrets<T>();
    }
    Configuration = builder.Build();
    return Configuration.GetSection(sectionName).Get<T>();
```

This code gets the environment variables for the .NET Core environment. It then gets the code to see if it is running in a software development environment or production environment. The configuration is then built for the environment it will be running in. So, if we are in debug mode, the configuration will be built for the development environment. And if we are in release mode, the configuration will be built for the production environment. If we are in development, then we add our `secrets` class as defined by the `T` variable.

10. Create a new folder, `Models`, and add the `Product` class using the following code:

```
using System.ComponentModel.DataAnnotations;
public class Product
{
        public Product() { }
        public Product(int id)
        {
            Id = id;
           Name = $"Item {Id} Name";
           Description = $"Item {Id} description.";
        }
        [Key]
        public int Id { get; private set; }
        public string Name { get; private set; }
        public string Description { get; private set; }
        public override string ToString()
        {
       return $"Id: {Id}, Name: {Name},
```

```
        Description: {Description}";
    }
}
```

Our `Product` class provides the model for our product data with `Id`, `Name`, and `Description` properties that are set via the constructor. We also override the `ToString` method to return a textual representation of the property values.

11. Add a `using` statement for `System.ComponentModel.DataAnnotations`. Change the struct to a class, and add the `[Key]` attribute to the `Id` property. We need these changes since we are using Entity Framework to connect to a database and extract data.

12. In the `CH07_LinqPerformance.Data` folder, add the `DatabaseContext` class:

```
using Microsoft.EntityFrameworkCore;
using CH07_LinqPerformance.Models;
public class DatabaseContext : DbContext
{
}
```

We have declared our `DatabaseContext` class, which inherits from the `DbContext` class. Now we'll need to add its internals.

13. Add the following items to the `DatabaseContext` class:

```
public DbSet<Product> Products { get; set; }
public DatabaseContext(string connectionString) :
    base(GetOptions(connectionString))
{
}
```

In this code, we have declared our `DbSet` of products property, which will hold a collect of our `Product` class, and a connection string member variable that will hold the string that connects us to our database. Our constructor is then declared, which takes in a connection string, which we pass into the `GetOptions` method that then gets passed into the base class constructor.

14. Add the `GetOptions` method to the `DatabaseContext` class:

```
private static DbContextOptions GetOptions(string
    connectionString)
{
    return SqlServerDbContextOptionsExtensions
```

```
            .UseSqlServer(
                new DbContextOptionsBuilder(),
                connectionString)
            .Options;
    }
```

This method returns the `DbContextOptions` for our SQL Server database connection. The connection string used is the one that is stored in our `secrets.json` file in development and in `appsettings.json` when in production.

15. Add the `OnModelCreating` method:

```
protected override void OnModelCreating(ModelBuilder
    modelBuilder)
{
        modelBuilder.Entity<Product>(entity =>
    {
                entity.HasKey(e => e.Id);
            entity.Property(e => e.Name)
                .HasMaxLength(50);
            Entity.Property(e => e.Description)
                .HasMaxLength(255);
        });
    }
```

Here, we are configuring our `Product` class that will be used in our `DbSet`. We are declaring that the `Id` field is our primary key and that the `Name` field has a maximum length of 50 while the `Description` field has a maximum length of 255.

16. Add the `DatabaseSettings` class to the `Configuration` folder:

```
public class DatabaseSettings
{
    public string ConnectionString { get; set; }
}
```

This class has a single property called `ConnectionString` that will hold your connection string to our `SampleData` database. Notice that the name of the class and property match the name of the JSON section and property!

17. Now, add `appsettings.json` to the root of your project with the following contents:

```
{
  "DatabaseSettings": {
    "ConnectionString": "Set in Azure. For
        development, set in User Secrets"
  }
}
```

This file has the same layout as the `secrets.json` file and the `DatabaseSettings` class. This file is used to store your connection string. In development, it is set in the secrets file, and in production, it is set in Azure. Now that you have your database configuration in place, you can add your benchmarking code.

18. Add a new class in the root of the project called `DatabaseQueryAndIteration` that implements `IDisposable` with the following code:

```
using BenchmarkDotNet.Attributes;
using BenchmarkDotNet.Order;
using CH07_Collections.Configuration;
using CH07_Collections.Data;
using CH07_Collections.Models;
using Microsoft.Extensions.Options;
using System;
using System.Collections.Generic;
using System.Diagnostics;
using System.Linq;
[MemoryDiagnoser]
[Orderer(SummaryOrderPolicy.Declared)]
[RankColumn]
public class DatabaseQueryAndIteration : IDisposable
{
}
```

This code declares our class and defines the fact that it implements `IDisposable`. It is also configured to be benchmarked.

19. Implement the `IDisposable` interface in our class:

```
private bool disposedValue;
protected virtual void Dispose(bool disposing)
{
    if (!disposedValue) {
        if (disposing)
            _context.Dispose();
        disposedValue = true;
    }
}
public void Dispose(){
    Dispose(disposing: true);
    GC.SuppressFinalize(this);
}
```

This code disposes of our managed resources and suppresses the call to the class finalizer method.

20. We have everything in place to benchmark the methods in this class, access database resources, and clean up after ourselves. Add the following code to the class:

```
private DatabaseContext _context;
[GlobalSetup]
public void GlobalSetup()
{
      var connectionString = SecretsManager
        .GetSecrets<DatabaseSettings>(nameof
          (DatabaseSettings)).ConnectionString;
  _context = new DatabaseContext(connectionString);
}
[GlobalCleanup]
public void GlobalCleanup()
{
      Dispose(true);
}
```

The _context variable provides us with our database access. The GlobalSetup() method gets our connection string from our secrets file, and creates a new DatabaseContext using the safely stored connection string. The GlobalSetup() method will run before our benchmarks. The GlobalCleanup() method calls the Dispose(disposing) method to clean up our managed resources after our benchmarks have finished running.

21. Next, add the QueryDb() method:

```
[Benchmark]
public void QueryDb()
{
        var products = (from p in _context.Products
                        where p.Id > 1select p);
foreach (var product in products)
    Debug.WriteLine(product.Name);
}
```

The QueryDb() method performs a simple LINQ query on the database by selecting products with an ID of greater than 1. It then iterates each product in the IQueryable<Product> list and writes the product name out to the debug window.

22. Now, add the QueryDbAsList() method:

```
[Benchmark]
public void QueryDbAsList()
{
List<Product> products = (from p in _context.Products
  where p.Id > 1select p).ToList<Product>();
foreach (var product in products)
Debug.WriteLine(product.Name);
}
```

QueryDbAsList() performs the same query as QueryDb(), except the processed type is of type List<Product>.

23. Add the `QueryDbAsIEnumerable()` method:

```
[Benchmark]
public void QueryDbAsIEnumerable()
{
var products = (from p in _context.Products
                where p.Id > 1
                select p).AsEnumerable<Product>();
foreach (var product in products)
    Debug.WriteLine(product.Name);
}
```

The `QueryDbAsIEnumerable()` method performs the same query as `QueryDbAsList`, but processes a type of `IEnumerable<Product>` instead.

24. Add the `QueryDbAsIEnumerator()` method:

```
[Benchmark]
public void QueryDbAsIEnumerator()
{
        var products = (from p in _context.Products
                        where p.Id > 1
                        select p).GetEnumerator();
    while (products.MoveNext())
    Debug.WriteLine(products.Current.Name);
}
```

`QueryDbAsIEnumerator()` does the same as the previous methods but operates on a type of `IEnumerator<Product>` and iterates using a `while` loop instead of a `foreach` loop.

25. The final method in this class that we need to add is the `QueryDbAsIQueryable()` method:

```
[Benchmark]
public void QueryDbAsIQueryable()
{
var products = (from p in _context.Products
                where p.Id > 1
                select p).AsQueryable<Product>();
foreach (var product in products)
```

```
        Debug.WriteLine(product.Name);
    }
```

This method is the same as `QueryDb` but explicitly operates on a type of `IQueryable<Product>`.

26. Replace the code in the `Main` method within the `Program` class with the following:

```
BenchmarkRunner.Run<DatabaseQueryAndIteration>();
```

This code runs your benchmarks. Do a release build of the code and run the executable from the command line. You should see a summary report similar to the following:

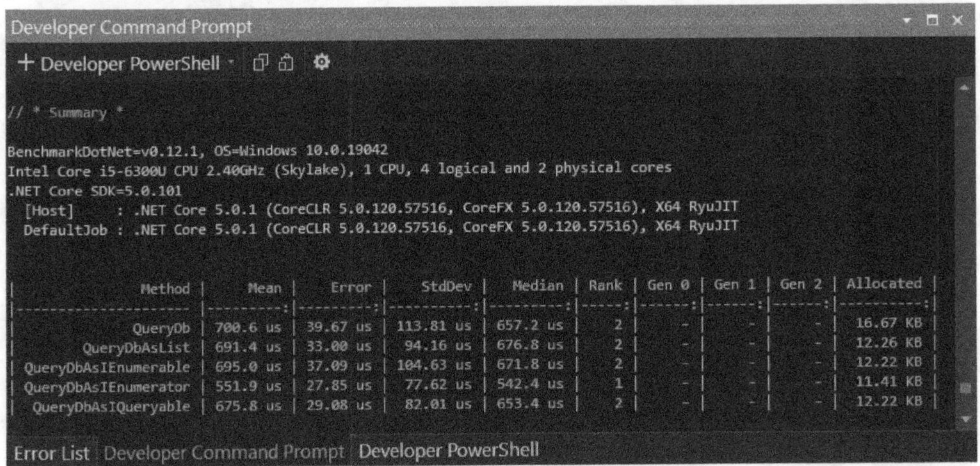

Method	Mean	Error	StdDev	Median	Rank	Gen 0	Gen 1	Gen 2	Allocated
QueryDb	700.6 us	39.67 us	113.81 us	657.2 us	2	-	-	-	16.67 KB
QueryDbAsList	691.4 us	33.00 us	94.16 us	676.8 us	2	-	-	-	12.26 KB
QueryDbAsIEnumerable	695.0 us	37.09 us	104.63 us	671.8 us	2	-	-	-	12.22 KB
QueryDbAsIEnumerator	551.9 us	27.85 us	77.62 us	542.4 us	1	-	-	-	11.41 KB
QueryDbAsIQueryable	675.8 us	29.08 us	82.01 us	653.4 us	2	-	-	-	12.22 KB

Figure 7.4: The different times and memory allocation of various database query types using LINQ

Let us summarize what we learn from the summary report after running our query benchmarks:

- In terms of memory usage, the worst performer is the `QueryDb()` method followed by the `QueryDbAsList()` method. `QueryDbAsIEnumerable()` and `QueryDbAsIQueryable()` are both slightly better than the previous two. But the best performing method in terms of memory allocation out of all five methods is the `QueryDbAsIEnumerator()` method.
- Speed wise, the `QueryDb()` method was the worst again. Followed by `QueryDbAsIEnumerable()`, then `QueryDbAsList()`, and then `QueryDbAsIQueryable()`. And again, the best performer in terms of speed is the `QueryDbAsIEnumerator()` method.

So, we can see that the best performing method for querying and iterating a database in both speed and memory usage terms is the `QueryDbAsIEnumerator()` method out of all the methods we've chosen to investigate.

In the next section, we will be investigating which is the fastest method for obtaining the last item in a collection.

Getting the last value of a collection

You are now going to see how the LINQ method that obtains the last element in the collection is really slow when compared to directly accessing the item by its index. This will be accomplished using benchmarking to measure the performance of different methods:

1. Update the `Main` method as follows:

```
static void Main(string[] args)
{
        BenchmarkRunner.Run<LinqPerformance>();
}
```

2. Open the `LinqPerformance` class.
3. Add the `GetLastPersonVersion1()` method:

```
[Benchmark]
public void GetLastPersonVersion1()
{
        var lastPerson = _people.Last();
}
```

   This method gets the last person in the collection using the LINQ-provided `Last()` method.

4. Add the `GetLastPersonVersion2()` method:

```
[Benchmark]
public void GetLastPersonVersion2()
{
        var lastPerson = _people[_people.Count - 1];
}
```

5. Here, we are using the index of the list to extract the last person in the list. At this point, it is worth noting that the difference between the two methods is that in the first method, this `Last()` method call is actually declared in `System.Linq.Enumerable`. The method signature is as follows:

```
public static TSource Last<TSource>(this
    IEnumerable<TSource> source);
```

So, the `Last()` call in the `GetLastPersonVersion1()` method performs various checks before the last value is returned. But the `GetLastPersonVersion2()` method does not perform these checks, and immediately returns the value at the last position. This explains why the method used in `GetLastPersonVersion1()` is much slower than accessing an element by its index in `GetLastPersonVersion2()`, as you will see in the following screenshot:

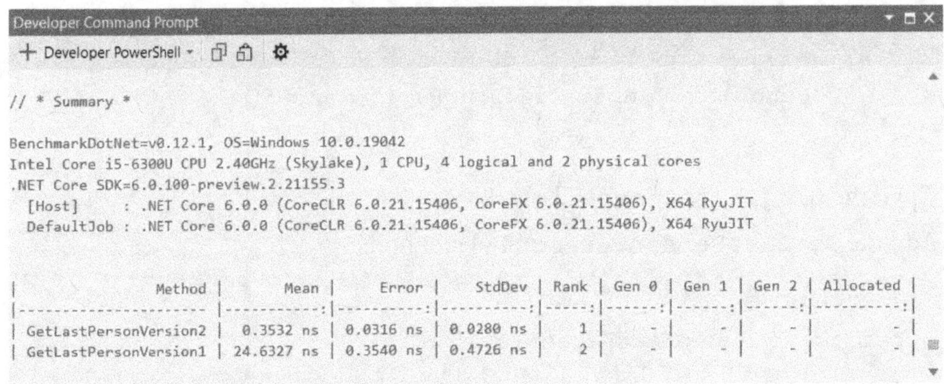

Method	Mean	Error	StdDev	Rank	Gen 0	Gen 1	Gen 2	Allocated
GetLastPersonVersion2	0.3532 ns	0.0316 ns	0.0280 ns	1	-	-	-	-
GetLastPersonVersion1	24.6327 ns	0.3540 ns	0.4726 ns	2	-	-	-	-

Figure 7.5: Get Last Person example performance using the Last() method and direct index access

Looking at the summary report of the benchmarks we have just run, it is evident that using the index for direct access is better than using the `Last()` method call in terms of improved performance.

We have seen how we can quickly access the last element in a collection. Let us now consider why we should avoid using the `let` keyword in LINQ queries.

Avoid using the let keyword in LINQ queries

You can use the `let` keyword to declare a variable and assign it a value to use in your LINQ query if the value is to be used several times within the query. At first glance, this may seem like you are improving performance since you only perform a single assignment, and then use the same variable several times. But this is not actually the case. Using the `let` keyword in your LINQ queries can actually decrease the performance of your LINQ query.

Let us work through some benchmark examples. In the `LinqPerformance` class, do the following:

1. Add the `ReadingDataWithoutUsingLet()` method:

```
[Benchmark]
public void ReadingDataWithoutUsingLet()
{
var result = from person in _people
    where person.LastName.Contains("Omega")
    && person.FirstName.Equals("Upsilon")
    select person;
}
```

   In this method, we are selecting people from the `_people` list with a last name of *Omega*, and a first name of *Upsilon* using LINQ without the `let` keyword.

2. Now, add the `ReadingDataUsingLet()` method:

```
[Benchmark]
public void ReadingDataUsingLet()
{
      var result = from person in _people
      let lastName = person.LastName.Contains("Omega")
      let firstName = person.FirstName.Equals("Upsilon")
      where lastName && firstName
      select person;
}
```

In this method, we are also selecting people from the `_people` list with a last name of *Omega* and a first name of *Upsilon*. But this time, we use the `let` keyword for both the filters and use them in the `where` clause.

3. Build the project and run the executable from the command line. You should see results similar to those shown in *Figure 7.6*:

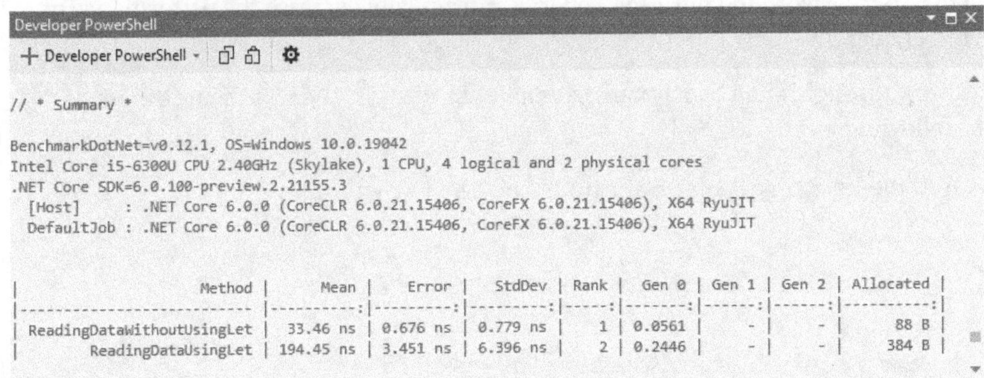

Figure 7.6: BenchmarkDotNet results for reading data with and without using the let keyword

As you can see from these results, the use of the `let` keyword in our query reduced the performance. The processing time increased and so did the memory allocation.

> **Note**
>
> You will see websites that promote the use of the `let` keyword in LINQ queries to improve performance and readability. But as you have seen in the example we have worked through, using the `let` keyword can seriously slow down the performance of your queries and increase memory usage. So, as a rule of thumb, take to measuring your performance for your particular queries and choosing the method that performs best for your query task.

In this section, we have seen how the use of the `let` keyword can increase the time taken and memory used to perform a simple `select` query using LINQ. This performance decrease can become a real problem when working with large volumes of data. In the next section, we will look at several methods for grouping data and see which method performs the best.

Increasing Group By performance in LINQ queries

In this section, we will look at three different ways of performing the same `Group By` operation. Each way provides a different performance level. You will see by the end of this section which method is best for performing fast `Group By` queries. The methods that we add in this section will be added to the `LinqPerformance` class.

For our scenario, we want to get a list of people from a collection that all share the same name. To extract those people, we will perform a `Group By` operation. Then, we will extract all those for whom the group count is greater than one, and then add them to a list of people.

Let us add our three methods that use the `GroupBy` clause to return a list of people:

1. Add the `GroupByVersion1()` method:

```
[Benchmark]
public void GroupByVersion1()
{
List<Person> People = _people.GroupBy(x => x.LastName)
                   .Where(x => x.Count() > 1)
                   .SelectMany(group => group)
                   .ToList();
}
```

   As you can see, we are grouping on the person's last name. We then filter the groups to include only those groups with a count greater than *1*. Those groups are then selected and then returned as a list of people.

2. Now, add the `GroupByVersion2()` method:

```
[Benchmark]
public void GroupByVersion2()
{
        IEnumerator<IGrouping<string, Person>> test =
             _people.GroupBy(p => p.LastName)
     .Where(p => p.Count() > 2).GetEnumerator();
List<Person> people = new List<Person>();
while (test.MoveNext())
{
```

```
        IGrouping<string, Person> current = test.Current;
        foreach (Person person in current)
        {
            people.Add(person);
        }
    }
}
```

In this method, we obtain an enumerator by grouping people by their last name and then filtering the groups to only include those groups with a count of *2* or more. Then we declare a new list of people. We then loop through the enumerator and obtain the current `IGrouping<string, Person>`. The grouping is then iterated through, and each person in the group is added to the list of people.

3. Add the `GroupByVersion3()` method:

```
[Benchmark]
public void GroupByVersion3()
{
        IEnumerator<IGrouping<string, Person>> test =
          _people.ToArray().GroupBy(p => p.LastName)
        .Where(p => p.Count() > 2).GetEnumerator();
        List<Person> people = new List<Person>();
while (test.MoveNext())
        {
        var current = test.Current;
        foreach (var person in current)
        {
            people.Add(person);
        }
    }
}
```

The `GroupByVersion3()` method is the same as and behaves the same as the `GroupByVersion2()` method, but with one main difference. We convert the list of people to an array before we perform the `Group By`.

4. Add the following annotations to the top of the `LinqPerformance` class:

```
[MemoryDiagnoser]
[Orderer(SummaryOrderPolicy.FastestToSlowest)]
[RankColumn]
```

These annotations will expand the data contained in the summary report as you will see shortly. Do a release build of the project and then run the project from the command line to benchmark these three methods. You should see the following benchmark summary report:

```
Developer Command Prompt
+ Developer PowerShell

// * Summary *

BenchmarkDotNet=v0.12.1, OS=Windows 10.0.19042
Intel Core i5-6300U CPU 2.40GHz (Skylake), 1 CPU, 4 logical and 2 physical cores
.NET Core SDK=6.0.100-preview.2.21155.3
  [Host]     : .NET Core 6.0.0 (CoreCLR 6.0.21.15406, CoreFX 6.0.21.15406), X64 RyuJIT
  DefaultJob : .NET Core 6.0.0 (CoreCLR 6.0.21.15406, CoreFX 6.0.21.15406), X64 RyuJIT

|          Method |     Mean |     Error |    StdDev | Rank |  Gen 0 | Gen 1 | Gen 2 | Allocated |
|---------------- |---------:|----------:|----------:|-----:|-------:|------:|------:|----------:|
| GroupByVersion3 | 1.961 us | 0.0270 us | 0.0252 us |    1 | 1.3390 |     - |     - |   2.05 KB |
| GroupByVersion2 | 2.011 us | 0.0265 us | 0.0248 us |    2 | 1.1826 |     - |     - |   1.81 KB |
| GroupByVersion1 | 2.204 us | 0.0351 us | 0.0595 us |    3 | 1.2817 |     - |     - |   1.97 KB |
```

Figure 7.7: The BenchmarkDotNet Group By summary report

As we can see, our first attempt at performing a `Group By` operation takes *2.204* microseconds, our second attempt takes *2.011* microseconds, and our third and final attempt takes *2.204* microseconds. So, we can see that converting our list to an array before performing a `Group By` speeds things up. Our final version is *0.243* microseconds faster than our original version, and that is despite the fact that more code is involved!

The section that follows will take you through the benchmarking of five different ways to provide filtering of lists. You will see how the different methods affect the performance of LINQ queries.

Filtering lists

In this section, we will look at various ways to filter a list using LINQ. We will see that the various ways all perform differently. By the end of this section, you will know the best way to filter a list for increased performance. You will be writing two different benchmarks that demonstrate query performance differences when using the `let` keyword and not using the `let` keyword. Let's begin writing our benchmarks:

1. Add the `FilterGroupsVersion1()` method:

```
[Benchmark]
public List<Person> FilterGroupsVersion1()
{
 return (from p in _people where
            _group1.Contains(p.LastName.ToLower())
            || _group2.Contains(p.LastName.ToLower())
            select p).ToList(
}
```

   The first of our benchmarks filters people that belong to `_group1` and `_group2`. Since the arrays are in lowercase, `LastName` is also converted to lowercase. The filtered people are then returned as a list of people.

2. Add the `FilterGroupsVersion2()` benchmark:

```
[Benchmark]
public List<Person> FilterGroupsVersion2()
{
      return (from p in _people
          let lastName = p.LastName.ToLower()
          where _group1.Contains(lastName)
          || _group2.Contains(lastName)
          select p).ToList();
}
```

This does the same as our first benchmark. The main difference is that we introduce the `lastName` variable using the `let` keyword, and assign it the lowercase `LastName` of the person.

3. Compile the project in release mode and run it from the command line. The benchmarks will be generated, and you should see a benchmark report similar to the one in *Figure 7.8*:

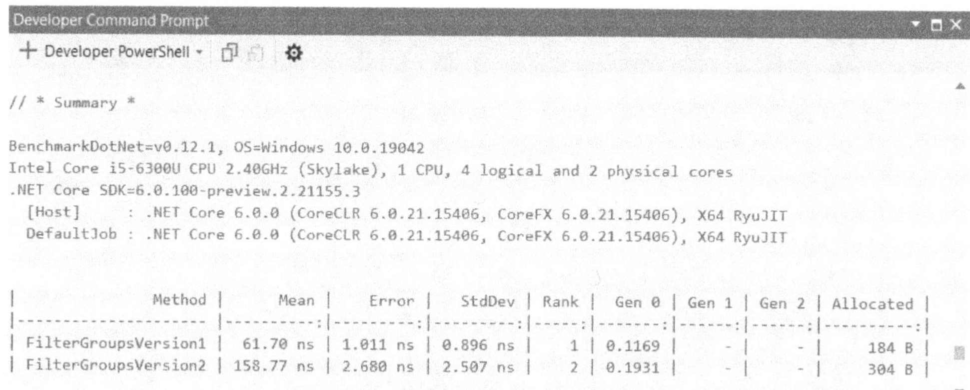

Figure 7.8: Benchmark Report for LINQ with and without using the let keyword

We can see in the summary report that using the `let` keyword slows things down considerably. And so, we will now investigate why the `let` keyword slows things down.

4. Open **ILDASM**, and load in `CH07_LinqPerformance.dll`.
5. Expand **CH07_LinqPerformance | CH07_ Linq.LinqPerformace. LinqPerformance**. You will see the two methods called `FilterGroupsVersion1` and `FilterGroupsVersion2`.
6. Double-click on the method `FilterGroupsVersion1` to reveal the intermediate language generated by the compiler.
7. Now, do the same with the `FilterGroupsVersion2` method. When you compare the IL for both methods, you will clearly see that the IL for `FilterGroupsVersion2` contains more lines of code than the IL for `FilterGroupsVersion1`.

And that explains why the `let` version of the code performs slower than the original code that does not use the `let` keyword. But can we do better than `FilterGroupsVersion1` in terms of performance? Well, it turns out that, yes, we can.

8. Add the `FilterGroupsVersion3` method:

```
[Benchmark]
public List<Person> FilterGroupsVersion3()
{
List<Person> people = new List<Person>();
for (int i = 0; i < _people.Count; i++)
{
    var person = _people[i];
    var lastName = person.LastName.ToLower();
    if (
        _group1.Contains(lastName)
        || _group2.Contains(lastName)
    )
    people.Add(person);
}
return people;
}
```

As you can see, we create a new people list. We then loop through the `_people` list. For each person, we get them from the `_people` list. We then assign the lowercase form of their name to a local variable. Using this variable, we check to see if either `_group1` or `_group2` contains the names. If they do, then the person is added to the `_people` list. Once the iteration has finished, the `_people` collection is returned.

9. Build and run the code again. You should see the following report:

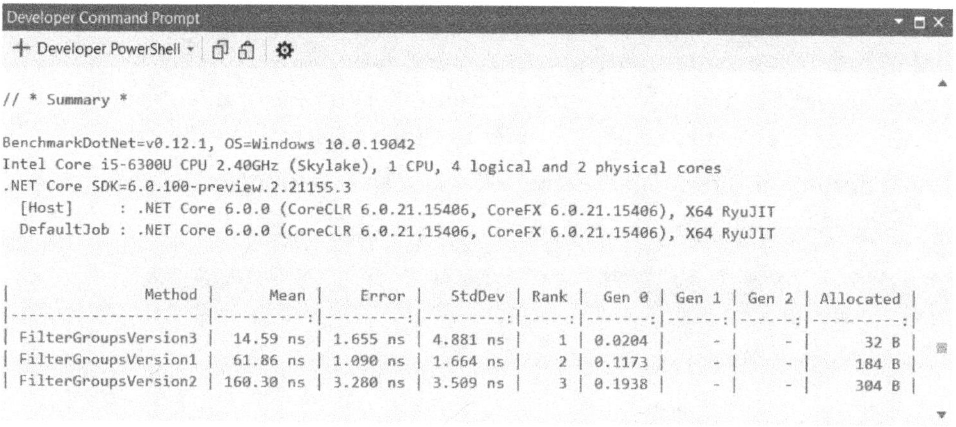

```
Developer Command Prompt
+ Developer PowerShell

// * Summary *

BenchmarkDotNet=v0.12.1, OS=Windows 10.0.19042
Intel Core i5-6300U CPU 2.40GHz (Skylake), 1 CPU, 4 logical and 2 physical cores
.NET Core SDK=6.0.100-preview.2.21155.3
  [Host]     : .NET Core 6.0.0 (CoreCLR 6.0.21.15406, CoreFX 6.0.21.15406), X64 RyuJIT
  DefaultJob : .NET Core 6.0.0 (CoreCLR 6.0.21.15406, CoreFX 6.0.21.15406), X64 RyuJIT

|               Method |      Mean |     Error |    StdDev | Rank |  Gen 0 | Gen 1 | Gen 2 | Allocated |
|--------------------- |----------:|----------:|----------:|-----:|-------:|------:|------:|----------:|
| FilterGroupsVersion3 |  14.59 ns | 1.655 ns  | 4.881 ns  |    1 | 0.0204 |     - |     - |      32 B |
| FilterGroupsVersion1 |  61.86 ns | 1.090 ns  | 1.664 ns  |    2 | 0.1173 |     - |     - |     184 B |
| FilterGroupsVersion2 | 160.30 ns | 3.280 ns  | 3.509 ns  |    3 | 0.1938 |     - |     - |     304 B |
```

Figure 7.9: The BenchmarkDotNet summary report showing FilterGroupsVersion3's performance

As you can see, we have three different versions of the code that produce the same output, and each one's execution time is different. Between these three different methods, `FilterGroupsVersion3` is by far the quickest method in achieving the desired result.

10. We will have another go at improving the performance of our LINQ filter query. Add the `FilterGroupsVersion4` method:

```
[Benchmark]
public List<Person> FilterGroupsVersion4()
{
    List<Person> people = new List<Person>();
for (int i = 0; i < _people.Count; i++)
{
    var person = _people[i];
    var lastName = person.LastName.ToLower();
    if (
        _group2.Contains(lastName)
        || _group1.Contains(lastName)
    )
    people.Add(person);
```

```
    }
    return people;
}
```

It can be seen that the only difference between `FilterGroupsVersion3` and `FilterGroupsVersion4` is the ordering of the `if` condition check.

11. Build the project and run the benchmark tests. *Figure 7.10* shows the performance summary:

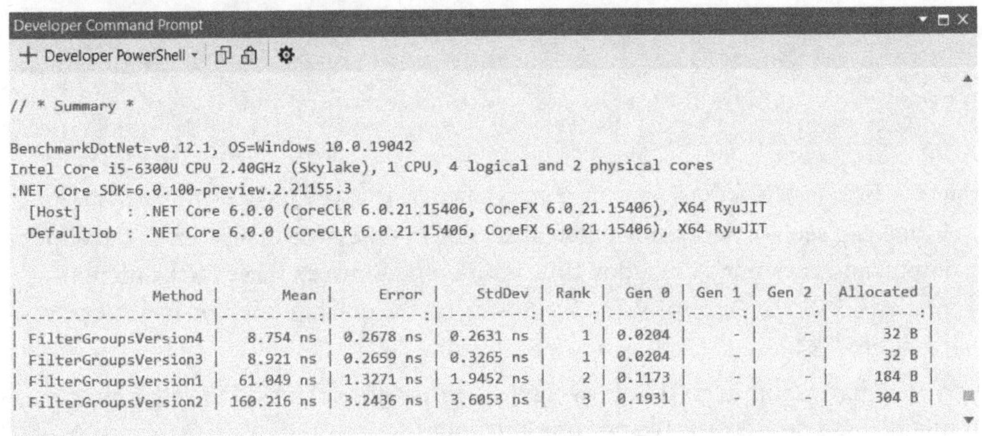


```
// * Summary *

BenchmarkDotNet=v0.12.1, OS=Windows 10.0.19042
Intel Core i5-6300U CPU 2.40GHz (Skylake), 1 CPU, 4 logical and 2 physical cores
.NET Core SDK=6.0.100-preview.2.21155.3
  [Host]     : .NET Core 6.0.0 (CoreCLR 6.0.21.15406, CoreFX 6.0.21.15406), X64 RyuJIT
  DefaultJob : .NET Core 6.0.0 (CoreCLR 6.0.21.15406, CoreFX 6.0.21.15406), X64 RyuJIT
```

Method	Mean	Error	StdDev	Rank	Gen 0	Gen 1	Gen 2	Allocated
FilterGroupsVersion4	8.754 ns	0.2678 ns	0.2631 ns	1	0.0204	-	-	32 B
FilterGroupsVersion3	8.921 ns	0.2659 ns	0.3265 ns	1	0.0204	-	-	32 B
FilterGroupsVersion1	61.049 ns	1.3271 ns	1.9452 ns	2	0.1173	-	-	184 B
FilterGroupsVersion2	160.216 ns	3.2436 ns	3.6053 ns	3	0.1931	-	-	304 B

Figure 7.10: The BenchmarkDotNet summary report showing FilterGroupsVersion4's performance

It is clear from the benchmark report that version 4 of our filter is the winning method in terms of performance. So, why is version 4 better than version 3? The `_group2` array contains fewer items than `_group1`. If you understand the business domain, you will be able to order the filter checks in such a way that the arrays with fewer items will be checked first.

You have seen how using the `let` keyword slows things down. But you have also seen how the ordering of checks in a conditional statement can also have an impact on performance. Placing the check with the least elements first within a conditional check statement will improve performance.

In the next section, we will look at closures in LINQ statements and how they affect query performance.

Understanding closures

In this section, we will understand closures from a C# perspective, and apply them to LINQ queries. Let's start with the definition of computer programming closures according to the content on Wikipedia.

> *Wikipedia: "In programming languages, a closure, also lexical closure or function closure, is a technique for implementing lexically scoped name binding in a language with first-class functions.*
>
> *Operationally, a closure is a record storing a function together with an environment.*
>
> *The environment is a mapping associating each free variable of the function (variables that are used locally but defined in an enclosing scope) with the value or reference to which the name was bound when the closure was created.*
>
> *Unlike a plain function, a closure allows the function to access those captured variables through the closure's copies of their values or references, even when the function is invoked outside their scope."*

To understand what's being said here, we will begin by understanding what first-class functions are.

A first-class function is a method that is treated by C# as a first-class data type. This means that you can assign a method to a variable and pass it around, and you can invoke it as you would a normal method. First-class functions can be created using anonymous methods and lambdas.

Free variables are variables that are not parameter variables to a method, and they are variables that are not local to that method, which, in plain English, means that they are variables that exist outside of a method, but are being referenced within a method's closing scope.

We are going to apply closures to a LINQ expression and benchmark them. The first one will be using LINQ with a closure that takes parameters, and the second one will be using LINQ with a closure that uses free variables. Follow these steps:

1. In the `LinqPerformance` class, comment out the current `[Benchmark]` annotated methods.

2. Add the `LinqClosureUsingParameters` method:

```
[Benchmark]
public void LinqClosureUsingParameters()
{           Func<string, char, char, bool> Between()
        {
            Func<string, char, char, bool> IsBetween
                  = delegate (
             string param1, char param2, char param3)
            {
                var character = param1[0];
                  return (
                        (character >= param2)
                        && (character <= param3)
                  );
            };
            return IsBetween;
        }
        var IsBetween = Between();
        var data = (from p in _people.ToList()
                where IsBetween(p.LastName, 'A', 'G')
                  select p).ToList();
}
```

In the `LinqClosureUsingParameters` method, we declare closure using a delegate with parameters. We declare a variable called `IsBetween` and assign the `Between` method to it. Then we perform a LINQ query and filter the results by calling `IsBetween`. The result is that we will have only those people whose last name's first letters are between A and G.

3. We can also use free variables. So, let us now look at a different example that uses free variables. Add the `LinqClosureUsingVariables` method:

```
[Benchmark]
public void LinqClosureUsingVariables()
{
```

```
Func<string, bool> Between()
        {
                    char first = 'A';
             char last = 'G';
             Func<string, bool> IsBetweenAG = delegate
                 (string param1)
             {
                    var character = param1[0];
                   return ((character >= first) &&
                       (character <= last));
               };
               return IsBetweenAG;
        }
        var IsBetweenAG = Between();
        var data = (from p in _people.ToList()
                     where IsBetweenAG(p.LastName)
                      select p).ToList();
}
```

In the LinqClosureUsingVariables method, we declare our closure using free variables to declare the first and last characters used for filtering the dataset. We then assign the Between method to the IsBetweenAG variable. Then, we perform a LINQ query and filter the results by passing in the last name of each individual into the IsBetweenAG method.

4. Add a method called NonLinqFilter:

```
[Benchmark]
public void NonLinqFilter()
{
        var data = _people.FindAll(
        x => x.LastName[0] >= 'A' && x.LastName[0]
            <= 'G');
}
```

In this method, we simply filter a list using its own FindAll method.

5. Make sure you are in Release mode and then run your project. You should end up with results similar to those in the following screenshot:

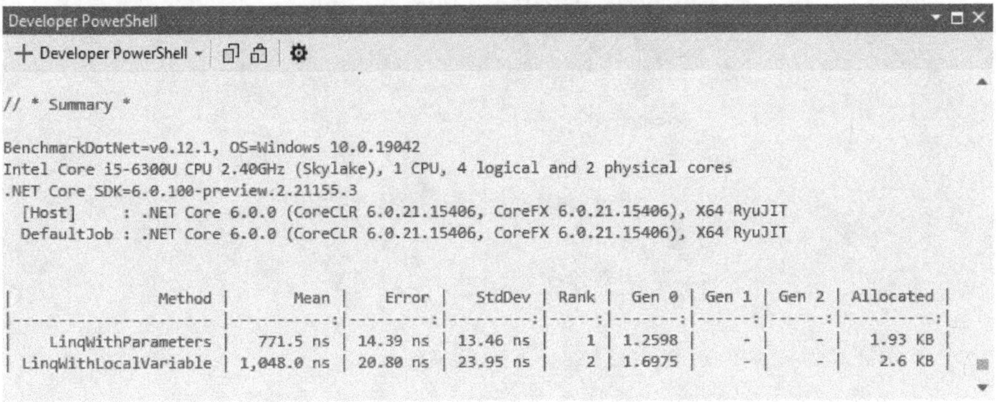

```
Developer PowerShell
+ Developer PowerShell

// * Summary *

BenchmarkDotNet=v0.12.1, OS=Windows 10.0.19042
Intel Core i5-6300U CPU 2.40GHz (Skylake), 1 CPU, 4 logical and 2 physical cores
.NET Core SDK=6.0.100-preview.2.21155.3
  [Host]     : .NET Core 6.0.0 (CoreCLR 6.0.21.15406, CoreFX 6.0.21.15406), X64 RyuJIT
  DefaultJob : .NET Core 6.0.0 (CoreCLR 6.0.21.15406, CoreFX 6.0.21.15406), X64 RyuJIT

|                 Method |       Mean |     Error |    StdDev | Rank |  Gen 0 | Gen 1 | Gen 2 | Allocated |
|----------------------- |-----------:|----------:|----------:|-----:|-------:|------:|------:|----------:|
|     LinqWithParameters |   771.5 ns | 14.39 ns  | 13.46 ns  |    1 | 1.2598 |     - |     - |   1.93 KB |
| LinqWithLocalVariable  | 1,048.0 ns | 20.80 ns  | 23.95 ns  |    2 | 1.6975 |     - |     - |    2.6 KB |
```

Figure 7.11: Closure benchmarks with and without parameters

As we can clearly see in the benchmarks of *Figure 7.11*, closures with parameters are faster and allocate less memory than closures without parameters. But it is far better to use a list's own `FindAll` method for filtering as it is faster and uses less allocated memory than LINQ and closures.

A situation when you may need to apply your own custom closures for use in LINQ queries is when you have complex data manipulation and query generation that cannot be dealt with easily with normal LINQ. In this case, closures would be of benefit to you. Having performed the benchmarking of closures, you now know to use closures with parameters for optimal performance when using LINQ. But if you don't need to use LINQ, then using a list's own methods may be more advantageous. And if you do have to work on lists, then it could pay to do the filtering of the dataset using non-LINQ methods first, then perform your LINQ queries on the filtered lists.

This chapter is now complete. But before we move on to *Chapter 8*, *File and Stream I/O*, let us summarize what we have learned in this chapter.

Summary

In this chapter, we studied LINQ performance by benchmarking a variety of ways to query, group, filter, and iterate data obtained from databases and in-memory collections. The most performant way to query a database was found to be using the `IEnumerator` interface. By disassembling code, we saw that the `let` keyword can degrade performance due to the extra lines of IL code produced by the compiler. We also saw how accessing the last element in a collection using its index is faster than calling the `Last()` method. And we also learned that filtering lists by filtering on objects with the least items first improves filter performance operations. Closures provided better overall performance when passing in parameters, compared to not passing in parameters.

In the next chapter, we will be looking at file and stream I/O performance. But for now, see if you can answer the following questions, and check out the further reading material to solidify what you have learned in this chapter.

Questions

1. Name some ways to improve LINQ performance.
2. What is wrong with using the `let` keyword in a LINQ query?
3. What is the best way to improve the performance of a `Group By` query?
4. What performs better, closures with parameters, or closures without parameters?

Further reading

- **Console User Secrets**: `https://github.com/jasonshave/ConsoleSecrets`.
- **Optimising LINQ**: `https://mattwarren.org/2016/09/29/Optimising-LINQ/`
- **Five Tips to Improve LINQ to SQL Performance**: `https://visualstudiomagazine.com/articles/2010/06/24/five-tips-linq-to-sql.aspx`.
- **Make your C# applications faster with LINQ joins**: `https://timdeschryver.dev/blog/make-your-csharp-applications-faster-with-linq-joins`.

- **LINQ Stinks – code smells in your LINQ**: `https://markheath.net/post/linq-stinks`.
- **How to get a value out of a Span<T> with Linq expression trees?**: `https://stackoverflow.com/questions/52112628/how-to-get-a-value-out-of-a-spant-with-linq-expression-trees`.
- **Linq ToLookup Method in C#**: `https://dotnettutorials.net/lesson/linq-tolookup-operator/`.
- **LINQ (C#) – ToLookup Operator Example And Tutorial**: `https://www.completecsharptutorial.com/linqtutorial/tolookup-operator-example-csharp-linq-tutorial.php`.
- **A Simple Explanation of C# Closures**: `https://www.simplethread.com/c-closures-explained/`.

8
File and Stream I/O

In this chapter, you will learn how to improve directory, file, and streaming performance. You will also learn how to efficiently enumerate directories, process small and large files, perform asynchronous operations, use local storage, handle exceptions, and work with memory efficiently.

We will cover the following topics in this chapter:

- **Understanding the various Windows file path formats**: This section provides information on the different file path formats that you will encounter on the Windows operating system. Also covered is the 256-character file path limit on Windows, and techniques that cover how to remove this limitation.
- **Considering improved I/O performance**: In this section, we will be benchmarking some code to see which method of coding performs fastest when it comes to calculating directory sizes and moving files. Plus, we will look at how to read and write files asynchronously.
- **Handling I/O operation exceptions**: We will cover how to handle I/O exceptions in this section. You will learn how to handle exceptions so that performance is not negatively impacted. You will also learn when to recover from exceptions, as well as when to exit them to preserve data integrity when exceptions cannot be graciously recovered from.
- **Performing memory tasks efficiently**: In this section, you will learn how to efficiently use memory when processing strings and dealing with objects. We will also discuss how to defragment the Large Object Heap.

- **Understanding local storage tasks**: In this section, we will discuss the various options for local file storage, some problems that can arise in networked environments, and when users install software just for themselves when multiple people use the same software on the same computer.

By the end of this chapter, you will be able to do the following:

- Understand the different Windows file path formats.
- Overcome the 256-character file path limit on Windows.
- Understand how hardware affects the performance of your code.
- Choose the best option for calculating directory sizes.
- Choose the best option for moving files.
- Read and write files asynchronously.
- Handle I/O and other exceptions effectively.
- Improve the performance of memory-based tasks.
- Understand what local file storage options are available to you.
- Understand the problems that can occur in networked environments, such as when applications that should be installed for all users on a single machine are installed only for the current user, and how to effectively resolve them.

Technical requirements

The following are the technical requirements for this chapter:

- Visual Studio 2022
- The source code for this book: `https://github.com/PacktPublishing/High-Performance-Programming-in-CSharp-and-.NET/tree/master/CH08`

Understanding the various Windows file path formats

You probably already know that .NET provides managed code that hides interaction with the Windows APIs from the end user. So, it will come as no surprise that the System.IO namespace passes file path information to the Windows APIs to handle. The Windows APIs perform the required task, and then control is handed back to .NET.

File paths in .NET can be absolute, relative, UNC paths, or DOS device paths. Non-Windows files and directories are case-sensitive. But on Windows, files and directories are case-insensitive. The following table provides examples of the different Windows file path formats:

Path	Description
C:\Work\Databases\DevDbDataDictionary.xls	Absolute path. The drive being used is C:\.
\temp\training.txt	Absolute path. This is relative to the current location.
2021\expenses.docx	The relative path to the current directory.
..\temp\training.txt	The relative path up a level from the current directory.
\\Staff\SEN\2020\Yr7\StudentWellbeing.txt	UNC path to the shared network resource.
\\.\C:\Admin\log.txt	DOS device path.
\\?\C:\Admin\log.txt	DOS device path.
\\.\Volume{GUID}\Admin\log.txt	DOS device path.

Table 7.1 – Windows path format examples

By default, Windows can only accept paths with a length of 256. As a programmer, you have probably encountered the `Destination Path Too Long` warning when backing up your files or moving them. A situation that can often lead to this warning is developing web projects using node modules via NPM. NPM packages can have particularly long file paths that exceed 256 characters in length, which will lead to this exception being raised.

You can remove the maximum path length limitation by either editing the registry or by editing the group policy. First, you will learn how to remove this limitation using the registry. Then, you will learn how to remove this limitation using the group policy.

Removing the maximum path length limitation using the registry

> **Note**
>
> Always exercise caution when making changes to the registry.

In this section, you will learn how to remove the file path limit of 260 characters by modifying the registry.

In terms of performance, the `MAX_PATH` issue on Windows can waste your time. Copying many gigabytes of data can be very time-consuming. This can be made worse if a file copy fails after 28 minutes of you moving files between locations on different disks.

So, with file management applications, for example, if a user is going to copy files between two locations that will raise a file length exception, it is best to warn the user and provide them with the option to restructure their files before they perform the copy, or offer to update the registry for them. This way, you can save the end user a lot of wasted time.

To manually remove the `MAX_PATH` file path limit, follow these steps:

1. Open **Registry editor**. You can do this by searching for `regedit`.
2. Once you have opened the registry editor, navigate to the following key:

```
HKEY_LOCAL_MACHINE\SYSTEM\CurrentControlSet\Control\
FileSystem
```

3. Identify the **LongPathsEnabled** key and set its value to `1`.
4. If the key does not exist, then add it as a **32-bit DWORD** with a value of `1`.
5. It may not be necessary, but it is a good idea to restart your computer for the changes to be picked up.

You should now be able to process files with paths with over 260 characters. If you experience permission issues after performing the preceding steps, then open the registry editor as an administrator. If you still have problems, then see your system administrator.

Now, let's learn how to do this using the local group policy editor.

Removing the maximum path length limitation using the group policy

You can also remove the file path limit of 260 characters by modifying the computer's policy. You can do this with the `gpedit.msc` tool. This may be unavailable on some versions of Windows, or may not be available because of enterprise group policies that have been put in place. If you find that to be the case, then see your system administrator. Otherwise, follow these steps:

1. Open **Local Group Policy Editor**. You can do this by searching for `gpedit.msc`.
2. Under **Computer Configuration**, navigate to **Administrative Templates | System | Filesystem**.

3. There will be a setting called **Enable Win32 long paths** set to `Not configured` by default. Edit this setting by setting it to `Enabled`.
4. It may not be necessary, but it is a good idea to restart your computer for the changes to be picked up.

With that, we've learned how to overcome the limiting path situation on Windows by editing the registry and local group policy.

> **Note**
> It is really important to remove the file path limitation. There have been instances where critical backups on client and server computers have failed due to this limitation being in place. It can also break your development project when you're working with third-party libraries.

We will now look at some considerations that will help boost I/O operations.

Considering improved I/O performance

There are several common I/O tasks that we do often, such as traversing directories searching for files, adding, renaming, moving and deleting directories, adding, renaming, moving, and deleting files, password protecting files and directories, encrypting and decrypting files and directories, and compressing files and directories. We also transmit and load files synchronously, asynchronously, and via streams such as file streams and memory streams. Then, there are all the NoSQL and SQL data operations, all of which will be happening frequently on corporate networks, and streaming data and audio/visual content at work and home.

When working with I/O, it is quite easy to completely slow a system down to the point that it becomes unusable while file reading and file writing is taking place. So, if you are going to be performing heavy I/O, you must keep the system where the work is being carried out fully operational and responsive for the end user and other processes.

If your hardware is poor, then no matter how good your software is, it will more than likely be slow!

> **Note**
> Before you consider optimizing your software to improve the speed and performance of I/O operations, you need to make sure that the hardware in place is suited to the type of I/O you will be performing. Otherwise, you could be wasting your time trying to improve your software!

When you're dealing with hardware to speed up input and output operations, things to consider include the speed of your network card, whether or not you are using SSD disks, the number of CPUs, and the amount of RAM in use.

You also need to consider what other software processes will be running on the target computers. Security software that's performing real-time scanning can often be overlooked when it comes to application slowdowns. When this is the case, you can have your application added as an exception to the antivirus software so that real-time scanning no longer slows down your software.

Another issue that's encountered in the wild is running one or more backups over the network during critical times of operation. No matter how efficient your program is, if it is running on a backup server, its performance can be severely impacted by the running backup software and process. This can also be the case if your software is not on the backup server, but requires the network to run and then send and receive files and data. The following are things to consider:

- Change backup schedules to run at non-critical times.
- Install your software on a different server with a better overall performance.
- Check your network for bottlenecks and alleviate those bottlenecks.
- Make sure your network cards are fast enough and configured appropriately.
- Make sure your Ethernet cables are up to date. Cat-5 cables are fine for typical internet traffic, but if you are doing a lot of file and data operations over your network, then you will want to upgrade to Cat-6a/Cat-7 cables for increased performance. However, with Cat-7 cables, you need to be careful not to damage the foil shielding when you bend the cable.

With web projects, it is important to reduce file size to speed up how files are transmitted and received over the internet. This helps reduce the overall page load time and results in happier customers. To improve the load performance of your web applications, enable the Windows Dynamic Content Compression feature. This will reduce the data's size, thus increasing the response time from the user's perspective. The need for data compression also applies to client/server applications, especially if the file and data sizes that are being transmitted are huge.

Employ caching to improve network performance. Caching will store resources locally or keep them in memory for a certain period. Should such resources be requested again, then the locally stored resources will be checked and used instead of the network resources. This increases the access and load times of resources, and it also reduces network traffic. Cached resources will be updated if the resources have been updated, if the cache period has expired, or if the user has cleared their cache.

The two most common data transfer mechanisms are XML and JSON. These are text files that store structured information. Parsers are required to extract information from such files so that the extracted data can be utilized in the applications. But not all XML and JSON parsers perform the same. It would be prudent to benchmark the performance of various XML and JSON parsers to help you choose the most efficient and performant one for your data processing needs.

When you're serializing and deserializing data, your objects and their hierarchies should match your JSON and XML formats so that processing is much faster.

Microsoft recommends that developers shouldn't use BinaryFormatter for transferring binary data as it is unsafe and can lead to **denial-of-service** (**DOS**) attacks. .NET offers several in-box serializers that can handle untrusted data safely:

- `XmlSerializer` and `DataContractSerializer` can serialize object graphs into and from XML. Do not confuse `DataContractSerializer` with `NetDataContractSerializer`.
- `BinaryReader` and `BinaryWriter` for XML and JSON.
- The `System.Text.Json` APIs can serialize object graphs into JSON.

Data types can vary in size as they can hold different data values, and data values can vary in length. Both number values and string values are variable in length. The bigger the number or string, the more bytes are saved to the file. The smaller the number or string, the fewer bytes are saved to the file. Likewise, with data type names, the longer the name, the more bytes are used, and the shorter the name, the fewer bytes are used.

While writing one or two files occasionally, the size of bytes may not be an issue to the end user or your application's performance. But when you move into the realms of batch file processing, the more bytes that have to be written per file, the longer batch processing will take to complete.

Depending on your OS version, drivers, disk, and networking hardware, it is possible that copying or moving small files is more performance-heavy than moving around large files. You can optimize file transfer at the OS level under the hood by leveraging burst copy or similar techniques.

As an example, you can have a lot of performance issues when moving around media files (photo/audio/video) or AI/ML datasets (usually text-based). If files are small (ranging from a few KBs to a few MBs), you can group them in ZIP files (without compression, if they're media files) so that it results in bigger files that can be transferred faster.

In the next section, we will be benchmarking three different methods for moving files. We will be using `File.Copy`, `FileInfo.MoveTo`, and obtaining `FileInfo` from the memory cache and using `FileInfo.MoveTo`. This will help us identify the quickest method to use in our applications, especially when large numbers of files need to be moved.

Moving files

A common function in various enterprise applications is the need to move around large numbers of files. For example, a reporting function may require the amalgamation of last month's sales figures from various teams to be entered into a data warehouse for report processing purposes. Those sales figures could reside in spreadsheets in various locations. Each spreadsheet would need to be moved to a central file storage location for further processing. The more files that you have in any file move operation, the more processing time will be required. So, it pays to know which method of moving a large number of files is the most performant in C#.

With that in mind, we will write a simple application to benchmark three different ways of moving files. Each method that we write will vary in performance. Our method of choice will be the method that performs the fastest, and this will be identified in our benchmark summary report once we have run our compiled executable. Let's start writing our benchmarks:

1. Start a new C# .NET 5 console application and name it `CH08_FileAndStreamIO`.
2. Install the `BenchmarkDotNet` NuGet package.
3. Add a new class called `MovingFiles` to the root of the project:

```
using BenchmarkDotNet.Attributes;
using BenchmarkDotNet.Order;
using System;
using System.Collections.Generic;
using System.IO;
using System.Text;

 [MemoryDiagnoser]
 [Orderer(SummaryOrderPolicy.Declared)]
 [RankColumn]
public class MovingFiles { }
```

   Our class is now set up to benchmark our methods and report on memory usage.

4. At the top of the class, add the following code (you can replace the `Moonshine-3.0.0.exe` file with a file of your own choosing):

```
private Dictionary<string, FileInfo> _cache;
private const string SOURCE_DIRECTORY =
@"C:\Temp\Source\";
private const string DESTINATION_DIRECTORY =
@"C:\Temp\Destination\";
private const string FILENAME = "Moonshine-3.0.0.exe";
```

   Here, we have declared a dictionary of `FileInfo` objects, which will act as our in-memory cache, and three constants for our source directory, destination directory, and filename. We will need these constants in the other methods we will be writing.

5. We need to have a procedure in place to prepare our code so that it can be benchmarked without exceptions being raised. If we don't, our benchmarks will fail to execute more than once because the file will have been moved. Each time a benchmark runs, the moved file needs to be moved back to its original location. So, we are going to need a `[GlobalSetup]` method and a `[GlobalCleanup]` method. First, add the `[GlobalSetup]` method to the `MovingFiles` class. We will call the `PreloadFilesAndCacheThem()` method here:

```
[GlobalSetup]
public void PreloadFilesAndCacheThem()
{
var files = new DirectoryInfo(SOURCE_DIRECTORY)
    .GetFileSystemInfos();
_cache = new Dictionary<string, FileInfo>();
foreach (var f in files)
{
    _cache.Add(f.FullName, f as FileInfo);
}
}
```

   This method is getting `FileSystemInfo` for each file in the source directory identified by the `SOURCE_DIRECTORY` string. Then, it instantiates `_cache` as a dictionary of `FileInfo` objects. After that, the list of files is iterated through, and the `FileInfo` object for the current file is added to `_cache`.

6. Add the `PreMoveCheck() [GlobalCleanup]` method:

```
[GlobalCleanup]
public void PreMoveCheck()
{
    if (File.Exists($"{SOURCE_DIRECTORY}{FILENAME}"))
    if (
        File.Exists(
            $"{DESTINATION_DIRECTORY}{FILENAME}")
    )
    {
        File.Delete(
            $"{DESTINATION_DIRECTORY}{FILENAME}");
    }
if (
    !File.Exists($"{SOURCE_DIRECTORY}{FILENAME}")
    && File.Exists(
        $"{DESTINATION_DIRECTORY}{ FILENAME}")
)
{
    FileInfo fileinfo =
        new FileInfo(
            $"{DESTINATION_DIRECTORY}{FILENAME}")
              fileinfo.MoveTo(
                $"{SOURCE_DIRECTORY}{FILENAME}");
}
}
```

7. The cleanup code checks whether the file already exists in `SOURCE_DIRECTORY`. If it does, then `DESTINATION_DIRECTORY` is checked for the file. If it exists, it is deleted. If the file does not exist in `SOURCE_DIRECTORY` but exists in `DESTINATION_DIRECTORY`, then the file is moved from `DESTINATION_DIRECTORY` back into `SOURCE_DIRECTORY`.

8. We need the `[GlobalSetup]` and `[GlobalCleanup]` methods because if they are not in place doing what they are doing, the benchmarks will fail because the file cannot be found.

9. Add the `FileCopy()` method to the `MovingFiles` class:

```
[Benchmark]
public void FileCopy()
{
    PreMoveCheck();
    File.Copy(
    $"{SOURCE_DIRECTORY}{FILENAME}"
    , $"{DESTINATION_DIRECTORY}{FILENAME}"
);
}
```

10. The `FileCopy()` method performs a `PreMoveCheck()` so that the file is in place, ready for the benchmark to run without failing. It then proceeds to copy the file from `SOURCE_DIRECTORY` to `DESTINATION_DIRECTORY`.
11. Now, add the `FileInfoMoveTo()` method:

```
[Benchmark]
public void FileInfoMoveTo()
{
    PreMoveCheck();
    FileInfo fileinfo = new FileInfo(
    $"{SOURCE_DIRECTORY}{FILENAME}"
);
fileinfo.MoveTo(
    $"{DESTINATION_DIRECTORY}{FILENAME}"
);
}
```

12. The `FileInfoMoveTo()` method also performs a `PreMoveCheck()`, ensuring that the file is in place, ready for the move. Then, it creates a `FileInfo` object for the specified file and uses the `MoveTo(string destinatation)` method to move the file from `SOURCE_DIRECTORY` to `DESTINATION_DIRECTORY`.
13. Add the `FileInfoReadCacheAndMoveTo()` method to the `MovingFiles` class:

```
[Benchmark]
public void FileInfoReadCacheAndMoveTo()
{
    PreMoveCheck();
```

```
        FileInfo fileInfo =
        _cache[$"{SOURCE_DIRECTORY}{FILENAME}"];
    if (fileInfo.Exists)
        fileInfo.MoveTo(
                    $"{DESTINATION_DIRECTORY}{FILENAME}"
        );
    }
```

14. The `FileInfoReadCacheAndMoveTo()` method performs a `PreMoveCheck()`. Then, it creates a `FileInfo` object from the `FileInfo` object stored in `_cache`. If the `FileInfo` object exists, it is then moved to `DESTINATION_DIRECTORY`.
15. Add the following line of code to the `Main` method in the `Program` class:

```
BenchmarkRunner.Run<MovingFiles>();
```

16. Build the project in `Release` mode, and then run the executable from the command line. You should see the following benchmark summary report:

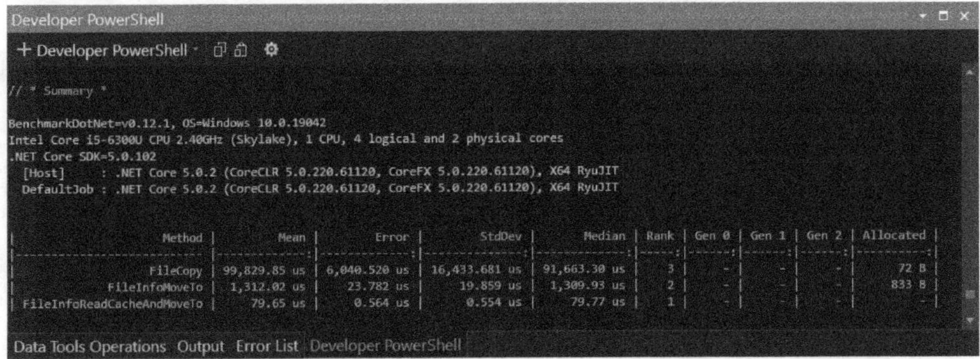

Figure 7.1 – The BenchmarkDotNet summary report for various file move operations

From the timings, we can see that the `File.Copy(string source, string destination)` method is the slowest method of moving files, followed by the `FileInfo.MoveTo(string destination)` method.

The fastest file move operation is to extract `FileInfo` from the in-memory cache and then use the `FileInfo.MoveTo(string destination)` method to perform the move operation.

In the next section, we will look at two different methods for calculating the size of all the files in a directory. We can then use the fastest method for when we need to calculate the size of directories, such as before doing a batch file move in an enterprise.

Calculating directory sizes

When you're batch processing files and directories, it can pay to know how large the sum of files is before moving them to a new location. This can help you determine the amount of time that it will take to copy the files, as well as whether the destination has space to store all the files.

An example of some dialog that pops up when you're copying or moving files is the Windows Explorer dialog. It traverses the files and directories to be moved or copied. As it does, it logs the total amount of bytes that are being used by the files and directories. Then, it provides a time estimate regarding how long it will take to move or copy those bytes. There are times when this process can take a very long time and be frustrating for the end user.

Another reason to know about directory sizes is when you have critical business needs that are time-sensitive. Prolonged file move operations can be detrimental to the business' time plan. In this section, we will calculate directory size by benchmarking two different methods. The method that performs the fastest is the one we would choose when calculating a directory's size. Let's begin:

1. Add a new class to the project called `GettingFileSizes` and configure it for benchmarking, as you did withthe `MovingFiles` class. Then, add the `DIRECTORY` constant to the top of the class:

```
public const string DIRECTORY = @"C:\Windows\System32\";
```

2. Add the `GetDirectorySizeUsingGetFileSystemInfos()` method:

```
[Benchmark]
public int GetDirectorySizeUsingGetFileSystemInfos()
{
DirectoryInfo directoryInfo =
    new DirectoryInfo(DIRECTORY);
FileSystemInfo[] fileSystemInfos =
    directoryInfo.GetFileSystemInfos();
int directorySize = 0;
for (int i = 0; i < fileSystemInfos.Length; i++)
{
    FileInfo fileInfo =
```

```
            fileSystemInfos[i] as FileInfo;
        if (fileInfo != null)
            directorySize += (int)fileInfo.Length;
    }
    return directorySize;
}
```

3. The `GetDirectorySizeUsingGetFileSystemInfos()` method creates a new `DirectoryInfo` object based on the directory defined in the `DIRECTORY` constant. Then, it gets an array of `FileSystemInfo` from the `DirectoryInfo` variable. The `FileSystemInfo` array is then iterated through and the `directorySize` variable is incremented. Once `directorySize` has been calculated, the value is returned to the caller.
4. Add the `GetDirectorySizeUsingArrayAndFileInfo()` method to the `MovingFiles` class:

```
[Benchmark]
public int GetDirectorySizeUsingArrayAndFileInfo()
{
    string[] files = Directory.GetFiles(DIRECTORY);
    int directorySize = 0;
for (int i = 0; i < files.Length; i++)
{
    directorySize +=
        (int)(new FileInfo(files[i]).Length);
}
return directorySize;
}
```

5. The `GetDirectorySizeUsingArrayAndFileInfo()` method gets a string array of filenames for the given directory. The array is then iterated and `directorySize` is incremented by the current file size. Once the iteration is complete, `directorySize` is returned.
6. Add the `benchmark` runner method to the `Main` method in the `Program` class, perform a `Release` build, and then run the executable from the command line. You will see the following report:

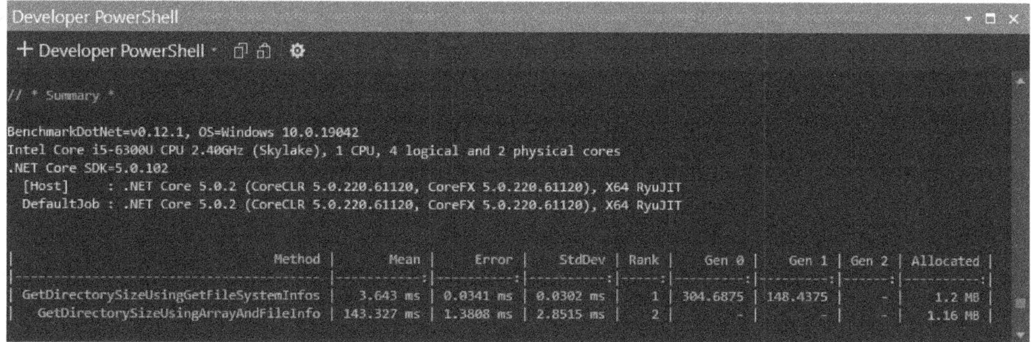

Method	Mean	Error	StdDev	Rank	Gen 0	Gen 1	Gen 2	Allocated
GetDirectorySizeUsingGetFileSystemInfos	3.643 ms	0.0341 ms	0.0302 ms	1	304.6875	148.4375	-	1.2 MB
GetDirectorySizeUsingArrayAndFileInfo	143.327 ms	1.3808 ms	2.8515 ms	2	-	-	-	1.16 MB

Figure 7.2 – The benchmark summary report for obtaining directory sizes

As you can see, we used two different methods to calculate the size of the `System32` directory. The slowest method of calculating a directory size was our second method. So, for performance reasons, the best method for calculating the size of a directory is to get `DirectoryInfo` for the directory in question. Then, you can call `GetFileSystemInfos()` and iterate through the result, summing the length of the `FileInfo` objects.

In the next section, we will look at asynchronous file operations.

Accessing files asynchronously

Why should you access files asynchronously? Well, here are a few reasons that you might consider when using asynchronous file access:

- Your user interface thread will be more responsive as the file operation won't block the user interaction if it takes a few seconds or longer to complete.
- An asynchronous process reduces the need for manually managed threads, making applications more scalable. ASP.NET and server-side applications are specific examples of applications that will benefit from asynchronous file processing.
- File access latency is also something you must consider. Computer resources such as the type of hard disk, network upload and download speeds, and real-time scanning by the security software, as well as file size, are all factors that can affect file access times.
- There is only a small overhead for using asynchronous tasks over threads.
- You can run asynchronous tasks in parallel.

The `FileStream` class gives you the most control over file access operations. You can configure the class to execute I/O operations at the operating system level. By doing this, you avoid blocking thread pool threads. To execute I/O operations at the operating system level, you must specify one of the following in the constructor call:

- `useAsync=true`
- `options=FileOptions.Asynchronous`

> **Note**
> This option can only be used with the `StreamReader` and `StreamWriter` classes when the stream that's provided to them is one that was opened by the `FileStream` class.

Now, let's look at a very simple example of performing asynchronous file writing and reading. Let's start by writing some text to a text file asynchronously. Then, we will read the text from the same file asynchronously.

Writing text to a file asynchronously

In this section, we will write some text to a text file asynchronously. There is a simpler way to perform this task but the method we will be using provides the most control and operates at the operating system level:

1. Add a new file to the `CH08_FileAndStreamIO` project called `AsyncFileAccess`.
2. Add a new method called `WriteTextToFileAsync(string text, string path)` to the `AsyncFileAccess` class:

```
public async Task WriteTextToFileAsync(
string text, string path
)
{
    byte[] encodeText =
      Encoding.Unicode.GetBytes(text);
    using var fileStream = new FileStream(
        path,
```

```
        FileMode.Create,
        FileAccess.Write,
        FileShare.None,
        bufferSize: 4096,
        useAsync: true
    );
    await fileStream.WriteAsync(
        encodeText, 0, encodeText.Length
    );
    }
```

Here, we pass a string of text in and the name of the file to write the text to. Then, we read all the text into a byte array. Next, we declare an asynchronous `FileStream` variable with a buffer size of 4,096 bytes, write the text asynchronously to the specified file, and wait for the operation to complete. The reason for using 4,096 bytes is that it is a power of two number and a memory page size. A page, memory page, or virtual page is a fixed-length contiguous block of virtual memory that's described by a single entry in the page table. So, when the system chooses to swap out a page to disk, it can do so in one go without any overhead involved.

3. Add the `ReadTextFromFileAsync(string path)` method to the `AsynFileAccess` class:

```
public async Task<string> ReadTextFromFileAsync(string
    path)
{
StringBuilder sb = new StringBuilder();
byte[] buffer = new byte[0x1000];
int numberOfBytesToDecode;

using var fileStream = new FileStream(
        path,
    FileMode.Open,
    FileAccess.Read,
    FileShare.Read,
```

```
        bufferSize: 4096,
        useAsync: true
    );
        while (
        (numberOfBytesToDecode = await fileStream.
          ReadAsync(buffer, 0, buffer.Length)) != 0
    )
        {
            sb.AppendLine(Encoding.Unicode.GetString(
            buffer, 0, numberOfBytesToDecode
        ));
    }
        return sb.ToString();
    }
```

In this method, we declare a `StringBuilder` for efficient string concatenation. Then, we declare and initialize a new byte array that will be our buffer and declare a `numberOfBytesToDecode` variable. A new `FileStream` object is instantiated.

The `numberOfBytesToDecode` variable is set by awaiting the call to the `ReadAsync` method. This variable is set for each iteration of the `For` loop. For each iteration of the loop, we obtain the number of bytes to be decoded. Then, we append a line to the output, with the items taken from the buffer. Finally, we return the resulting string.

4. Add the `DemonstrateAsyncFileOps()` method to the `AsyncFileAccess` class:

```
public async Task DemonstrateAsyncFileOps()
{
await WriteTextToFileAsync(
        "Supercalifragilisticexpialidocious",
        @"C:\Temp\File\film.txt"
);
 string text = await ReadTextFromFileAsync(
    @"C:\Temp\File\film.txt"
```

```
);
Console.WriteLine($"The Text written was: {text}");
}
```

The `DemonstrateAsynFileOps()` method writes some text to a file asynchronously by calling the asynchronous write operation. Then, it reads the text back asynchronously by calling the asynchronous read operation. The result is then printed to the console window.

5. Modify your `Program` class's `Main` method as follows:

```
static async Task Main(string[] args)
{
    AsyncFileAccess afa = new AsyncFileAccess();
    await afa.DemonstrateAsyncFileOps();
}
```

This code creates a new instance of our `AsyncFileAccess` class, and then calls the `DemonstrateAsyncFileOps()` method.

6. Build and run your code. In your console window, you should see the following line printed out:

```
The Text written was: Supercalifragilisticexpialidocious
```

As can be seen from our simple example, asynchronous file access is fairly straightforward. In the next section, we will look at how to handle I/O exceptions.

Handling I/O operation exceptions

When working with I/O operations, you can encounter several different exceptions. The base I/O exception is `IOException`. It pays to differentiate between the different I/O exceptions and to log them as this can help expedite problem resolution.

The following table provides a breakdown of the various I/O exceptions that can be raised by your I/O operations. By trapping these specific exceptions, you can provide a more detailed exception log entry that helps with identifying the root source of the problem more easily:

Exception	Description
`IOException`	The base I/O exception that all the other I/O exceptions derive from.
`FileNotFoundException`	This exception is raised when an attempt to read a file fails.
`DirectoryNotFoundException`	This exception is raised when an attempt to read a directory fails.
`DriveNotFoundException`	This exception is raised when an attempt to read a drive fails.
`PathTooLongException`	This exception is raised when the file path exceeds the 256 Windows path length limit.
`OperationCancelledException`	This exception is raised when an I/O operation is canceled.
`UnauthorizedAccessException`	This exception is raised when read/write access has been attempted on a file or directory that the user does not have access to.

Table 7.2 – Microsoft .NET I/O exceptions

Now that you know about the kind of I/O exceptions that can be raised, you also need to know about the correct way to handle, log, and display such exceptions.

As programmers, we need to write code that can detect malfunctioning code. Code that malfunctions leaves a computer program in an undefined state. This can lead to side effects that are unexpected and unpredictable. A computer program that is in an unpredictable state can lead to all manner of issues such as performance slowing down, application hangs, and invalid data, leading to incorrect information. This can lead to serious business and consumer issues, and that is not good.

Therefore, your code needs to be fault-tolerant and should be able to handle faults appropriately. Exceptions should be handled so that data integrity remains intact. You should also bear in mind that there are two categories of exceptions that your computer program should be aware of:

- **Expected exceptions** are exceptions that your computer program can recover from.
- **Unexpected exceptions** are exceptions that your computer program is unable to recover from.

The expected exceptions need to be handled silently. You know what has the potential to fail and why, so you can put defensive code in place to act against such code-raising exceptions in the first place. This is important, as you don't want bubbling exceptions since this reduces application performance. In turn, a reduction in application performance impacts the user experience.

Allowing exceptions to propagate through your computer program is expensive in terms of performance. With this in mind, best practice stipulates that it is better to handle exceptions at the point where they occur within your code for improved application performance.

When you're trapping for errors using a `try/catch` block, it is also a good practice to have multiple `catch` blocks. The only exceptions that would form the `catch` blocks are those that can be thrown by the current method. You would put the exception `catch` blocks in an order where the most specific exception is at the top, and then reduce to the least specific, which would be your bottom `catch` block. This helps make your code more readable to fellow programmers, and it also makes debugging your code for specific exceptions much easier.

You can use exception filters to handle an exception when a specific condition is present. If the exception filter returns true, then the exception is handled. But if it returns false, the search for an exception handler continues. It is preferable to use exception filters instead of catching and rethrowing because filters leave the stack unharmed. If a later handler dumps the stack, you can see where the exception originally came from, rather than just the last place it was rethrown.

When an unexpected exception occurs, it must be thrown because it can have a seriously detrimental effect on the predictability of your computer program. When unexpected exceptions occur, you should log the exception and exit to protect the integrity of your data.

This is why using `System.Exception` is a bad idea in that it swallows all exceptions. Your methods should only trap for the exceptions that they expect to be raised. All unexpected exceptions should be handled by the application in such a way that the exception is logged and the program is exited. It is in the main application's `try/catch` block that you would have your `System.Exception` catch block to catch unexpected exceptions. This block would handle all unexpected exceptions that are allowed to bubble up back to the main application code.

When unexpected exceptions propagate back to the main application code's exception `catch` block, you can extract the underlying base exception by calling `Exception.GetBaseException()`. This will get the original exception that was raised, causing any subsequent exceptions to also be raised.

In my experience, I have found that IT professionals will often neglect to review the event log and application logs when troubleshooting. However, when they have drawn blanks and have asked for my help, this has usually been my first port of call. It may be that nothing gets logged in **Event Viewer**, and nothing gets logged by the application. But there are times when valuable information does get logged, and it can be a time-saver in terms of problem-solving and getting the application working again in a more stable manner.

There are essentially three different locations where an exception can be logged:

- **Application log files**: When an exception is encountered, it will be logged by the application to a text file, JSON file, or XML file.
- **Event Viewer**: When an expected exception is encountered, this will be logged by the application to a named event log. When an unexpected exception is encountered such as an application hang, the system will log this exception in either the Windows Application Log or the Windows System Log.
- **The database**: When an application is encountered, the application will log the exception to a database table.

Whichever mechanism or mechanisms you choose is down to you and your application needs. However, you must make sure that the logs are well-formatted and that the data that's provided is meaningful. Logs are no good if they are hard to read and contain lots of noise!

> **Note**
>
> Use a best practice that dictates managed and unmanaged resources should be correctly disposed of, especially if an application does crash. When providing tech support, I have often come across situations where applications have crashed and locked resources, and where resources have been kept alive in memory. This leads to bad user experiences and can lead to files, directories, and other resources not being accessible, and the application itself not being able to start up. Often, in these cases, the only options are to kill the application using Task Manager or restart the computer.

Performing memory tasks efficiently

When benchmarking C# programs, you will see that sometimes, the objects that allocate the most memory will be faster than the methods that allocate fewer objects. A case in point is strings. Using formatted strings can allocate fewer memory interpolated strings. However, formatted strings can be slower than using interpolated strings. We are going to demonstrate this with a really simple piece of code:

1. Add a class to the **CH08_FileAndStreamIO** project called `Memory` and configure it for using *BenchmarkDotNet*.
2. Add the `ReturnFormattedString()` method:

```
[Benchmark]
public string ReturnFormattedString()
{
return string.Format("{0} {1} {2} {3} {4} {5} {6}
    {7} {8} {9}", "The", "quick", "brown", "fox",
    "jumped", "over", "the", "lazy", "dog", "."
);
}
```

   This method returns a formatted string. It is essentially one line and contains no named variables.

3. Add the `ReturnInterpolatedString()` method to the `Memory` class:

```
[Benchmark]
public string ReturnInterpolatedString()
{
    string thep = "The";
    string quick = "quick";
    string brown = "brown";
    string fox = "fox";
    string jumped = "jumped";
    string over = "over";
    string thel = "the";
    string lazy = "lazy";
    string dog = "dog";
```

```
        string period = ".";
    return $"{thep} { quick } { brown } { fox }
    {jumped} {over} {thel} {lazy} {dog} {period}";
    }
```

This method declares several strings and assigns values to them. It then returns the interpolated string. This method covers multiple lines and looks like it will be slower and use the most memory. However, the only way to know for sure is to run the benchmarks.

4. Add the `BenchmarkRunner.Run<Memory>();` call to your `Main` method, do a `Release` build, and then run the executable from the command line. The following screenshot shows the memory that was allocated and the time it took to perform each method:

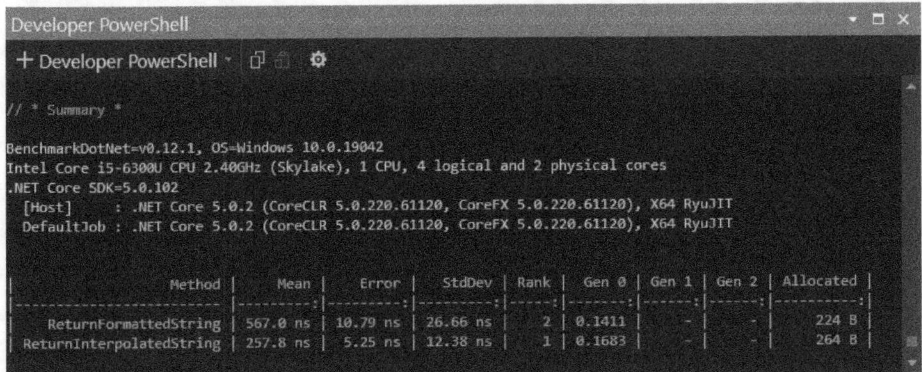

Figure 7.3 – The Benchmark summary report comparing String.Format against interoperable strings

As you can see, even though we can declare multiple variables and allocate the most memory using our string interoperability method, it is much faster than doing the same thing with `String.Format`. If you have a lot of string processing to do, such as in batch report generation or document processing, then you can almost halve the time it takes to perform your string manipulations using string interoperability. The memory also never reaches generation 1, so it is dealt with efficiently by the garbage collector.

Also, you need to reduce the amount of boxing and unboxing that you do. Every time you convert a value type into a reference type, it will be stored on the heap. And every time you convert a reference type into a value type, you place it on the stack. So, what are the performance implications for doing this? Boxing and unboxing are computationally expensive processes. The more computations that are required to perform a function, the slower the process will be. So, by eliminating unnecessary computations caused by boxing and unboxing, you speed up your application and can end up using less memory. So, when you can, try and use value types on the stack instead of reference types on the heap.

Avoid code duplication in your objects. If you have multiple constructor overrides, then place the common code in the common constructor and do the same with your methods. A class with duplicate code will use more memory than the same class correctly coded to have no duplication. You should always look for ways to refactor your objects to reduce code bloat, and removing code duplication and reusing code is an easy way to do this.

Memory fragmentation can be a major cause of performance issues for C# programs. Memory fragmentation occurs when objects are added to the heap, garbage is collected, and then other objects fill the available space. If you end up with free space between the objects in memory, then your memory has become fragmented. The GC will perform a compacting collection when it is most efficient to do so. Doing this manually should only be done after carefully investigating the scenario in question.

In C#, you can defragment the **Large Object Heap** (**LOH**) using the garbage collection settings that are available, as follows:

```
GCSettings.LargeObjectHeapCompactionMode =
GCLargeObjectHeapCompactionMode.CompactOnce;
GC.Collect();
```

This code ensures that the objects on the LOH occupy a contiguous area of memory. All the free space that is located between objects in memory is removed and placed at the end of the allocated memory.

You should also consider not using finalizers. An object will remain in memory longer if it uses finalizers. This will cause a build-up of memory usage. And a build-up of memory usage will lead to reduced performance by your applications.

It is a best practice to dispose of objects and resources when you have finished with them. This helps prevent objects remaining in memory that are not being used, and also releases locks on resources such as files and directories.

When utilizing disposable objects, you should always try and use the `using` statement. This is because when the block of code finishes, the object will automatically be disposed of. When you write a class that uses various disposable resources, even if it does not own those disposable resources, you should implement the disposable pattern.

So far, we have looked at file and memory operations and how performance can be impacted. Now, let's turn our attention to local storage tasks.

Understanding local storage tasks

On Windows 10, there are several locations that you can use to store data locally. These are as follows:

- **Local**: Located in the user's `AppData` folder, this folder can contain settings, files, and folders. This folder is used for data that is not that easy to recreate or download. If you have backup applications that can back up a user's `AppData` folder, then anything stored in the `Local` folder will be backed up.
- **Local Cache**: Only files created using the `ApplicationData.LocalCacheFolder` property can be stored in the local cache. Items stored using the local cache will be persisted across sessions.
- **Roaming**: Roaming profiles can be used by network users to store their local data on the server. This has the advantage that prudent network managers will ensure profiles are backed up regularly, so users will always have a restore point if they happen to lose data.
- **Temporary**: Use the `AppData\Temp` folder for temporary data. It is a good idea to clean data in the `Temp` folder when you have finished with it. Application initialization and shutdown are good points to perform system housekeeping.
- **C:\ProgramData**: This location is a best practice location for storing application data. However, this location does not always get backed up. So, it is always a good idea to provide an in-application way to ensure data is regularly backed up and stored in a safe location in case your computer dies, which does happen!

It's down to you regarding how and where you store your data. From my extensive experience providing IT support to schools, they can have some extremely complicated and very hardened systems security-wise. You cannot assume your application will be installed on the `C:\` drive, and you cannot assume you will have access to the `C:\ProgramData` folder.

Many business and assessment hours have been lost by schools trying to install and run educational vendor software on such complicated systems. Often, this leads to remote technical support sessions.

Another problem that can often arise is the use of the Microsoft VirtualStore. When a user installs software and they are presented with the question, `Install for anyone who uses this computer` or `Install for Just Me`, they tend to select the latter. On Windows 10 computers, `Install for Just Me` puts the stored data for the installed application into the user's virtual store. But selecting `Install for anyone who uses this computer` will normally store application data in the `C:\ProgramData\YOUR_APPLICATION` folder.

A telltale sign that a user has installed the software for only themselves to use is when multiple people log onto an office computer, and each person has a copy of the data. When this happens, multiple copies of the data exist. These copies can be found in each person's virtual store.

This is exactly what happened to me and my colleagues. We develop educational software that comes in standalone, network, and online formats. For our standalone customers, we offer a single-user license. The data for the application is stored in a Microsoft Access database. Originally a problem on Windows 7, which remains a potential problem on Windows 10, is users being given the prompt to install for just them or all users. When they install for all users, the Microsoft Access database can be found under `C:\ProgramData\CompanyName\ProductName`. All users who log onto the computer to use our software will see the same datasets. But should a user select to install only for themselves, then our software's data will be stored under the user profile's VirtualStore

The location of the Virtual Store is `C:\Users\%USERNAME%\AppData\Local\VirtualStore`. This is useful to know because it reduces your time locating the data for the various users under their profiles. The difficulty arises when the customer demands that the data be merged and stored in a central location. When this situation arises, uninstall the software and reinstall it, making sure that you select the option to `Install for all users`. Then, request the users stop using the software until you have provided them with the merged data. Information such as this may not increase the performance of your C# and .NET programs, but it certainly improves your performance when you're providing technical support. And that can be a feather in your cap, as I have found to my benefit! And as programmers/technical support staff/software developers, we all go through personal performance reviews to see how well we are doing in our roles.

Now that we have concluded the material for this chapter, let's summarize what we have learned.

Summary

In this chapter, we started by looking at various file paths. There are four different types of file paths – absolute paths, relative paths, UNC paths, and DOS device paths.

After discussing the various types of paths, we learned that, by default, Windows and Windows Server are limited to a complete file path length of 256 characters. In today's world of open source and web-based software working across platforms, this maximum standard length on Windows computers can be very limiting. This can cause backup issues when you're performing disk-to-disk backups, and deeply nested projects can blow the maximum file path length. To overcome this limitation, we learned how to remove the limit by accessing and modifying the registry.

The next thing we looked at was the various considerations for improving disk I/O. We started looking at I/O performance considerations by considering the different hardware devices that can affect performance. Then, we benchmarked some code to find the most efficient ways of calculating directory sizes, moving files, and performing asynchronous file manipulation.

The next thing we looked at was exception handling. We came to understand that bubbling up exceptions unnecessarily affects performance and that they should be caught and dealt with at the source. We also came to understand that we should not swallow exceptions by catching generic exceptions. Generic exceptions should only be a last resource for logging purposes before you close the application down due to encountering a non-recoverable exception.

We then looked at memory tasks. After benchmarking `string.Format` and interpolated strings, where we learned how using interpolated strings almost doubled our `performane.Next`, we considered memory fragmentation, which can occur when we're adding and removing objects of various sizes. We also learned how to compact fragmented memory to make it run more efficiently.

Finally, we looked at local storage tasks. We discussed the various types of local storage available and their uses. Plus, we discussed the end user installation of our products, which can result in different logged-on users having their own sets of data. This problem arises when users choose to install for themselves instead of all users. Thus, each user has their copy of the application data stored against the profile in `C:\Users\%USERNAME%\AppData\Local\VirtualStore`.

In the next chapter, we will look at networking. But before we do, see if you can answer the following questions. Then, improve your knowledge on the topic of I/O performance by looking at the *Further reading* section.

Questions

Answer the following questions to test your knowledge of this chapter:

1. What are the various Windows file path formats that you need to be aware of?
2. How do you remove the 256-character limit for Windows file paths?
3. Which method is the most efficient for calculating directory sizes?
4. Which method is the most efficient for moving files?
5. When should you catch exceptions using the `Exception` class?
6. What is the base I/O `Exception` class?

7. What file location options do you have for local storage?
8. What is one of the potential pitfalls that may be encountered when users install your software?
9. What is the Microsoft Virtual Store?
10. Where is the Microsoft Virtual Store located?

Further reading

For more information regarding the topics that were covered in this chapter, take a look at the following resources:

- *File and Stream I/O*: `https://docs.microsoft.com/dotnet/standard/io/`.
- *Pipes*: `https://docs.microsoft.com/dotnet/standard/io/pipe-operations`.
- *Faster file move method other than File.Move*: `https://stackoverflow.com/questions/18968830/faster-file-move-method-other-than-file-move`.
- *C# GetFileSystemInfos can get file sizes quickly*: `https://thedeveloperblog.com/getfilesysteminfos`.
- *Performance of writing to a file in C#*: `https://stackoverflow.com/questions/9437265/performance-of-writing-to-file-c-sharp`.
- *Asynchronous File Processing*: `https://docs.microsoft.com/dotnet/csharp/programming-guide/concepts/async/using-async-for-file-access#:~:text=%20Asynchronous%20file%20access%20(C#)%20%201%20Use,writing%2010%20text%20files.%20For%20each...%20More`.
- *How to iterate file directories with PLINQ*: `https://docs.microsoft.com/bs-cyrl-ba/dotnet/standard/parallel-programming/how-to-iterate-file-directories-with-plinq?view=dynamics-usd-3`.
- *Handling I/O exceptions in .NET*: `https://docs.microsoft.com/dotnet/standard/io/handling-io-errors`.
- *Calling Windows 10 APIs from a desktop application*: `https://blogs.windows.com/windowsdeveloper/2017/01/25/calling-windows-10-apis-desktop-application/#vZiZ96PlZUqTduts.97`.

- *Performance Improvements in .NET 6*: `https://devblogs.microsoft.com/dotnet/performance-improvements-in-net-6/`.
- *Page (Computer Memory)*: `https://en.wikipedia.org/wiki/Page_(computer_memory)`.

9
Enhancing the Performance of Networked Applications

You will be very hard pressed when you turn your computer on to not have a single application using some kind of network application. Your operating system connects to a network to download and install Windows updates. Installed applications will poll (check at regular intervals) application servers over the internet to see if there are newer versions available for download.

Browsers download audio and visual data over the internet, and websites allow you to upload and download files. Business applications communicate with database servers. Communication applications send large volumes of textual, audio, and visual data over networks – often with multiple people from various parts of the world involved in online video meetings and training sessions. Your fintech applications communicate with your financial providers over the internet. This is only just scratching the surface.

Our world is very interconnected via technology, and it is networks that make all this possible. I am sure that you have felt some frustration as a user of a website or application when it experiences a slowdown, an application hang, or the application temporarily freezes while some other tasks block the UI until they have been completed, preventing you from doing any work.

Due to this, having applications that are highly performant over a network is crucial in today's fast-paced world. And that is why Microsoft is busy working to always improve the efficiency and speed of their software. One such piece of software that is relatively new on the scene is **Google Remote Procedural Calls (gRPCs)**. A software framework for making **Remote Procedural Calls (RPCs)**, gRPC/gRPC-Web has received a performance boost.

In this chapter, you will learn how to speed up the performance of network applications. You will also learn how to communicate over a network using the **Transmission Control Protocol (TCP)** and **User Datagram Protocol (UDP)** network protocols. Then, you will learn how to perform **network tracing** processes with the OSI network layer reference model and a selection of TCP and UDP networking protocols. **Cache** management will also be covered so that you can improve the efficiency of resource retrieval. Then, you will learn how to make requests and handle responses over the internet, as well as how to use `System.IO.Pipelines` to provide performant streaming capabilities.

The following topics will be covered in this chapter:

- **Understanding the network layers and protocols**: To produce working network software, you don't necessarily need to know anything about networks and how they work – that is, unless you are writing low-level software to improve the network performance of networked applications. In this section, we will start looking at improving the network performance of software by looking at the different layers of a network and the protocols that live in those layers.
- **Improving web-based network traffic**: Many of us use the internet daily during our work, family, education, and leisure time. The internet works over a web-based network that covers the globe. This network is made up of very slow copper wire networks to ultra-fast fiber-optic networks, and many computers with varying degrees of processing power. In this section, we will learn how to improve traffic over the internet to improve internet resource transfer. You will also learn how to monitor web application performance using Microsoft Edge.
- **High-performance communication using gRPC**: In this section, we will learn how to perform high-speed network inter-process communication using gRPC and gRPC-Web. When it comes to gRPC-Web, we will be using Blazor Server for the server-side code and Blazor WebAssembly for the client-side code.

- **Optimizing internet resources**: To improve resource upload and download times, it pays for you to spend time performing the right kinds of resource optimization. In this section, we will learn how to optimize images, text characters, and data transmission.
- **Using pipelines for content streaming**: In this section, you will learn how to break down the data processing, data transmission, and data reception phases into several atomic tasks that work together using pipelines.
- **Caching resources in memory**: In this section, you will learn how to cache resources in memory to reduce page transfer and display times. This can help reduce network load for other users and prevent bottlenecks and throttling.

Upon completing this chapter, you will be able to do the following:

- Understand and apply UDP-based and TCP-based network protocols
- Monitor and identify problems with network traffic
- Improve the network retrieval performance of resources using caching
- Issue web requests and process responses securely
- Efficiently stream content over a network such as the internet using pipelines

> **Note**
> As with all performance-sensitive work, all the techniques and examples in this chapter, as well as throughout this book, should be measured in the context of your application. The overhead of certain techniques mentioned may not be necessary, depending on the scale your networking applications need to handle.

Technical requirements

To follow along with the contents of this chapter, you will need the following:

- Visual Studio 2022 or later
- Microsoft Edge
- This book's source code: `https://github.com/PacktPublishing/High-Performance-Programming-in-CSharp-and-.NET/tree/master/CH09`

Understanding the network layers and protocols

When writing applications that interact with networks, it is very useful to know what **network protocols** are used. A network protocol is a defined set of rules that defines how data is formatted, transmitted, and received between different devices and applications over a network. Different network protocols are used for different tasks. Some protocols are secure by nature, while some protocols are insecure by nature. The **OSI network layer reference model** is a good place to start in understanding the layers of a device's networking capabilities and the associated protocols for those layers.

OSI stands for **Open Standards Institute**. The OSI network layer reference model is a conceptual model that defines and standardizes the communication between computers and telecommunication devices. It is independent of the technology that underpins such communication, so it is a technology-agnostic representation of the layers and protocols of a device's network layers:

OSI Reference Model	Protocol Suite
Application	HTTP, HTTPS, SSL, FTP, TFTP, Telnet, NTP, NNTP
Presentation	Email: SMTP, POP, IMAP
Session	TCP, UDP
Transport	
Network	IP, IGMP, ICMP, ARP, RARP
Data Link	Ethernet, PPP
Physical	

Table 9.1 – The OSI network layer reference model

As you can see, a network has seven layers of operation. These layers are as follows:

- **Application layer**: The application layer provides a user interface that allows users to send and receive data over a network. This layer contains all the applications you use and that operate behind the scenes to interact with the presentation level. For example, your internet browser employs the HTTP, HTTPS, and FTP protocols for transmitting and receiving files over the internet, while email clients use POP3, SMTP, and IMAP for sending and receiving email data.
- **Presentation layer**: The presentation layer encrypts, formats, and compresses the data ready for network transmission. The protocols that are employed in this layer include POP/SMTP, Usenet, HTTP, FTP, Telnet, DNS, SNMP, and NFS.

- **Session layer**: The session layer initiates and terminates sessions with remote systems. This is the layer where ports are assigned for network communication. Examples include port 25 for POP/SMTP, port 532 for Usenet, port 80 for HTTP, port 443 for HTTPS, ports 20 and 21 for FTP, port 23 for Telnet, port 53 for DNS, ports 161 and 162 for SNMP, and the use of an RPC Portmapper for NFS.
- **Transport layer**: The transport layer breaks down data streams into smaller data delivery segments using either TCP or UDP.
- **Network layer**: The network layer provides logical addressing using either the original IPv4 or the newer IPv6.
- **Data Link layer**: The data link layer prepares the data for transmission. This layer translates information from the network layer into a format that can be easily transmitted over the required network type using the SLIP, PPP, 802.2 SNAP, and Ethernet II protocols.
- **Physical layer**: The physical layer is responsible for moving data between device locations. The types of physical network connections that this layer can make available include RS-X, CAT1 to CAT8, ISDN, ADSL, ATM, FDDI, and Coaxial Cable.

The following protocols are used in these various levels:

- **Domain Name Service (DNS)**: The purpose of the DNS protocol is to translate host names into IP addresses and vice versa using what is known as domain name resolution. Port 53 is the default port for DNS.
- **Dynamic Host Configuration Protocol (DHCP)**: The purpose of DHCP is to assign IP address-related information dynamically to networked devices. Ports 67 and 68 are the default ports for DHCP.
- **Hypertext Transfer Protocol (HTTP)**: The purpose of HTTP is to enable the transfer of web pages and supporting material over the internet. Port 80 is the default port for the HTTP protocol.
- **Hypertext Transfer Protocol Secure (HTTPS)**: The purpose of HTTPS is to enable the safe transfer of web pages and their supporting material securely over the internet. Port 443 is the default port for HTTPS.
- **Secure Shell (SSH)**: The purpose of the SSH protocol is to securely connect to a remote computer and move files around and execute various commands. The default port for SSH is port 22.

- **Secure Socket Layer** (**SSL**): The purpose of the SSL protocol is to secure the data that is transferred between a server and a web browser. Port 443 is the default port for SSL.
- **File Transfer Protocol** (**FTP**): The purpose of FTP is to transfer files over the internet. Ports 20 and 21 are the default ports for FTP.
- **Telnet**: Telnet provides insecure two-way interactive text-based communication between two computers using virtual terminal connections. Port 23 is the default port for Telnet.
- **Trivial File Transfer Protocol** (**TFTP**): The default port for TFTP is port 69.
- **Simple Mail Transfer Protocol** (**SMTP**): The purpose of SMTP is to ensure the safe transmission of emails over the network. Port 25 is the default for the SMTP protocol.
- **Post Office Protocol version 3** (**POP3**): The purpose of POP3 is to download and read emails from email servers. Port 110 is the default port for POP3.
- **Internet Message Access Protocol 4** (**IMAP4**): The purpose of IMAP is to access emails on a remote email server without the need to download them. Port 143 is the default port for IMAP.
- **Remote Desktop Protocol** (**RDP**): The purpose of RDP is to establish a remote connection to a computer and take control of it. Port 3389 is the default for RDP.
- **Transmission Control Protocol** (**TCP**): The purpose of TCP is to provide trustworthy assurance that transmitted data will be received. TCP enables data sending and receiving. Different protocols come under the banner of TCP and each TCP protocol has a default port number.
- **User Datagram Protocol** (**UDP**): The purpose of UDP is to provide untrusted data transmission without the assurance that the data will be received. UDP only allows data to be transmitted. Different protocols come under the banner of UDP and each UDP protocol has a default port number.
- **Internet Protocol** (**IP**): The purpose of the IP is to address how data packets are routed between host computers on a TCP/IP network.
- **Ethernet**: The purpose of the Ethernet protocol is to provide control over how data is transmitted over a LAN per the IEEE 802.3 protocol.
- **Point-to-Point** (**PPP**): The purpose of the PPP protocol is to establish a data link connection between two routers using authentication, transmission encryption, and data compression.

- **Network Time Protocol** (**NTP**): The purpose of NTP is to provide clock synchronization between computer systems over packet-switched data networks with variable latency.
- **Network News Transport Protocol** (**NNTP**): The purpose of NNTP is to transport Usenet articles (*netnews*) between news servers. It is also used by end user client applications to read and post articles.

These are only a small subset of the various network protocols that are used in today's world. You are encouraged to further research the various protocols in use if you do a lot of programming that requires network access. You can find some useful articles to assist your advancement in the *Further reading* section.

Once you understand what network protocols are used for, you can select the protocol that best suits your need. This helps reduce overhead. For example, if you only want to transmit data and do not wish to receive it or care whether it is received or not, then you would use the UDP network protocol. However, if you must guarantee that data is sent and received, then you must use TCP instead.

The Internet Engineering Task Force (**IETF**) has defined two **Request For Comments** (**RFCs**) network transport protocols that have become internet standards. RFC 768 (UDP) defines UDP, whereas RFC 793 (TCP) defines TCP. Here are the official links for these RFCs for you to look at:

- RFC 768 (UDP): `https://tools.ietf.org/html/rfc768`
- RFC 793 (TCP): `https://tools.ietf.org/html/rfc793`

TCP is a connection-oriented protocol responsible for ensuring that data is transferred reliably across networks via sessions. The sender and receiver agree on what data will be transferred. Packet error checking is performed on the received data. If there are errors, then a request is submitted to re-transmit the packet that failed. TCP is often used with **IP**. Packets are made aware of where to go and how to get there by IP. The combination of the TCP and IP protocols, when they work together, is defined as TCP/IP.

UDP differs from TCP as it is connectionless. UDP receivers listen for UDP packets with sessions being established. No error checking is performed with UDP. Therefore, packets may be lost with the receiver being unaware of the loss of those packets. UDP does not acknowledge the sender when data is received or when packets are lost.

With TCP establishing connections for communication sessions and performing error checking and resubmitting lost or corrupted packets, it is generally considered slower than UDP. UPD is faster than TCP because it does not establish connections for sessions or perform error checking. Therefore, TCP is the best option when data must be received without errors, such as with financial transactions. However, UDP is the best option when it comes to streaming live images, such as when you are watching a movie. That's why movies can sometimes appear a bit grainy at times.

In the real world, the OSI model does not exist in all practicality. Instead, the universally accepted network model that is tangible in a practical way is the TCP/IP model.

The TCP/IP model

The TCP/IP model differs from the OSI model in that there are only four layers that make up the TCP/IP model. These layers are as follows:

- Application layer
- Transport layer
- Internet layer
- Network interface layer

So, how do the layers of the TCP/IP model map to the OSI model? The following table presents both models and their layers side by side for comparison:

TCP/IP Model	**OSI Model**
Application Layer	Application Layer
	Presentation Layer
	Session Layer
Transport Layer	Transport Layer
Internet Layer	Network Layer
Network Interface Layer	Data Link Layer
	Physical Layer

Table 9.2 – Comparison between the TCP/IP model and the OSI model

Let's describe each layer in the TCP/IP model:

- The **Application layer** enables users to initiate communication between applications and systems over a network. This can be sending an email, opening a web page, running an application over a network, accessing application information from a database, and performing file transfers over a network.
- The **Transport layer** resolves host-to-host communication.
- The **Internet layer** connects different networks.
- The **Network Interface layer** is the physical hardware that enables network communication between a server and its hosts.

Now that we have learned about the TCP/IP model, in the next section, we will write a simple email application and discuss how it relates to the TCP/IP model.

Writing an example email application with the TCP/IP model

In this section, we are going to write a simple console application that sends an email using SMTP. Then, we will discuss how this email is sent through the TCP/IP model. To write a simple console application, follow these steps:

1. Start a new .NET 6.0 Console Application and call it `CH09_OsiReferenceModel`.
2. Add a new class called `EmailServer` with the following `using` statements:

```
using System;
using System.Net.Mail;
```

   We need these two namespaces for handling exceptions and sending emails.

3. Add the following method:

```
public static void SendEmail(
string from, string to, string title, string message
)
{
    try
    {
    MailMessage mailMessage = new MailMessage();
    mailMessage.From = new MailAddress(from);
```

```
        mailMessage.To.Add(to);
        mailMessage.Subject = title;
        mailMessage.Body = message;
        SmtpClient smtpServer = new SmtpClient();
        smtpServer.DeliveryMethod =
            SmtpDeliveryMethod.Network;
        smtpServer.Host = "smtp-mail.outlook.com";
        smtpServer.Port = 587;
        smtpServer.UseDefaultCredentials = false;
        smtpServer.Credentials = new
            System.Net.NetworkCredential("EMAIL_ADDRESS",
                "PASSWORD");
        smtpServer.EnableSsl = true;
        smtpServer.Send(mailMessage);
    }
    catch (Exception ex)
    {
                throw ex.GetBaseException();
    }
    }
```

The preceding code takes the necessary parameters for sending our email programmatically. A `MailMessage` is built up from those parameters. Then, we initialize and configure a `SmtpClient` to connect to a networked-host email server that sends our email.

4. Update the `Program` class, as follows:

```
using CH09_OsiReferenceModel;
Console.WriteLine("Hello World!");
SendMail();
Console.WriteLine("Email has been sent.");
```

Here, we are writing a greeting to the console window. Then, we are calling `SendMail()` to send our email, and then finishing with a message.

5. Now, add the `SendMail()` method:

```
static void SendMail()
{
EmailServer.SendEmail(
    "FROM_EMAIL"
    , "TO_EMAIL"
    , "Test Message"
    , "Test Body. You can delete!"
);
}
```

Replace the email addresses with valid ones. This method calls the `SendMail` method in the `EmailServer` class.

Run the program; you should have an email in your email account.

With your project working, it is time to discuss how your project links in with the TCP/IP network model. Let's start by looking at the following diagram:

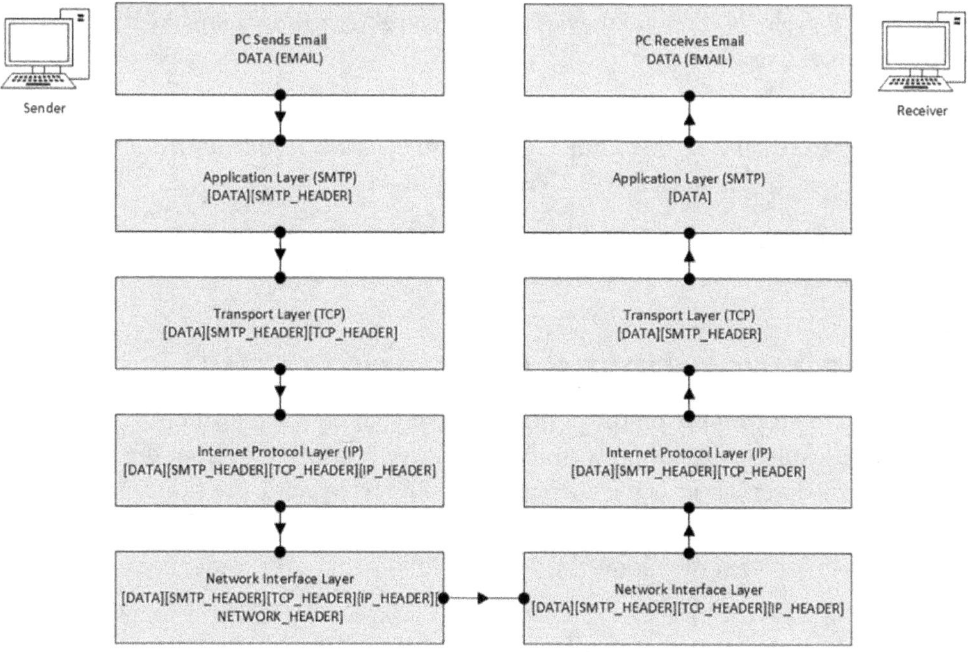

Figure 9.1 – Sending and receiving an email over a network via SMTP using the TCP/IP protocol

First, start with your email client putting together an email, and the user clicking **Send**. When the data hits the Application layer, this is where the SMTP protocol comes into play. In this layer, the recipient is contacted, and the data is formatted and prefixed with an SMTP header.

The email is then passed to the Transport layer. TCP is employed in this layer and is used to break down messages into smaller packets prefixed with a TCP header.

From the transport layer, the email is passed to the Internet layer. IP formats the email packets so that they're ready to be transmitted over the internet and prefixes them with an IP header. These formatted TCP/IP packets are then passed to the Network interface layer.

At the network interface layer, the sender and receiver IP addresses are added to the header that is prefixed to the email. The email is then sent to the receiver.

When the email packet reaches the receiver, it first hits the network layer. The header for the network layer is removed, and the email packet is passed to the Internet layer. The IP header is removed, and the email packet is passed to the transport layer.

At the transport layer, the email packets are then reassembled. Once all the packets have been assembled with the TCP headers removed, they are passed to the application layer, where the SMTP protocol removes the SMTP header, passes the pure email data to the client, and closes the session.

With that, we have covered the conceptual OSI model and the practical four-layer TCP/IP model. Sending an email was the example we used to discuss the journey from the sender to the receiver over the four-layer TCP/IP layer.

Now that you understand the different layers that make up a network and some of the different network protocols and their uses, let's look at network tracing.

Improving web-based network traffic

It is a good idea to keep an eye on the performance of your web applications. This helps you see how well your application transmits and receives information from the network we all know as the internet or, as it is increasingly being referred to, the cloud. You can even track down those calls that are taking a long time to complete, enabling you to improve the responsive performance of your application.

There are various ways that you can accomplish this task. But we will only focus on one way, and that way is to record your application's performance in the web browser using the in-built development tools performance analyzer. Specifically, we will be looking at using Microsoft Edge's **development tools**. This will be the topic of the next section.

Recording your web-applications performance using Microsoft Edge

In this section, you will be using the Microsoft Edge web browser to analyze the performance of your web applications. Internet is the name given to the **Wide Area Network** (**WAN**) that we use every day to browse the web. Sometimes, web applications can be slow, and they are often much slower than their desktop counterparts. That is where the developer tools provided by various browsers come in.

With the browser developer tools, you have some powerful capabilities for seeing what your application is doing behind the scenes. The main features that are provided by various browsers are as follows:

- The ability to navigate the elements of the currently loaded website to view the HTML structure, styles employed, computed styles, layout, event listeners, DOM breakpoints, properties, and accessibility.
- You can view console messages, including any error messages raised.
- You can view all the resources that make up a page with sources, synchronize changes with the local filesystem, override page assets with files from a local folder, view content scripts served by extensions, and create and save code snippets for later reuse.
- You can record and view the network traffic generated by a page, including information such as name, status, type, initiator, size, time, and waterfall with the **Network** tab.
- You can record a process. This information can be extremely detailed and you can save screenshots, record memory usage, and view the web vitals for the page with the **Performance** tab.
- You can profile memory usage and have the option to record a heap snapshot, allocate instrumentation on time, and allocate a sample.
- You can see and debug the background services for your applications on the **Application** tab, including their storage and caches.
- Security, which enables you to view the main origin and secure origins of your application, along with its security information, such as whether it has a valid **SSL certificate**.

Each browser from various vendors works in subtly different ways. Developers each have their preferences as to which browser and set of developer tools they like to use. In this section, we will be using the **Microsoft Edge Network** and **Performance** tabs to analyze the performance of a web page. To do so, follow these steps:

1. Open **Microsoft Edge** and press *F12* to open the developer tools. The following screen should appear:

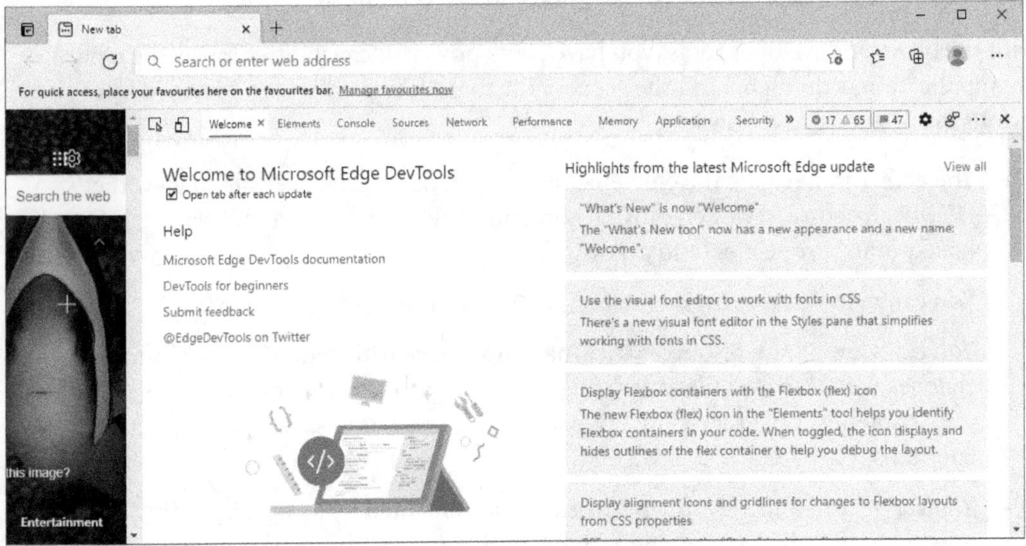

Figure 9.2 – Microsoft Edge developer tools displaying the default tab

2. Click on the **Network** tab.
3. In the address box, type `docs.microsoft.com`.

   The website will now load. As it does, you will see the network traffic being generated and logged. The following screenshot shows a portion of the data that's been ordered by the resources that took the longest time to process:

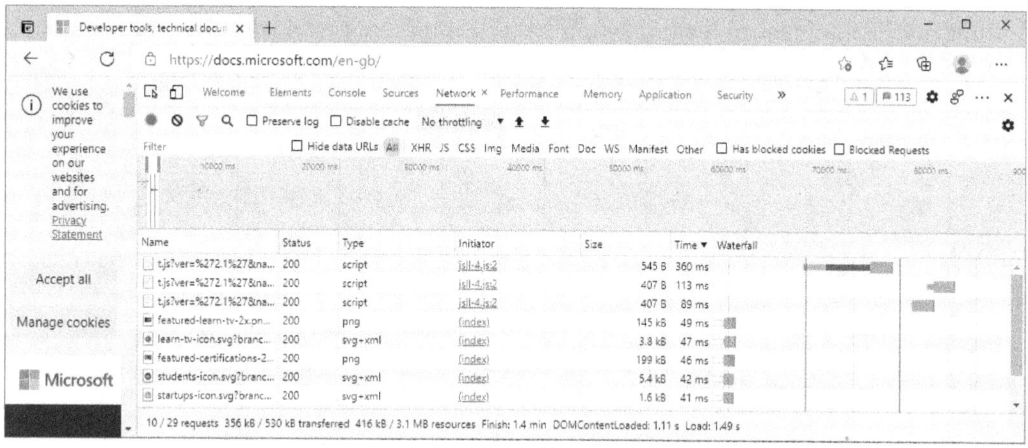

Figure 9.3 – The Microsoft Edge developer tools' Network tab displaying network traffic data

As you can see, the **Network** tab is useful for seeing what resource has been requested (**name**), the **status** and **type** values of the request, what initiated the request (**initiator**), the request **size** and **time** to process, and its visual representation on the **Waterfall** chart. This information can be applied to your pages and their resources to reduce the overall size of a complete request and reduce the time it takes to complete the request.

Now that we've seen the **Network** tab in action, let's look at the **Performance** tab in action. To do so, follow these steps:

4. Click on the **Performance** tab, and then click on the **record** button.
5. Type `docs.microsoft.com` into the address bar and press *Enter*.
6. Once the page has fully loaded, stop the recording by clicking on the popup dialog's **Stop** button.

The profile that has just been captured will now be loaded and presented to you. How long this process will take varies based on how long you were recording and how much traffic was generated.

Once the profile has finished loading, you should be presented with the following screen:

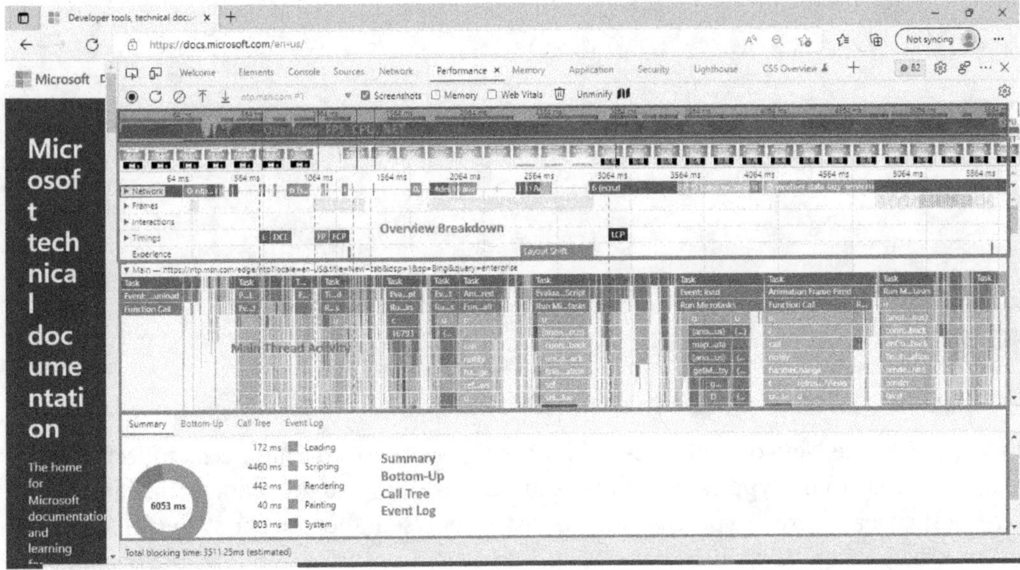

Figure 9.4 – The Microsoft Edge performance profile for docs.microsoft.com

You may be unable to read the contents of the preceding screenshot. That's okay – this screenshot just represents the amount of data that you can glean using the performance profiler. You have screenshots, a Waterfall chart, a breakdown of all the methods and properties that were utilized to load the URL, and a summary of the types of traffic by time, such as loading time, scripting time, rendering time, painting time, system time, and idle time.

You can use this information to find where most of the time is being taken up for a request and identify the method where the time is being consumed. This will help you identify the areas of your web projects that may be candidates for performance improvement.

There is a wealth of information that can be gathered regarding the performance of your application using browser tools. And not all that information has been covered here. For instance, we have not even touched on the memory profiling tab in Microsoft Edge Developer Tools due to this chapter's page length restrictions. However, you are actively encouraged to try out all the different features available in the web browser's development tools for yourself to help you profile and improve the performance of your web applications and their network utilization.

Now that we have learned how to use browser development tools to profile the internet traffic that's produced by our application requests and responses, let's look at the performance-enhanced **gRPC Remote Procedural Call** (**gRPC**) framework for high-speed network data transfer and communication.

High-performance communication using gRPC

What is gRPC? It is an open source **RPC** framework. Applications use RPC to talk to each other. gRPC is built upon the modern technologies of HTTP/2 for the **transport protocol layer** and **protocol buffers** (**Protobuf**) for serializing technology for messages. Protobuf also provides a language-neutral contract language.

gRPC has been designed with modern high-performance and cross-platform applications in mind. There are implementations for all manner of programming languages. This enables applications developed on different operating systems and in different programming languages to talk to each other.

gRPC is in an opinionated contract-first framework with the contract being defined in a **proto file**. This proto file contains your API definitions and the messages they will send and receive. Code generation is then used to generate strongly typed clients and messages for your language and platforms, which in our case will be C# and .NET. The language of gRPC is binary and designed for computers. This makes gRPC perform better than text-based HTTP APIs. The complexity of remoting is hidden from the programmer in the gRPC framework. Much of the work you would normally have to do by hand as a programmer is done for you by code generation tools. And so, all you must do is call methods on your clients and await the results. For increased developer productivity and application performance, you are better off using gRPC over HTTP APIs.

HTTP APIs are content-first and consider the shape of URLs, HTTP methods, JSON and XML, and more. REST APIs are code-first. Normally, you would write your code and then generate Swagger or RAML contracts afterward. REST APIs are human-readable as they are text-based. This makes them easy to debug with the right tools, but these APIs perform slower than gRPC. REST APIs deal with low-level HTTP, so you have more to think about in terms of HTTP requests, responses, and routing. This is more complicated than using gRPC, but you do end up with a high degree of control. So, even though HTTP APIs are not big on performance, they will appeal to the widest developer audience. They can be easier to get started with. However, they can become incredibly complex and deep-routed when you're working on complicated enterprise software.

Now that you know about gRPC and HTTP, you will appreciate that the fastest network and inter-application communication will be carried out by gRPC and not HTTP. And since this book is about performance, we will now demonstrate gRPC at work with a simple demonstration.

Programming a simple gRPC client/server application

In this section, we will be building a gRPC service that returns a single message. Then, we will write a client to call the gRPC service and update our client and server so that we can stream messages. Let's begin by writing our gRPC service.

Building a gRPC service

In this section, we are going to build a gRPC service in **Visual Studio**. Later in this chapter, we will consume this service. To build a gRPC service in Visual Studio, follow these steps:

1. Open Visual Studio and select **Start a new project**.
2. Search for and select the ASP.NET Core **gRPC Service** template and click **Next**.
3. On the **Configure your new project** page, change the location to where you would like, name the project `CH09_GrpcService`, and click **Create**.
4. You will then be presented with the **Additional information** page. Ensure the latest version of .NET Framework is selected from the drop-down; this should be .NET 6.0.
5. Click on the **Create** button. An ASP.NET project will be scaffolded. The proto files for your service will be placed in the **Protos** folder, and your services will be placed in the **Services** folder. Configuration settings will be stored in the `appsettings.json` file.

6. Make sure the project is set as the startup project, and then run it. You should be presented with a **Trust ASP.NET Core SSL Certificate** dialog. Click on **Yes**.
7. You will now be presented with a security dialog, informing you that you are about to install a security certificate. Click **Yes** to install it. Once the certificate has been installed, your service should be running. The gRPC service URLs are `http://localhost:5000` and `https://localhost:5001`.

> **Note**
>
> Ports 5000 and 5001 may be different on your system if they are already in use.

8. Enter `https://localhost:5001` in a browser; you should get the following message: **Communication with gRPC endpoints must be made through a gRPC client.** To learn how to create a client, visit `https://go.microsoft.com/fwlink/?linkid=2086909`. This message informs us that the next step is for us to write a client that will be able to communicate with the service.

And that is how easy it is to get started with a gRPC service. Open the `greet.proto` file in the `Proto` folder and enter the following code:

```
syntax = "proto3";
option csharp_namespace = "CH09_GrpcService";
package greet;
// The greeting service definition.
service Greeter {
  // Sends a greeting
  rpc SayHello (HelloRequest) returns (HelloReply);
}
// The request message containing the user's name.
message HelloRequest {
  string name = 1;
}
// The response message containing the greetings.
message HelloReply {
  string message = 1;
}
```

As you can see, the proto language is straightforward. In this file, we stated the language's syntax, assembly namespace, and the name of the package. We then provided a service definition, which defines the RPC request and response, followed by the request and response messages.

> **Note**
>
> There is a lot of code generation that goes on under the hood. So, in case you are wondering where certain files are located, you will find them hiding away in your `Obj\Debug\net6.0\Protos` folder.

Since we are using gRPC for our service, we need a client. So, in the next section, we will build our client.

Building a gRPC client

In this section, we are going to add a gRPC client project that will consume our gRPC service. Also, for our client project, we will write a simple console application. To add a client project, follow these steps:

1. Start a new .NET 6.0 Console Application project named `CH09_GrpcServiceClient` and change the target framework to `.NET 6.0`.
2. Right-click on the project's **Service dependences** node in the Solution Explorer and select the **Add Connected Service** menu option. This will present you with the following tab:

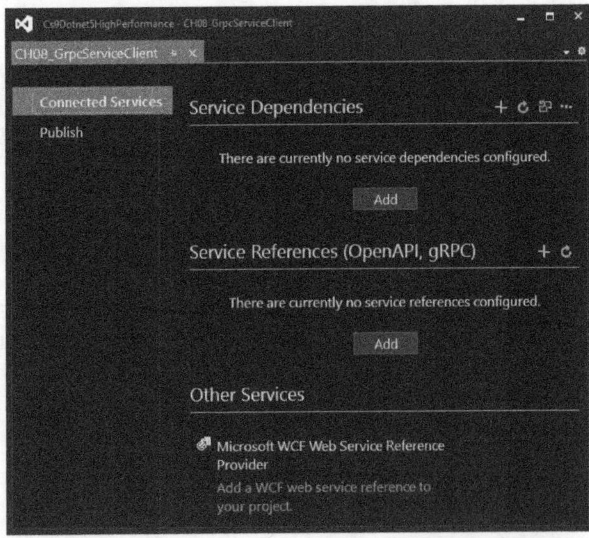

Figure 9.5 – The Connected Services tab in Visual Studio

3. Click on the **Add** button under the **Service References (OpenAPI, gRPC)** section. This will bring up the **Add service reference** dialog, as shown in the following screenshot:

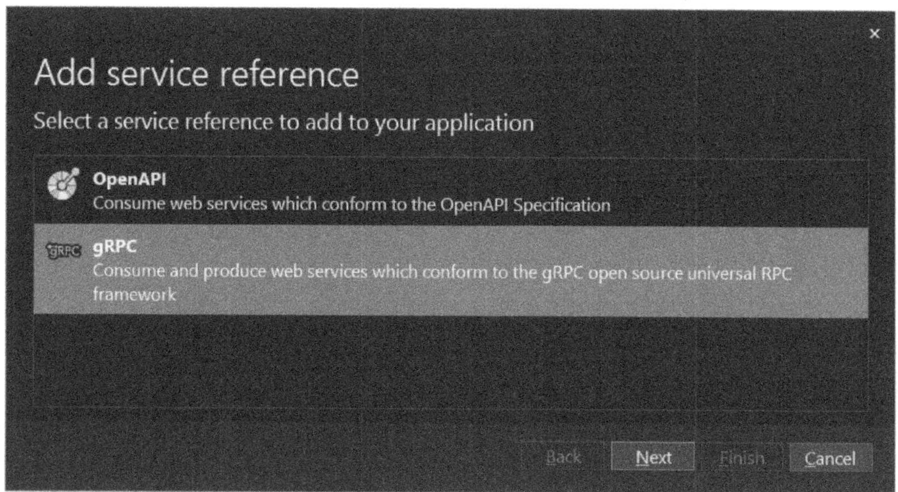

Figure 9.6 – The Add service reference dialog in Visual Studio

4. Click on the **gRPC** option and then click on the **Next** button. The wizard dialog presented in the preceding screenshot will move to the **Add a new gRPC service reference** page, as shown here:

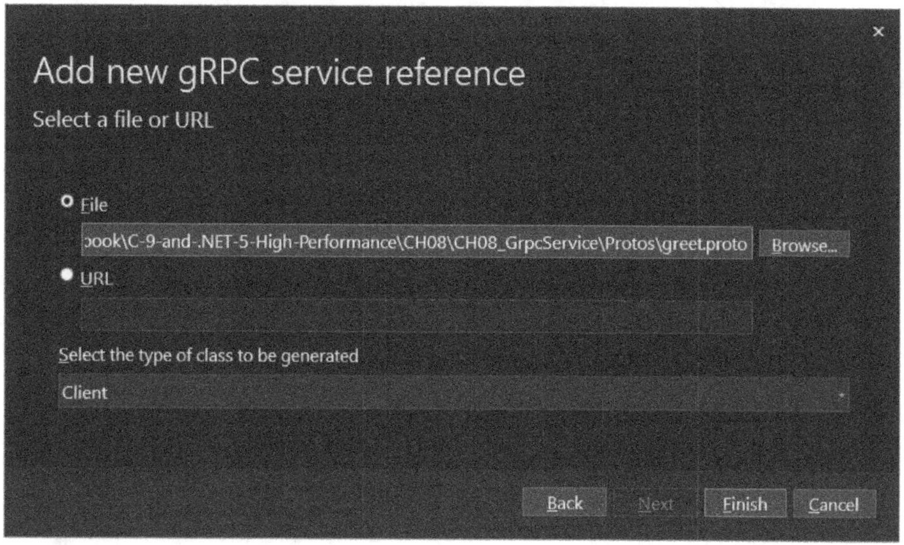

Figure 9.7 – The Add a new gRPC service reference page of the Add service reference dialog

5. Click on the **Browse** button, navigate to the `greet.proto` file in your gRPC service project, and select it. Ensure the **client** option is selected from the dropdown list. Then, click **Finish**.
6. The dialog will change to **Service reference configuration progress**. When you get a message stating **Successfully added service reference(s)**, click the **Close** button. Your gRPC connected service will now appear in the **Service References** section of the **Connected Services** tab, as shown here:

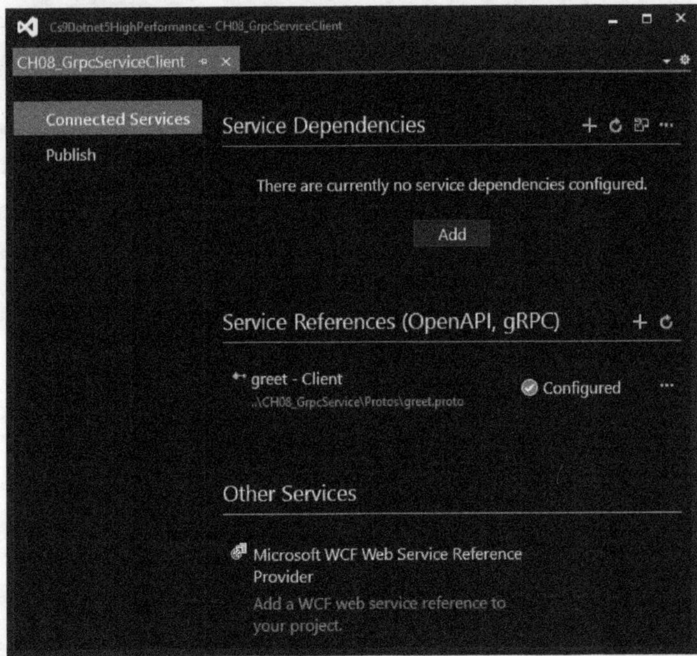

Figure 9.8 – The Connected Services tab displaying our connected gRPC service

With that, you have added a client project to your gRPC service. With the client projected added, we can now write the console applications. Follow these steps:

7. Open the `CH09_GrpcServiceClient.csproj` file by selecting it in the Solution Explorer. You should see the following XML:

```
<Project Sdk="Microsoft.NET.Sdk">
<PropertyGroup>
    <OutputType>Exe</OutputType>
    <TargetFramework>net6.0</TargetFramework>
</PropertyGroup>
```

```
<ItemGroup>
    <PackageReference Include="Google.Protobuf"
        Version="3.13.0" />
    <PackageReference Include="Grpc.Net.ClientFactory"
        Version="2.32.0" />
    <PackageReference Include="Grpc.Tools"
        Version="2.32.0">
        <PrivateAssets>all</PrivateAssets>
        <IncludeAssets>runtime; build; native;
            contentfiles; analyzers;
            buildtransitive</IncludeAssets>
    </PackageReference>
</ItemGroup>
<ItemGroup>
    <Protobuf
        Include="..\CH09_GrpcService\Protos\greet.proto"
        GrpcServices="Client">
        <Link>Protos\greet.proto</Link>
    </Protobuf>
</ItemGroup>
</Project>
```

In the preceding XML code, you can see the references to Google's `Protobuf` library and the `gRPC` libraries. You will also see a Protobuf link to your proto file in the gRPC service, which indicates that your project is taking on the role of the client.

8. Open the `Program` class in your client project.
9. Update the `Main(string[] args)` method, as follows:

```
static async void Main(string[] args)
{
    await ExecuteGrpcClient();}
```

In our entry point method, we call the asynchronous `ExecuteGrpClient()` method. However, because we cannot mark our main method as async, we have to call `Wait()` on the `ExecuteGrpcClient()` method:

```
tatic async Task ExecuteGrpcClient()
{
```

```
    GrpcChannel grpcChannel =
        GrpcChannel.ForAddress("https://localhost:5001");
    Greeter.GreeterClient greeterClient =
        new Greeter.GreeterClient(grpcChannel);
    HelloReply helloReply =
        await greeterClient.SayHelloAsync(new HelloRequest
        {
            Name = "gRPC Demonstration!"
        });
    Console.WriteLine(
        $"Message From gRPC Server: {helloReply.Message}");
}
```

Because we will be awaiting an asynchronous call, we must make the `ExecuteGrpcClient()` method asynchronous with the async modifier. This method does not return anything. However, it cannot be declared void, so we must provide `Task` as the return type. Then, we must declare our gRPC channel by pointing it to our gRPC HTTPS address. Then, we must declare our client by passing in the gRPC channel we have just declared and initialized. Next, we must obtain a reply by awaiting our asynchronous call to our server method and passing in a message request where we set the properties as necessary. Finally, we must print the response from the server to the console window.

10. Open the server project in the terminal and type `dotnet run`. The server will be running locally on port 5001.
11. Then, open the client project in the terminal window and type `dotnet run`. It will print the following message in the console window:

```
Message From gRPC Server: Hello gRPC Demonstration!
```

With that, you have successfully written a gRPC server and consumed its message by writing and running a gRPC client. So what? What does this mean for you? It means that you now have a cross-platform way of communication between different applications using a common protocol. And what is the big deal in this regard? Well, say that you have several legacy applications written in various languages and you want to migrate them all to a common platform and programming language such as .NET or C# – you now have a straightforward way to accomplish this.

By using gRPC, you can provide a phased migration from legacy platforms to the .NET 5 and higher platforms and C# 9 and higher programming language. You would accomplish this by writing gRPC clients for your .NET clients and legacy clients. This would enable you to start using .NET and C# as you incrementally replace older systems. Then, gradually, as the older systems are replaced by one modern system, you can fully utilize .NET and C# and benefit from all the performance improvements the Microsoft teams have made to the language and framework. Plus, you can leverage all the business and performance benefits of using the Microsoft ecosystem, which includes the Microsoft Azure Cloud services that have been built with security, scalability, and performance in mind.

At this point, it is worth noting the various languages that are officially supported by gRPC. The officially supported languages, operating systems, compilers, and SDKs are shown in the following table:

Language	**Operating System**	**Compiler/SDK**
C/C++	Linux, macOS	GCC 4.9+, Clang 3.4+
C/C++	Windows 7+	Visual Studio 2015+
C#	Linux, macOS	.NET Core, Mono 4+
C#	Windows 7+	.NET Core, .NET 4.5+
Dart	Windows, Linux, macOS	Dart 2.2+
Go	Windows, Linux, macOS	Go 1.13+
Java	Windows, Linux, macOS	JDK 8 recommended (Jelly Bean+ for Android)
Kotlin/JVM	Windows, Linux, macOS	Kotlin 1.3+
Node.js	Windows, Linux, macOS	Node v8+
Objective-C	macOS 10.10+, iOS 9.0+	Xcode 7.2+
PHP	Linux, macOS	PHP 7.0+
Python	Windows, Linux, macOS	Python 3.5+
Ruby	Windows, Linux, macOS	Ruby 2.3+

Table 9.3 – Officially supported languages by gRPC

As we can see, gRPC is well supported across languages, operating systems, SDKs, and compilers. So, gRPC is the perfect networking technology to bring disparate systems together using one harmonious messaging framework.

So far, you have consumed a unary request and know that gRPC can be used with various operating systems and programming languages. But what if you need to handle a whole batch of gRPC requests? How do we do that? Good question. We'll learn how to do this in the next section.

Streaming multiple gRPC requests

In this section, we will be modifying our client and server gRPC projects to send and process message streams. By the end of this project, you will be sending 10 messages from the server to the client. On the client, you will process each message as it comes in and write it to the console window. To do so, follow these steps:

1. Update the `greet.proto` file in the `CH09_GrpcService` project, as follows:

```
// The greeting service definition.
service Greeter {
// Sends a greeting
rpc SayHello (HelloRequest) returns (HelloReply);
rpc SayHelloStream(HelloRequest)
    returns (stream HelloReply);
}
```

   You will see that you have added a new message stream to our service definition. Instead of returning a single `HelloReply` message, the message stream returns a stream of messages of the `HelloReply` type.

2. In the `GreeterService` class of the `CH09_GrpcServer` project, add the following method:

```
public override async Task SayHelloStream(HelloRequest
    request, IServerStreamWriter<HelloReply>
        responseStream, ServerCallContext context)
{
    for (int i = 0; i < 10; i++)
    {
    await responseStream.WriteAsync(new HelloReply
    {
        Message = $"Response Stream Message: {i}"
    });
```

```
        await Task.Delay(TimeSpan.FromSeconds(1));
    }
}
```

In this method, you are iterating 10 times. For each iteration, you await the `responseStream.WriteAsync(HelloReply)` method. In this asynchronous call, you set the message on the `HelloReply` object. With each iteration taking only milliseconds, you will purposefully slow down the processing of the task for human eyes to see each method being written one after the other. This delay slows down your task by 10 seconds – a second for each iteration. In a normal application, you would normally not have such a delay in place.

3. Now that you have updated your server project, rebuild both projects to see the changes and move to your `CH09_GrpcServiceClient` project.
4. In the `Program` class, move the code inside the `ExecuteGrpcClient()` method into its own method called `SingleGrpcMessageClient()`. Then, add the following two lines of code to the `ExecuteGrpcClient()` method:

```
await SingleGrpcMessageResponse();
await GrpcMessageResponseStream();
```

The preceding code contains two asynchronous calls: one for a single message, and one for streaming multiple messages.

5. Add the `GrpcMessageResponseStream()` method:

```
static async Task GrpcMessageResponseStream()
{
    GrpcChannel grpcChannel =
        GrpcChannel.ForAddress("https://localhost:5001");
    Greeter.GreeterClient greeterClient =
        new Greeter.GreeterClient(grpcChannel);
    AsyncServerStreamingCall<HelloReply> helloReply =
        greeterClient.SayHelloStream(new HelloRequest
        {
            Name = "gRPC Streaming Demonstration!"
        });
    await foreach (HelloReply item in
        helloReply.ResponseStream.ReadAllAsync())
```

```
            {
                Console.WriteLine(item.Message);
            }
        }
```

`GrpcMessageResponseStream()` creates a `GrpcChannel` and assigns it to a new client. A call is then made to a gRPC stream. This iterates through all the items in the stream that have been sent back to the client from the server, and then prints the message for each item to the console window.

6. Open each of the projects in their own terminals and type the `dotnet run` command. This will start the server and run the client. You should see the following console window output:

```
Message From gRPC Server: Hello gRPC Demonstration!
Response Stream Message: 0
Response Stream Message: 1
Response Stream Message: 2
Response Stream Message: 3
Response Stream Message: 4
Response Stream Message: 5
Response Stream Message: 6
Response Stream Message: 7
Response Stream Message: 8
Response Stream Message: 9
```

You now know how to use gRPC with desktop applications. In the next section, you will learn how to use gRPC with Blazor.

Programming a simple gRPC Blazor application

Blazor is a web programming model. With Blazor, you can have server-side Blazor projects that you would write when you have sensitive information that you need to keep secret. You can have client-side Blazor projects when application performance is of the utmost importance. As part of an organization's enterprise application, you have many different Blazor server-side and client-side applications working together as one.

For gRPC to work with web projects, a wrapper called gRPC-Web has been developed. This enables you to have both gRPC-Web services and gRPC-Web clients. With gRPC-Web, it is possible to build end-to-end pipelines that are compatible with the HTTP/1.1 and HTTP/2 protocols. This provides a competitive edge over browser APIs that are unable to call gRPC HTTP/2, especially when you consider that not all .NET platforms have support for HTTP/2 via the `HttpClient` class. Another benefit of gRPC-Web is that you don't have to use just TCP for **Inter-Process Communication** (**IPC**). For IPC, you can also use named pipes (UDP) and **Unix domain sockets** (**UDS**).

> **Note**
>
> The default template app for Blazor has a fetch data page that uses JSON for its data backend. The data size for this JSON file is 627 bytes. But when JSON is replaced with gRPC, the size of the data is reduced to 309 bytes. This example shows that data transfer is quicker using gRPC-Web than it is using JSON, as there is not so much data to transmit and receive over the network. The reduced size of the data transmission using gRPC-Web means that more requests can be made over the network before the requests need to be throttled.

In .NET 6.0, applications are made smaller via aggressive trimming. You can aggressively trim gRPC-based applications to reduce their size and increase their performance, especially when it comes to sending data over networks. This is because of the in-built code generation that is part of gRPC.

In web projects, gRPC cannot be directly accessed. Therefore, a proxy project known as gRPC-Web was introduced to enable the use of gRPC with web projects.

In the following sections, we will be writing a Blazor client and Server gRPC application consisting of a Blazor Server Application and a Blazor WebAssembly Application.
Let's begin.

The blank solution

We need to start with a blank solution:

1. Open Visual Studio and search for `Blank Solution`.
2. Create the blank solution and name it `CH09_BlazorGrpc`.

This will provide a blank solution to which we can add our client and server Blazor applications. Next, we will work on our client project.

The Blazor client project

In this section, we will build our Blazor client gRPC application. Follow these steps:

1. Add a new **Blazor WebAssembly** app called `CH09_BlazorGrpc.Client`.
2. Add the following NuGet packages:

   A. `Google.Protobuf`

   B. `Grpc.Net.Client`

   C. `Grpc.Net.Client.Web`

   D. `Grpc.Tools`

3. Add a folder called `Protos` and a file to that folder called `person.proto`.
4. Open the `person.proto` file and add the following code:

```
syntax = "proto3";
option csharp_namespace = "CH09_BlazorGrpc.Client";
package grpcpeople;
service Person {
  rpc GetPeople (PeopleRequest) returns (PeopleResponse);
}

message PeopleRequest {
}

message PeopleResponse{
        repeated PersonResponse people = 1;
}

message PersonResponse {
     string name = 1;
}
```

   Our proto file defines the proto definition version as proto3. So, the `proto3` syntax will be used. The namespace for our service definition is `CH09_BlazorGrpc.Client`. The name that's been given to our package is `grpcpeople`. There are three messages called `PeopleRequest`, `PeopleResponse`, and `PersonResponse`. Finally, we define our service as `Person` with an RPC called `GetPeople` that takes a `PeopleRequest` and returns a `PeopleResponse`.

5. Add the following imports to the `_Imports.razor` file:

```
@using CH09_BlazorGrpc.Client
@using CH09_BlazorGrpc.Client.Shared
@using Grpc.Net.Client;
@using Grpc.Net.Client.Web;
```

These imports will be available to all our files.

6. Locate the `Pages/Index.razor` page and replace its contents with the following code:

```
@page "/"
@using CH09_BlazorGrpc.Client
<PageTitle>Index</PageTitle>
<h1>People from Grpc Service</h1>
@foreach(var person in model.People)
{
    <p>Name : @person.Name</p>
}
@code{
    private PeopleResponse model = new PeopleResponse();
    protected override async Task OnInitializedAsync()
    {
            using var channel = GrpcChannel.ForAddress
                ("https://localhost:7272/", new
                    GrpcChannelOptions
    {
        HttpHandler = new GrpcWebHandler(new
            HttpClientHandler())
    });
            var client = new Person.PersonClient
                (channel);
            model = await client.GetPeopleAsync(
                                new PeopleRequest { });
    }
}
```

The preceding code will call the gRPC service that has been located by the service app and list the people that have been returned.

That's our client application completed. Now, let's write our server application.

The Blazor server project

In this section, we will write our server application, which will contain our service responsible for returning the requested data to the client. Let's begin:

1. Add a new Blazor Server app called `CH09_BlazorGrpc.Server`.
2. Add the `Grpc.AspNetCore` and `Grpc.AspNetCore.Web` NuGet packages.
3. Copy the `Protos` folder and its contents from the client project and paste it into the server project.
4. Add the `PeopleService` class to the root of the server project.
5. Replace the contents of the `PeopleService` class with the following code:

```
namespace CH09_BlazorGrpc.Server;
using Grpc.Core;
using CH09_BlazorGrpc.Client;
public class PeopleService : Person.PersonBase
{
    public override async Task<PeopleResponse>
        GetPeople(PeopleRequest request,
            ServerCallContext context)
    {
        PeopleResponse response = new PeopleResponse();
        response.People.Add(new PersonResponse { Name =
            "Person One" });
        response.People.Add(new PersonResponse { Name =
            "Person Two" });
        response.People.Add(new PersonResponse { Name =
            "Person Three" });
        return response;
    }
}
```

This service has a single method that returns a list of people.

6. Replace the code in the `Program.cs` file with the following:

```
using CH09_BlazorGrpc.Server;
var builder = WebApplication.CreateBuilder(args);
builder.Services.AddGrpc(options =>
{
    options.EnableDetailedErrors = true;
    options.MaxReceiveMessageSize = 2 * 1024 * 1024;
    // 2 MB
    options.MaxSendMessageSize = 5 * 1024 * 1024; // 5 MB
});
builder.Services.AddCors(setupAction =>
{
    setupAction.AddDefaultPolicy(policy =>
    {
        policy.AllowAnyHeader().AllowAnyOrigin()
            .AllowAnyMethod()
          .WithExposedHeaders("Grpc-Status",
               "Grpc-Message", "Grpc-Encoding",
                   "Grpc-Accept-Encoding");
    });
});

var app = builder.Build();
app.UseCors();
app.UseRouting();
app.UseGrpcWeb(new GrpcWebOptions { DefaultEnabled =
    true });
app.UseEndpoints(endpoints =>
{
    endpoints.MapGrpcService<PeopleService>();
});

app.Run();
```

The preceding code configures our Blazor application to use gRPC and sets up our `PeopleService` class so that our client application can use it. We also configured `Cors` so that our gRPC requests and responses don't get blocked.

7. Right-click on the solution and select **Properties**.
8. Under **Startup Project**, select **Multiple startup projects** and change **Action** to **Start** for both the client and the server projects.
9. Click on **OK** to close the property dialog.

Run the project. Two browser and two console windows should open. If all goes well, then you should see the following browser window:

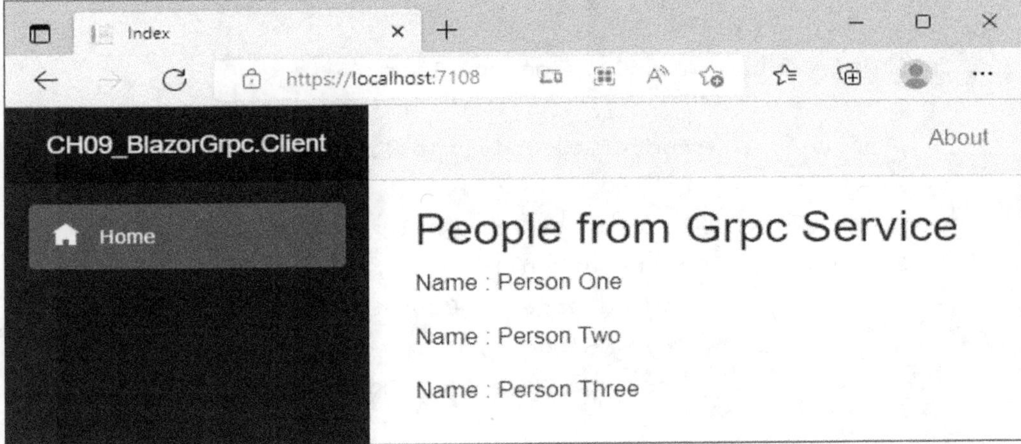

Figure 9.9 – The client Blazor application showing the response from the gRPC service in the server app

> **Note**
>
> The port numbers depend on the ports that are available on a system. So, alternative ports will be used if ports 5000 and 5001 are already in use. This has happened here, in which port 7272 is being used for the server app and port 7108 is being used for the client app.

With that, you have learned about desktop and web-based network data transmission and communication using gRPC and gRPC-Web, both of which have received several performance enhancements along with C# and .NET Framework. You have also used the Blazor server and Blazor WebAssembly to perform web data transmissions and receive data.

You can use this information to replace your code that uses the JSON data format with the binary format of gRPC. This should cut down the size of your data transmissions and reduce the time it takes for the data to be transmitted and received, thus improving the performance of your networked applications – especially those applications that deal with huge volumes of data.

Optimizing internet resources

The best web page is a web page that does the minimum it needs to present the necessary data that you want your users to view. Noisy web pages take longer to load and can be a source of irritation to your end users.

When you use advert services and analytical and health monitoring services, these can produce unnecessary network traffic and an increase in page load time. So, you need to be concise in what data you gather regarding the page that is loading. You also need to reduce the number of resources that your page is downloading. Some of these resources will be explained here.

Images

Images are one of the resources that can significantly increase the time it takes to load a page. Therefore, it is important to use the right image format and compression for your images. It is often necessary to reduce the file size of images. Images usually come in three file formats: **JPEG/JPG**, **PNG**, and **GIF/animated GIF**. When it comes to image optimization, you are best off experimenting based on your website requirements. This is because you will need to factor in the tradeoff between image quality and image size, depending on your specific requirements.

An example of a tool that you can use for PNG optimization is PNGGauntlet by Ben Hollis: `https://pnggauntlet.com/`. This tool creates small PBGs by combining PNGOUT, OptiPNG, and DeflOpt with no loss of image quality. It can also convert the JPG, GIF, TIFF, and BMP file formats into PNG. You can configure the tool to your liking.

Text characters

When transmitting text over the internet, the more characters you have, the larger the file you will have. As a page grows, the time to load that page increases. You can reduce the size of each request and response by enabling deflate or `gzip` compression. Most, if not all, web servers provide web compression. You will have to look at how to enable web compression in the web server that you are using.

You can also reduce the size of your HTML, CSS, and JavaScript files in production by using minification. During development, when you have reached the stage where you are ready to deploy your application, you can employ tools such as webpack that will condense your files by removing unnecessary whitespace, comments, and unused code. Tools such as webpack can drastically reduce the size of your files.

This size reduction results in less data being transmitted over a network, meaning that the files a user has requested get downloaded on their device much faster. The quicker that requested files are downloaded to a user's device, the quicker the requested page will be rendered for them to view.

Data transmission

Transmitting data over a network takes time. That time can vary based on several different factors, such as the amount of network traffic and the route taken. Not all networks use fiber optic, and there are still locations over the internet that are still on slow copper wire connections.

One way to reduce network traffic and load time for networked resources is to **cache** them on the user's computer that requested the resource. When a network resource is requested, the application will check if it exists in the cache. If it does, then the item will be retrieved from the cache on the user's computer. But if the item is not in the cache, it will be downloaded over the network and stored in the user's cache. When an item is being retrieved from the cache, the expiration date and time will be checked for the resource. If the expiry date and time have been reached, then the resource will be downloaded from over the network.

Also, when working with large volumes of data, it is best to filter the data on the server and only return the subset of the data that you require. If the amount of data that you require is quite large, then employ data paging, whereby the data is divided into pages. Then, you only need to download a page as it is requested. This reduces the time it takes to receive the data once the request has been made.

Using pipelines for content streaming

`System.IO.Pipelines` is a high-performance I/O .NET library that was first shipped with .NET Core 2.1 and was born from performance work carried out by the Kestrel team. The purpose behind pipelines is to reduce the complexity of correctly parsing stream and socket data.

In this section, we will learn how to use pipelines with sockets. We will write to small console applications. The first console application will listen for incoming requests on port `7000` and output the contents to the console window. The second console application will listen for the newline key. When it is detected, it will send the contents of the command line to the server on port `7000`. By completing this project, you will see how easy it is to write a network communication application with a minimal number of lines of code using pipes and sockets.

Let's start by writing our server console app.

Writing and running a TCP server console application

In this section, we will use sockets and pipelines to write a console application that listens for incoming data on port `7000`. When data is received, it is processed and output to the console window. To write a TCP server console application, follow these steps:

1. Start a new .NET 6.0 Console Application called `CH09_TcpServer`.
2. Add the `System.IO.Pipelines` NuGet package.
3. Add a new class called `SocketExtensions`:

```
using System;
using System.Net.Sockets;
using System.Runtime.InteropServices;
using System.Text;
using System.Threading.Tasks;
internal static class SocketExtensions
{
}
```

This is our `SocketExtensions` class, which we will build up with extension methods to simplify our socket code.

4. Add the `ReceiveAsync` extension method:

```
public static Task<int> ReceiveAsync(this Socket socket,
    Memory<byte> memory, SocketFlags socketFlags)
{
    ArraySegment<byte> arraySegment = GetArray(memory);
return SocketTaskExtensions.ReceiveAsync(socket,
    arraySegment, socketFlags);
}
```

This method extends a socket to delimit a section of a one-dimensional array. It receives data from a connected socket and returns a `Task` that represents the asynchronous receive operation.

5. Add the `GetString` extension method:

```
public static string GetString(this Encoding encoding,
     ReadOnlyMemory<byte> memory)
{
     ArraySegment<byte> arraySegment = GetArray(memory);
      return encoding.GetString(arraySegment.Array,
          arraySegment.Offset, arraySegment.Count);
}
```

This method extends a socket to delimit a section of a one-dimensional array. Then, it decodes a sequence of bytes into a string and returns the decoded string.

6. Add the `GetArray` method:

```
private static ArraySegment<byte> GetArray(Memory<byte>
     memory)
{
return GetArray((ReadOnlyMemory<byte>)memory);
}
```

This method gets contiguous memory and returns a delimited section of a one-dimensional array.

7. Add the final extension method – that is, `GetArray`:

```
private static ArraySegment<byte> GetArray
     (ReadOnlyMemory<byte> memory)
{
if (!MemoryMarshal.TryGetArray(memory, out var result))
{
     throw new InvalidOperationException("Buffer backed by
          array was expected");
}
return result;
}
```

This method tries to get a segment from the underlying memory buffer. The return value indicates the success of the operation. A delimited segment of a one-dimensional array is returned.

8. Switch to the `Program` class.
9. Replace the `Program.cs` file's source code with the following code:

```
using CH09_TcpServer;

using System;
using System.Buffers;
using System.IO.Pipelines;
using System.Net;
using System.Net.Sockets;
using System.Text;
using System.Threading.tasks;
Socket listenSocket = new Socket(SocketType.Stream,
    ProtocolType.Tcp);
listenSocket.Bind(new IPEndPoint(IPAddress.Loopback,
    7000));
Console.WriteLine("Listening on port 7000");
listenSocket.Listen(120);
while (true)
{
    Socket socket = await listenSocket.AcceptAsync();
    _ = ProcessLinesAsync(socket);
}
```

Our top-level code creates a socket on port `7000`. Then, it listens for incoming data on port `7000` and processes the data.

10. Add the `ProcessLinesAsync` method:

```
tatic async Task ProcessLinesAsync(Socket socket)
{
Console.WriteLine($"[{socket.RemoteEndPoint}]:
    connected");
NetworkStream stream = new NetworkStream(socket);
PipeReader reader = PipeReader.Create(stream);
```

```
while (true)
{
    ReadResult result = await reader.ReadAsync();
    ReadOnlySequence<byte> buffer = result.Buffer;

    while (TryReadLine(ref buffer, out
        ReadOnlySequence<byte> line))
        ProcessLine(line);     reader.AdvanceTo
            (buffer.Start, buffer.End);
    if (result.IsCompleted)
        break;
}
    await reader.CompleteAsync();
        Console.WriteLine($"[{socket.RemoteEndPoint}]:
            disconnected");
}
```

With this method, we pass in a socket. The socket is assigned to a new `NetworkStream` object. Then, the new `NetworkStream` object is passed into a new `PipeReader` object. While there is data to be read, we read and process each line in the stream in turn. Once the stream has been completely read from start to finish, we mark the reader as complete so that no more data will be read from it.

11. Now, add the `TryReadLine` method:

```
static bool TryReadLine(ref ReadOnlySequence<byte>
    buffer, out ReadOnlySequence<byte> line)
{
SequencePosition? position = buffer.PositionOf
    ((byte)'\n');
if (position == null)
{
    line = default;
    return false;
}
            line = buffer.Slice(0, position.Value);
            buffer = buffer.Slice(buffer.GetPosition
                (1, position.Value));
```

```
        return true;
    }
```

This method attempts to read a line of a `ReadOnlySequence` of bytes. If is unable to, it will return false. But if it can, it will set the line it can read as a `ReadOnlySequence` of bytes and return true.

12. Add our final method for our TCP Server called `ProcessLine`:

```
static void ProcessLine(in ReadOnlySequence<byte> buffer)
{
foreach (ReadOnlyMemory<byte> segment in buffer)
{
    Console.Write(Encoding.UTF8.GetString(segment.Span));
}
Console.WriteLine();
}
```

All we are doing here is printing the contents of the stream to the console window line by line.

13. Run the program. You should see something similar to the following:

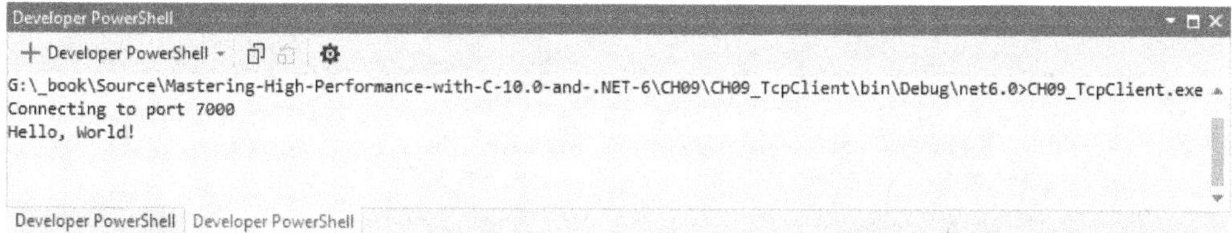

Figure 9.10 – The TCP Server in a running state ready to receive connections on port 7000

You now have your TCP Server project up and running. The next step in developing your understanding of pipelines is to write our TCP Client project. We will do this in the next section.

Writing and running a TCP client console application

In this section, we will continue looking at pipelines by writing the TCP client console application that will be sending console input to the TCP Server. The data received by the TCP Server from our TCP client will be displayed in the TCP Server's window:

1. Start a new .NET 6.0 Console Application called `CH09_TcpClient`.
2. In the `Program.cs` file, you will need to include the following namespaces:

   ```
   using System;
   using System.IO;
   using System.IO.Pipelines;
   using System.Net;
   using System.Net.Sockets;
   using System.Threading.Tasks;
   ```

   We will need these namespaces for our TCP Client to read the console input and send it to the TCP Server for processing.

3. Add the following top-level statements:

   ```
   Socket clientSocket = new Socket(SocketType.Stream,
       ProtocolType.Tcp);
   Console.WriteLine("Connecting to port 7000");
   clientSocket.Connect(new IPEndPoint(IPAddress.Loopback,
       7000));
   NetworkStream networkStream = new NetworkStream
       (clientSocket);
   await Console.OpenStandardInput().CopyToAsync
       (networkStream);
   ```

   For our TCP Client, we simply open a new TCP socket on port `7000` and connect using the `IPAddress.Loopback` address. Then, we pass the socket into a new `NetworkStream`. Finally, we listen for input from the console window's standard input and copy that input to the network stream that transmits the data to our TCP Server for processing.

4. Run the program. You should see the following:

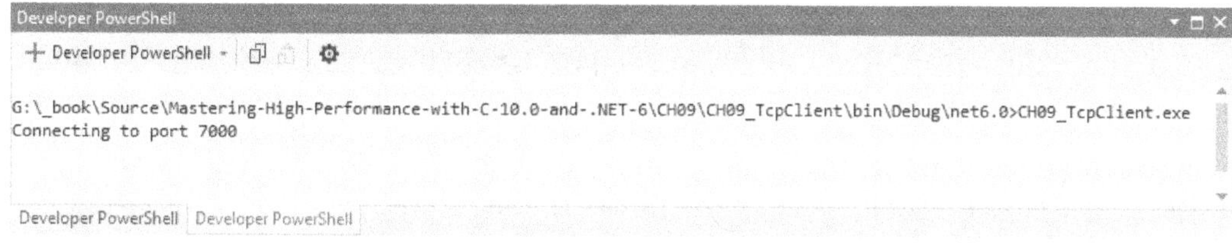

Figure 9.11 – The TCP Client listening on port 7000

5. Type `Hello, World!` and press *Enter*. Your TCP Client console application should look as follows:

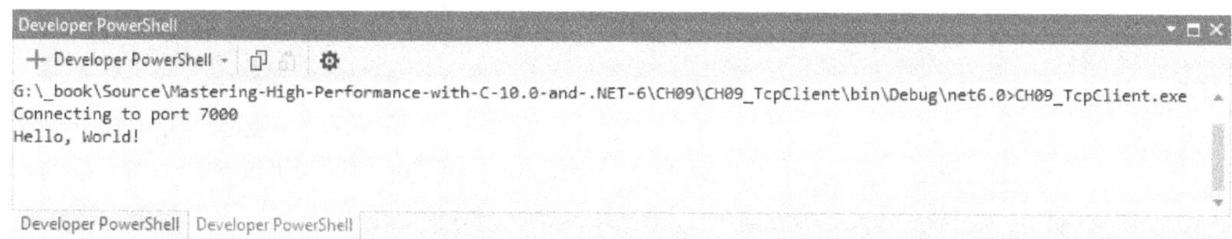

Figure 9.12 – The TCP Client console window displaying user input

6. Observe the TCP Server console window. You will see that the message **Hello, World!** has appeared since you typed the same message in the TCP Client window and pressed *Enter*, as shown here:

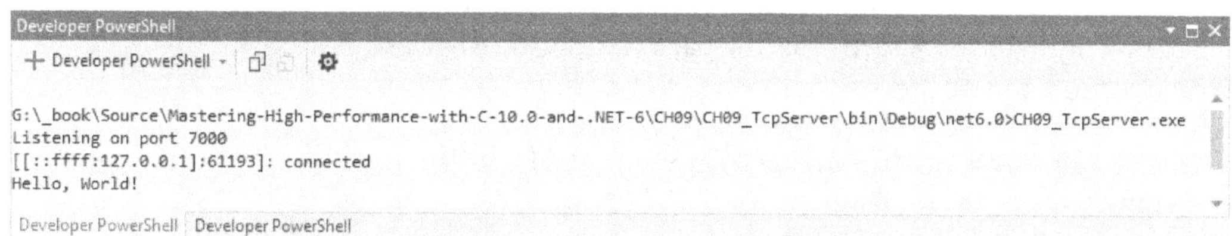

Figure 9.13 – The TCP Server console window displaying the response from the TCP Client

With that, you have finished writing and running the TCP client and server console applications, and you have seen just how simple it is to write a console application with sockets and pipelines. The code is very minimal and you can chain multiple pipelines together. For example, on the client end, a chained pipeline could be the serialization of an object followed by its encryption. Then, at the server end, the data could be decrypted and deserialized, and the resulting object could then be passed to LINQ, which would save the data contained in the object to a database. We can use sockets and pipelines with most C# project types, and you are encouraged to experiment with your own little projects to further your knowledge.

Caching resources in memory

Caching items in memory requires RAM to be allocated so that they can be stored and retrieved efficiently. Storing frequently accessed resources in memory greatly improves the performance of applications.

A typical application that benefits from caching is a website. A traditional website will consist of HTML pages that define the structure of the visual web page that's displayed to end users, CSS, which styles the page and makes it look nice, and JavaScript, which makes websites dynamic and interactive.

Many pages of a website can use the same resources, such as data, images, sounds, files, and objects. Caching – temporarily storing some item so that it can be retrieved efficiently – can be done with a database, filesystem, or memory.

In this section, we will learn how to store items in memory. Microsoft recommends the use of their `Microsoft.Extensions.Caching.Memory` NuGet package for caching items in memory. Therefore, we will follow their guidance and use this library in our example project.

We will be creating a very simple ASP.NET Core website that displays the current time and the cached time. When the cached time has expired, we will reset the cache. Each time the home view is called, we will output some text to the immediate window that displays the current time, the cached time, and the time difference in seconds.

After each specified period has elapsed, you will see that the cache is reset, along with the time that's output to the screen after the page refresh. To write our ASP.NET Core MVC web application, follow these steps:

1. Start a new empty ASP.NET Core MVC Web Application, ensuring that your target framework is `net6.0` and called `CH09_AspNetCoreCaching`.
2. Add the `Microsoft.Extensions.Caching.Memory` NuGet package, and then add the `using` statement for this package to the `HomeController` class.
3. Add an `IMemoryCache` member variable and update the `HomeController` constructor, as follows:

```
private IMemoryCache _memoryCache;
public HomeController(ILogger<HomeController> logger,
    IMemoryCache memoryCache)
{
    _logger = logger;
    _memoryCache = memoryCache;
}
```

Our `_memoryCache` variable will hold our cache in memory. The object that's being used as our memory cache is injected into the `HomeController` constructor as a parameter and assigned to our variable.

4. Next, add the `SetCache` method:

```
private void SetCache(string key, object value)
{
    var cachedEntryOptions =
     new MemoryCacheEntryOptions()
         .SetSlidingExpiration(TimeSpan.FromSeconds(20));
     _memoryCache.Set(key, value, cachedEntryOptions);
}
```

This method accepts a key and a value. We set our `MemoryCacheEntryOptions` with a sliding expiration of 20 seconds and then set the cached entry's value, which will expire in 20 seconds.

5. The next thing we need to do is update the `HomeController` constructor's `Index` method, as shown here:

```
public IActionResult Index()
{
     DateTime whenCached;
    bool exists = _memoryCache.TryGetValue("WhenCached",
       out whenCached);
    if (!exists)
    {
       Debug.WriteLine("Creating cached entry...");
       whenCached = DateTime.Now;
       SetCache("WhenCached", whenCached);
    }
    else
    {
        DateTime now = DateTime.Now;
       double differenceInSeconds =
           now.Subtract(whenCached).TotalSeconds;
       if (differenceInSeconds < 20)
       {
           Debug.WriteLine($"Now: {now}, When Cached:
```

```
                    {whenCached}, Time Difference (Seconds):
                        {differenceInSeconds}");
                return View(whenCached);
            }
            else
            {
                Debug.WriteLine("Resetting cache...");
                whenCached = DateTime.Now;
                SetCache("WhenCached", whenCached);
            }
        }
        return View(whenCached);
    }
```

The preceding code declares a `DateTime` variable called `whenCached`. It checks if the value exists. If it does, its value will be set to the time when the variable was cached. If the variable does not exist, then it will be added to the cache. If it does exist, then the difference in time between now and when the variable was cached is calculated, and the results will be output to the debug window if the cache has not expired. If the cache has expired, then the cached variable will be updated with the current time.

6. Now, we need to update our Home view's HTML code, as follows:

```
@model DateTime?
@{
    ViewData["Title"] = "Index";
}
<h1>Index</h1>
<div class="row">
    <span>
        When Cached: @Model.Value.ToString();
    </span>
    <span>
        Current Time: @DateTime.Now.ToString();
    </span>
</div>
```

The preceding code defines our model for the Razor page. The title of our page is set to `Index`. Our main page title is `Index`. Finally, we have a row that defines when the variable was cached and the current time.

7. Now, we need to update our `Program.cs file` to inform our website to use memory caching:

```
builder.Services.AddControllersWithViews();
builder.Services.AddMemoryCache();
```

With that, our services have been configured to use the memory cache.

With that, we have configured our MVC application to use memory caching with sliding expiration. This means that we are now ready to run our project. Run the project and refresh a few times within 20 seconds, and then watch what happens. You will see that the cached and current times start the same. Then, when you refresh the page, you will see that the cached time remains the same, but the current time is ahead of the cached time. Then, when 20 seconds is over, the cached time will be updated in sync with the current time, as shown here:

Figure 9.14 – ASP.NET Core MVC memory caching example in action

As you can see from the preceding screenshot and by running the code, we now have a way of storing items in a computer's memory cache, and we can determine when its cache value expires and has to be updated. This is a really simple way to improve a networked application's network performance. It also reduces the amount of data that is transmitted over a network. This, in turn, helps reduce bandwidth problems and reduces transaction and network traffic costs for cloud-hosted operations.

That concludes this chapter. Now, let's summarize what we have learned from working through this chapter.

Summary

In this chapter, you studied the OSI reference model to understand the different layers of a network and the various protocols available for each layer. You also learned that the various protocols can be grouped into two main groups: TCP and UDP.

Then, you learned about web browser development tools, which allow you to monitor your website's activities, such as memory usage and network traffic. You also saw the errors it raises via the console window. This can help identify problems and resolve them.

From there, you learned how to add gRPC for desktop clients and servers, and gRPC-Web for web-based clients and servers. You learned that gRPC helps reduce the size of data compared to the JSON data format, thus reducing page load time.

After that, you learned how to optimize internet resources. This includes using the correct file format and reducing the size of images, caching items to reduce network traffic and load times, reducing the number of background services that are running, and limiting the number of resources that your page loads. You also considered filtering data on the server and dividing it into pages that are returned as requested.

Finally, you learned how to write and run TCP client and server console applications before looking at memory caching, in which you can use ASP.NET Core MVC as your host project.

In the next chapter, we will be working with data by benchmarking different methods for inserting, updating, and deleting data efficiently. This will help us choose the best method for data operations based on our benchmark results. But before we do that, take some time to go through the *Further reading* section to further your knowledge on improving network performance. Also, try your hand at the questions to see how much knowledge you have retained.

Questions

Answer the following questions to test your knowledge of this chapter:

1. Name the seven layers of the OSI reference model.
2. Name some network protocols.
3. What is the difference between TCP/IP and UDP?
4. How can you see what errors are produced by your web page, what network traffic it produces, and how much memory it uses?
5. What are gRPC and gRPC-Web?
6. How you can optimize internet resources?

Further reading

To learn more about the topics that were covered in this chapter, take a look at the following resources:

- OSI seven layers model explained with examples: `https://www.computernetworkingnotes.com/ccna-study-guide/osi-seven-layers-model-explained-with-examples.html#:~:text=The%20OSI%20(Open%20System%20Interconnection)%20Reference%20Model%20is,and%20software%20applications%20which%20work%20in%20dissimilar%20environments`
- TCP/IP model: `https://ipcisco.com/lesson/tcp-ip-model/`
- Overview of common TCP and UDP default ports: `https://www.examcollection.com/certification-training/network-plus-overview-of-common-tcp-and-udp-default-ports.html#:~:text=%20Overview%20of%20common%20TCP%20and%20UDP%20default,FTP%20is%20to%20transfer%20files%20over...%20More`
- List of TCP and UDP port numbers: `https://en.wikipedia.org/wiki/List_of_TCP_and_UDP_port_numbers`
- Internet protocol suite: `https://en.wikipedia.org/wiki/Internet_protocol_suite`
- .NET network tracing: `https://www.shanebart.com/dotnet-network-tracing/`
- dotnet-trace instructions: `https://github.com/dotnet/diagnostics/blob/master/documentation/dotnet-trace-instructions.md`
- How to view and kill processes using the Terminal in Mac OS X: `https://www.chriswrites.com/how-to-view-and-kill-processes-using-the-terminal-in-mac-os-x/`
- How to find a process name using a PID number in Linux: `https://www.tecmint.com/find-process-name-pid-number-linux/`
- High-performance services with gRPC – what's new in .NET 5: `https://www.youtube.com/watch?v=EJ8M2Em5Zzc`
- gRPC-Web with .NET: `https://www.youtube.com/watch?v=UV-VnlcpDhU`

- .NET Conf 2021 new Blazor WebAssembly capabilities in .NET 6: `https://www.youtube.com/watch?v=kesUNeBZ1Os&list=PLdo4fOcmZ0oVFtp9MDEBNbA2sSqYvXSXO&index=20`
- .NET Conf 2021 high-performance services with gRPC – what's new in .NET 6: `https://www.youtube.com/watch?v=CXH_jEa8dUw&list=PLdo4fOcmZ0oVFtp9MDEBNbA2sSqYvXSXO&index=31`
- Everything about Blazor: `https://codewithmukesh.com/blog/category/dotnet/blazor/`

10
Setting Up Our Database Project

In this and the following two chapters, we will be improving the performance of your database-based applications. In this chapter, we will be setting up our relational database and the code to access that database. In the next chapter, we will write benchmarks to test the performance of the different frameworks, which consist of Entity Framework, Dapper, and ADO.NET. Finally, in *Chapter 12, Responsive User Interfaces*, we will learn how to improve the performance of SQL Server and Cosmos DB.

Data is extensively used in all aspects of our daily lives. In today's world of big data, the volume of data being collected and stored for all kinds of analysis is phenomenal. When working with data, performance can slow down exponentially as the size of your data grows. And depending on how much data you have to process, time is often critical.

In this chapter, we will create a database and populate it, and we will write the code to access the database and perform insert, update, select, and delete operations. Our database access code will consist of Entity Framework, Dapper.NET, and ADO.NET.

> **Note**
>
> No code performance improvements will be discussed in this chapter. We are only concerned with setting up our database and source code in preparation for the benchmarking that we will be doing in the next chapter.

In this chapter, we will cover the following topics:

- Creating and populating a SQL Server database
- Writing code to access the database using Entity Framework
- Writing code to access the database using Dapper.NET
- Writing code to access the database using ADO.NET

After completing this chapter, you will be able to do the following:

- Log on to SQL Server Management Studio and execute database creation and seeding scripts
- Store secrets in `secrets.json` when developing so that secrets don't get stored in version control
- Access SQL Server databases and perform **Create/Insert, Read/Select, Update, and Delete** (**CRUD**) operations using Entity Framework
- Access SQL Server databases and perform CRUD operations using Dapper.NET
- Access SQL Server databases and perform CRUD operations using ADO.NET

Technical requirements

To follow along with this chapter, you will need to ensure that you have the following:

- SQL Server 2019 Express Edition or higher
- SQL Server Management Studio
- Visual Studio 2022
- This book's source code: `https://github.com/PacktPublishing/High-Performance-Programming-in-CSharp-and-.NET/tree/master/CH10`

Setting up our database

In this section, we will set up our database and get our project ready for benchmarking. We will be benchmarking different methods of inserting, updating, selecting, and deleting data. Let's start with setting up our database:

1. Visit `https://github.com/Microsoft/sql-server-samples/tree/master/samples/databases/northwind-pubs`.
2. Download the `instnwnd.sql` file.

3. Once the file has been downloaded, open it in SQL Server Management Studio.
4. Execute the file. This will install the database.
5. Open a new query window and enter the following SQL code:

```
USE [Northwind]
GO
SET ANSI_NULLS ON
GO
SET QUOTED_IDENTIFIER ON
GO
CREATE PROCEDURE [dbo].[InsertProduct]
    @ProductName NVARCHAR(40),
    @CategoryID INT,
    @SupplierID INT,
    @Discontinued BIT
AS
BEGIN
SET NOCOUNT ON;
INSERT INTO
        Products (
            ProductName,
            CategoryID,
            SupplierID,
            Discontinued,
             QuantityPerUnit
        )
    VALUES (
        @ProductName,
        @CategoryID,
        @SupplierID,
        @Discontinued,
         '1'
    )
END
GO
```

Once the code has been entered, execute the script. This code generates the `InsertProduct` stored procedure. This stored procedure inserts a product into the `Products` table of the `Northwind` database.

6. Replace the existing SQL with the following SQL:

```
USE [Northwind]
GO
SET ANSI_NULLS ON
GO
SET QUOTED_IDENTIFIER ON
GO
CREATE PROCEDURE [dbo].[GetProductName]
    @ProductName NVARCHAR(40)
AS
BEGIN
    SET NOCOUNT ON;
    SELECT
        Top 1 ProductName
    FROM
        Products
    WHERE
        ProductName LIKE @ProductName
END
GO
```

Execute the SQL to generate the `GetProductName` stored procedure. A product name can have different variations. This stored procedure gets the top 1 name for the given product.

7. Replace the existing SQL code with the following SQL:

```
USE [Northwind]
GO
SET ANSI_NULLS ON
GO
SET QUOTED_IDENTIFIER ON
GO
CREATE PROCEDURE [dbo].[FilterProducts]
    @ProductName NVARCHAR(40)
```

```
AS
BEGIN
        SET NOCOUNT ON;
        SELECT
            *
        FROM
            Products
        WHERE
            ProductName LIKE @ProductName
END
GO
```

Execute the SQL to generate the `FilterProducts` stored procedure. The stored procedure returns all the products whose names contain the search term.

8. Now, replace the existing SQL with this SQL:

```
USE [Northwind]
GO
SET ANSI_NULLS ON
GO
SET QUOTED_IDENTIFIER ON
GO
CREATE PROCEDURE [dbo].[UpdateProductName]
        @OldProductName NVARCHAR(40),
        @NewProductName NVARCHAR(40)
AS
BEGIN
    SET NOCOUNT ON;
     UPDATE
         Products
         SET
            ProductName = @NewProductName
         WHERE
            ProductName = @OldProductName
END
GO
```

Execute this SQL to generate the `UpdateProductName` stored procedure. This procedure updates a product name from its current name to a new name.

9. Replace the existing SQL with the following:

```
USE [Northwind]
GO
SET ANSI_NULLS ON
GO
SET QUOTED_IDENTIFIER ON
GO
CREATE PROCEDURE [dbo].[DeleteProduct]
    @ProductName NVARCHAR(40)
AS
BEGIN
    SET NOCOUNT ON;
     DELETE FROM
            Products
     WHERE
            ProductName = @ProductName
END
GO
```

Execute this code to generate the `DeleteProduct` stored procedure. This deletes products from the database that match the given product name.

10. Once the database has been installed and all the procedures have been written and executed, you can close SQL Server Management Studio.

Now that we have set up our database, we will set up our database access project.

Setting up our database access project

In this section, we will be creating our database access project and classes. In the next chapter, we will be writing some benchmarks that reference the classes that we will write in this chapter. Create the project as follows:

1. Open Visual Studio and create a new .NET 6.0 console application called `CH10_DataAccessBenchmarks`.
2. Add the latest version of the `Microsoft.EntityFrameworkCore.SqlServer` NuGet package.

3. Add the latest version of the `Dapper` NuGet package.
4. Add the latest version of the `System.Data.SqlClient` NuGet package.
5. Add a new folder called `Configuration`, and add two classes called `DatabaseSettings` and `SecretsManager`.
6. Add a folder called `Data`, and add three classes called `AdoDotNetData`, `DapperDotNet`, and `EntityFrameworkCoreData`.
7. Add a folder called `Models`, and add three classes called `Product`, `SqlCommandModel`, and `SqlCommandParameterModel`.
8. Add a folder called `Reflection`, and add a class called `Properties`.
9. On the main root, add a class called `BenchmarkTests`.
10. Save the project.

With that, we have created and updated our database with the stored procedures we will be calling, and we have also put in place the project, folders, and class files that we will be using to benchmark the various types of data operations we normally perform on a database from code. Let's start by writing the `Properties` class.

Writing the Properties class

As part of our benchmarking, we need to obtain the `FieldCount` value of a `DbDataRecord`. But the property is not readily accessible without using reflection. Therefore, to make our lives easier, we will write a class called `Properties` that helps us get the values of properties using reflection easily. Follow these steps:

1. Open the `Properties` class and add the following `using` statements:

```
using System.Data.Common;
using System.Reflection;
internal class Properties
{
}
```

We need both these namespaces to be imported as we are using reflection and need access to the `DbDataRecord` class.

2. Add the `GetProperty` method:

```
public static PropertyInfo GetProperty<T>(string name)
{
      return typeof(T).GetProperty(name);
}
```

This method takes a generic type and a property name. Then, it obtains the property and returns it as a `PropertyInfo` instance.

3. Now, add the `GetValue` method:

```
public static T GetValue<T, U>(U source, string name)
{
        return (T)GetProperty<U>(name).GetValue(source);
}
```

   This method takes a generic object type, return type, and property name. Then, it calls the `GetProperty` method by passing in the generic object type and property name. The `GetValue` method is then called, passing in the source object. The result is cast to the generic return type and returned to the caller.

4. Add the `GetFieldCount` method:

```
        public static int GetFieldCount(DbDataRecord
            record)
        {
            return GetValue<int, DbDataRecord>(
            record, "FieldCount"
        );
    }
```

   This method accepts a `DbDataRecord` object. It calls our `GetValue` method by passing in the return type, our `DbDataRecord`, and our `FieldCount` property name. An integer is returned that contains the number of fields that our `DbDataRecord` object has.

With that, we have created our `Properties` class. As part of our benchmarking, we will be inserting, reading, editing, and deleting data from a SQL Server database. And so, in the next section, we will update our `DatabaseSettings` class.

Writing the DatabaseSettings class

Our `DatabaseSettings` class is really simple: it contains a single property. Open the database and add the following property:

```
public string ConnectionString { get; set; }
```

This property holds our connection string for the SQL Server database. We will be setting this property in each of our benchmark methods. Then, it will be passed to the constructors of our data access classes.

Because database connection strings are a sensitive form of data that should be kept very private, we will be storing our database connection strings in a `secrets.json` file during the development process. But in production, we will obtain the connection string from an `appsettings.json` file. And so, in the next section, we will be writing a `SecretsManager` class.

Writing the SecretsManager

In this section, we are going to update our `SecretsManager` class so that we can safely obtain secrets.

> **Note**
>
> Our development environment will use a `secrets.json` file. This is very serious as private credentials have been found and accessed on source code hosting sites such as GitHub before now, and we don't want to be the ones responsible for checking in code that contains secrets that should be kept private.

Follow these steps:

1. Add the following NuGet packages:

```
Microsoft.Extensions.Configuration
Microsoft.Extensions.Configuration.JsonFile
Microsoft.Extensions.Configuration.EnvironmentVariables
Microsoft.Extensions.Configuration.UserSecrets
```

   We need these packages so that we can configure the project for user secrets and `appsettings.json`.

2. Open the `SecretsManager` class and add the following `using` statements:

```
using Microsoft.Extensions.Configuration;
using System;
using System.IO;
```

We need these `using` statements for our property, filesystem, and environment variable access, and for access to the Microsoft `IConfiguration` interface.

3. Add the `Configuration` property:

```
public static IConfiguration Configuration
{
        get; private set;
}
```

This property will hold the correct configuration object, which depends on whether we are in development or production mode.

4. Now, add the `GetSecrets` method:

```
public static string GetSecrets<T>(string sectionName)
where T : class
{
var devEnvironmentVariable =
    Environment
        .GetEnvironmentVariable("NETCORE_ENVIRONMENT");
var isDevelopment =
    string.IsNullOrEmpty(devEnvironmentVariable)
    || devEnvironmentVariable.ToLower() == "development";
var builder = new ConfigurationBuilder()
    .SetBasePath(Directory.GetCurrentDirectory())
        .AddJsonFile(
            "appsettings.json",
            optional: true,
            reloadOnChange: true
    )
    .AddEnvironmentVariables();
//only add secrets in development
if (isDevelopment)
{
    builder.AddUserSecrets<T>();
}
Configuration = builder.Build();
```

```
return Configuration.GetSection($"{typeof(T).Name}
    :{sectionName}").Value;
}
```

This method determines whether we are in development or non-development mode. If we are in development mode, then we use the secrets configuration mode. Otherwise, we obtain secrets from the `appsettings.json` file. The method accepts a section name, which is the name of the secret we want to retrieve, and it returns the value of that secret.

With that, we have finished writing our `secrets` class. For our data manipulation benchmarks, we will be focusing on a single table – the `Products` table of the `Northwind` database. We will need a class that will act as a model for the data. So, in the next section, we'll write the `Product` class.

Writing the Product class

In this section, we will update our `Product` class. It is a simple object that is used for data manipulation benchmarks and contains properties that match the `Products` table in the `Northwind` database. Follow these steps:

1. Open the `Product` class and update it as follows:

```
using System;
using System.ComponentModel.DataAnnotations;
using System.ComponentModel.DataAnnotations.Schema;
[Table("Products")]
public class Product
{
}
```

   Here, we annotated our class with the `Table` annotation, passing the name of the table in the `Northwind` database that this class maps to into the annotation.

2. Add the following properties and annotations:

```
[Key]
public int ProductID { get; set; }
public string ProductName { get; set; }
[ForeignKey("Suppliers")]
public int SupplierID { get; set; }
```

```
[ForeignKey("Categories")]
public int CategoryID { get; set; }
public string QuantityPerUnit { get; set; } = "1"
public decimal UnitPrice { get; set; }
public Int16 UnitsInStock { get; set; }
public Int16 UnitsOnOrder { get; set; }
public Int16 ReorderLevel { get; set; }
public bool Discontinued { get; set; }
```

These properties match the columns of the `Product` table in the `Northwind` database. The `[Key]` annotation identifies the `ProductID` property as the table's primary key. Two foreign keys are identified by the `[ForeignKey]` annotation. We pass the name of the table into this annotation, which contains the primary key.

And that's it – we've finished writing our `Product` class. We will be using several commands and parameters when accessing data. To make life easy, we will have a `SqlCommandModel` class for defining our commands, and a `SqlCommandParameterModel` class for defining our command parameters. Let's begin by writing the `SqlCommandModel` class.

Writing the SqlCommandModel class

In this section, we write a simple class that models a SQL command. Follow these steps:

1. Open the `SqlCommandModel` class, define the class as public, and add the `System.Data` namespace.
2. Now, add the following three properties:

```
public string CommandText { get; set; }
public CommandType CommandType { get; set; }
public SqlCommandParameterModel[] CommandParameters {
    get; set; }
```

The `CommandText` property holds our SQL command. This may be the name of a stored procedure or a SQL statement. The `CommandType` property determines whether the command is a `Text` command or a `StoredProcedure` command, while the `CommandParameters` property contains an array of SQL command parameters.

Now that we have written `SqlCommandModel`, let's write the `SqlCommandParameterModel` class.

Writing the SqlCommandParameterModel class

In this section, we'll write our `SqlCommandParameterModel` class. This class is simply a SQL parameter definition model.

Open the `SqlCommandParameterModel` class, make the class public, and add the `System.Data` namespace.

Now, add the following three parameters:

```
public string ParameterName { get; set; }
public DbType DataType { get; set; }
public dynamic Value { get; set; }
```

This class models a standard parameter that consists of the name of the parameter, its database type, and its value.

With that, we have created the core functionality that we need in place for our data access classes. In the following sections, we will be writing data access classes to access data using Entity Framework, Dapper, and ADO.NET.

The reason behind choosing SQL Server for the database server is that it is one of the most common database servers and is used in many business scenarios the world over. In professional environments where SQL Server is employed, the three most common data access methods are Entity Framework, Dapper, and ADO.NET. That is why we will be benchmarking them in this chapter. Let's start by writing our ADO.NET data access class.

Writing the AdoDotNet class

In this section, we will be writing our data insertion methods. However, we will not be running our benchmarks, which will be performed in the next chapter as we analyze our results. Follow these steps:

1. Update the `AdoDotNetData` class, as follows:

```
using CH10_DataAccessBenchmarks.Models;
using CH10_DataAccessBenchmarks.Reflection;
using System;
using System.Collections;
using System.Collections.Generic;
using System.Data.Common;
```

```
using System.Data.SqlClient;
using System.Reflection;
internal class AdoDotNetData : IDisposable
{
private readonly SqlConnection _sqlConnection;
private bool _isDisposed;
public AdoDotNetData(string connectionString)
{
        _sqlConnection =
         new SqlConnection(connectionString);
}
public void Dispose()
{
    Dispose(_isDisposed);
}
public void Dispose(bool disposing)
{
        if (disposing)
        {
        _sqlConnection.Dispose();
        _isDisposed = true;
    }
}
}
```

In the preceding code, we implemented the `IDisposable` pattern. When we have finished with our class, we dispose of our class, which also disposes of disposable objects that it holds in memory.

2. Add the `ExecuteNonQuery` method:

```
    internal void ExecuteNonQuery(SqlCommandModel
        model)
    {
        SqlCommand sqlCommand
         = new (model.CommandText, _sqlConnection);
     sqlCommand.CommandType = model.CommandType;
     foreach (SqlCommandParameterModel parameter in
```

```
        model.CommandParameters)
    sqlCommand.Parameters.Add(new SqlParameter()
        {
            ParameterName = parameter.ParameterName,
            DbType = parameter.DataType,
            Value = parameter.Value
    });
    _sqlConnection.Open();
    sqlCommand.ExecuteNonQuery();
    _sqlConnection.Close();
}
```

This method takes a `SqlCommandModel` object. A new instance of the `SqlCommand` object is created. We pass the SQL command and SQL connection into the constructor during instantiation. Then, we loop through the command parameters, instantiating and adding a `SqlParameter` for each `model.CommandParameter` to the `sqlCommand` object. Next, we open a connection to the database, execute the query, and close the connection.

3. Add the following code:

```
internal int ExecuteNonQuery(string sql)
{
try
{
_sqlConnection.Open();
return new SqlCommand(sql, _sqlConnection)
    .ExecuteNonQuery();
}
finally
{
_sqlConnection.Close();
}
}
```

The preceding code executes the non-query SQL code that's been passed in via the `sql` string.

4. Add the following generic scalar method:

```
internal T ExecuteScalar<T>(string sql)
{
        try
        {
            _sqlConnection.Open();
            return (T)new SqlCommand(sql, _sqlConnection)
                .ExecuteScalar();
}
        finally
        {
            _sqlConnection.Close();
}
}
```

This method takes a SQL command as a string. A connection to the database is opened, and a new `SqlCommand` is instantiated. The `ExecuteScalar` command is executed, which returns a single value from the database. Before the value is returned, it is cast to the generic type specified by the caller and returned as that type. The connection is then closed.

5. Add the following scalar method:

```
internal T ExecuteScalar<T>(SqlCommandModel model)
{
SqlCommand sqlCommand = new(
        model.CommandText, _sqlConnection);
sqlCommand.CommandType = model.CommandType;
      foreach (SqlCommandParameterModel parameter in
            model.CommandParameters)
            sqlCommand.Parameters.Add(new SqlParameter()
            {
                ParameterName = parameter.ParameterName,
                DbType = parameter.DataType,
                Value = parameter.Value
        });
    _sqlConnection.Open();
      T data = (T)sqlCommand.ExecuteScalar();
```

```
        _sqlConnection.Close();
        return data;
    }
```

This method takes a `SqlCommandModel` and uses it to build up a `SqlCommand`. The `SqlCommand` class is executed by calling the `ExecuteScalar` method and is cast to the generic type before being returned.

6. Add the following reader method:

```
    internal IEnumerator<T> ExecuteReader<T>(string sql)
    {
        Type TypeT = typeof(T);
        ConstructorInfo ctor =
          TypeT.GetConstructor(Type.EmptyTypes);
    if (ctor == null)
        {
    throw new InvalidOperationException($"Type
        {TypeT.Name} does not have a default
            constructor.");
    }
        _sqlConnection.Open();
    IEnumerator data = new SqlCommand(sql, _sqlConnection)
      .ExecuteReader().GetEnumerator();
    while (data.MoveNext())
        {
            T newInst = (T)ctor.Invoke(null);
            DbDataRecord record = (DbDataRecord)
                    data.Current;
            int fieldCount = Properties
                 .GetFieldCount((DbDataRecord)
                    data.Current);
         for (int i = 0; i < fieldCount; i++)
            {
                string propertyName = record.GetName(i);
                PropertyInfo propertyInfo = TypeT
                    .GetProperty(propertyName);
                 if (propertyInfo != null)
```

```
            {
                object value = record[i];
                if (value == DBNull.Value)
                    propertyInfo
                        .SetValue(newInst, null);
                else
                    propertyInfo
                        .SetValue(newInst, value);
        }
    }
        yield return newInst;
}
}
```

This method takes a SQL statement and executes it by calling the `ExecuteReader` method. Once the method has been executed, we obtain the reader's enumerator. Then, we iterate through the enumerator and build up an object for the current iteration and yield the result.

7. Add the following reader method:

```
internal IEnumerator<T> ExecuteReader<T>
    (SqlCommandModel model) {
Type TypeT = typeof(T);
ConstructorInfo ctor
        = TypeT.GetConstructor(Type.EmptyTypes);
if (ctor == null) {
throw new InvalidOperationException($"Type
    {TypeT.Name} does not have a default
        constructor.");
}
SqlCommand sqlCommand
    = new(model.CommandText, _sqlConnection);
sqlCommand.CommandType = model.CommandType;
foreach (SqlCommandParameterModel parameter in
```

```
    model.CommandParameters)
sqlCommand.Parameters.Add(new SqlParameter() {
ParameterName = parameter.ParameterName,
DbType = parameter.DataType, Value =
    parameter.Value});
_ sqlConnection.Open();
SqlDataReader reader = sqlCommand.ExecuteReader();
if (reader.HasRows) {
while (reader.Read()) {
T newInst = (T)ctor.Invoke(null);
for (int i = 0; i < reader.FieldCount; i++) {
     string propertyName = reader.GetName(i);
     PropertyInfo propertyInfo
         = TypeT.GetProperty(propertyName);
     if (propertyInfo != null) {
         object value = reader[i];
         if (value == DBNull.Value)
               propertyInfo.SetValue(newInst, null);
         else
               propertyInfo.SetValue(newInst, value);
     }
}
    yield return newInst;
}
}
  _ sqlConnection.Close();
}
```

This reader method takes a `SqlCommandModel` and builds up a `SqlCommand`. It executes the reader and obtains `SqlDataReader`. It iterates through the reader and builds up an instance of the generic type that is then yielded to the user.

That's our ADO.NET data access class completed. Now, let's learn how to write the Entity Framework data access class.

Writing the EntityFrameworkCoreData class

In this section, we will be writing the methods for our Entity Framework data access class. The code we will write in this section will be executed in the next chapter. Follow these steps:

1. Open the `EntityFrameworkCoreData` class and edit it as follows:

```
using CH10_DataAccessBenchmarks.Models;
using Microsoft.EntityFrameworkCore;
using System.Collections.Generic;
using Microsoft.Data.SqlClient;
using System.Linq;
using Microsoft.EntityFrameworkCore.SqlServer
    .Infrastructure.Internal;
public class EntityFrameworkCoreData : DbContext
{
    private string _connectionString = string.Empty;
    public DbSet<Product> Products { get; set; }
    public EntityFrameworkCoreData(string
        connectionString) : base(GetOptions
            (connectionString))
    {
        _connectionString = connectionString;
    }

    private static DbContextOptions GetOptions(string
        connectionString)
    {
    return SqlServerDbContextOptionsExtensions
        .UseSqlServer(new DbContextOptionsBuilder(),
            connectionString).Options;
    }
```

Our class inherits from the `DbContext` class of the `Microsoft.EntityFrameworkCore` library. We declare a variable to hold our database connection string, and a variable to hold a collection of `Products`. In our constructor, we set the connection string and call the base constructor.

2. Add the `OnConfiguring` method:

```
protected override void OnConfiguring
    (DbContextOptionsBuilder optionsBuilder)
{
    optionsBuilder.UseSqlServer(_connectionString);
}
```

This method determines that we will be using SQL Server and passes in the SQL Server connection string that we will be using.

3. Add the following method, which executes raw SQL:

```
public int ExecuteSQL(string sql)
    {
        return Database.ExecuteSqlRaw(sql, null);
}
```

This method takes a SQL statement and executes it against the database as raw SQL. The returned value is the number of records affected by the execution of the statement.

4. Add the following method for executing a stored procedure as a non-query:

```
public int ExecuteNonQuerySP(SqlCommandModel model)
    {
        SqlParameter[] parameters
        = new SqlParameter[model.CommandParameters
          .Length];
        for (int i = 0; i < parameters.Length; i++)
        {
            parameters[i] = new SqlParameter(
            model.CommandParameters[i].ParameterName,
             model.CommandParameters[i].Value
             );
    }
        if (parameters.Length == 4)
            return Database.ExecuteSqlRaw(
             model.CommandText, parameters[0],
             parameters[1], parameters[2],
             parameters[3]
```

```
            );
        else if (parameters.Length == 2)
            return Database.ExecuteSqlRaw(
                model.CommandText, parameters[0],
                parameters[1]
            );
        else
            return Database.ExecuteSqlRaw(
                model.CommandText, parameters[0]
            );
}
```

In this method, we build up a `SqlParameter` array from our `SqlCommandModel`. Then, we execute raw SQL by passing in each of the parameters to the stored procedure. This execution is a non-query and returns the number of rows affected by running the procedure.

5. The following method will execute and return a scalar value of the `string` type:

```
public string ExecuteScalarSP(string productName)
    {
        return Products.FromSqlRaw(
            "EXEC FilterProducts @ProductName={0}",
            new SqlParameter() {

        ParameterName = "@ProductName", Value =
                productName })
            .AsEnumerable().FirstOrDefault()
              .ProductName;

    }
```

This method executes a stored procedure with a single parameter. We obtain the enumerable return object and filter it to get the first record. The name of the product is then returned as a string.

6. Add the final method to our class, which returns an enumerator:

```
public IEnumerator<Product> ExecuteReaderSP(string
    productName)
```

```
{
return Products.FromSqlRaw(
        "EXEC FilterProducts @ProductName={0}",
        new SqlParameter() {
            ParameterName = "@ProductName",
            Value = productName
        }
    ).GetEnumerator();
}
```

This executes a stored procedure with a single parameter and returns an enumerator full of filtered products.

With that, we have written all our Entity Framework classes. Now, it's time to write our Dapper.NET methods.

Writing the DapperDotNet class

In this section, we'll write our Dapper.NET methods. This is the last section before we write our benchmarking methods. We will run the code we write in this section in the next chapter. Follow these steps:

1. Open the `DapperDotNet` class, add the `SimpleCRUD` package, and modify it as follows:

```
public class DapperDotNet : IDisposable
{
    private bool isDisposed = false;
     private IDbConnection _dbConnection;
     public DapperDotNet(string connection)
    {
     SimpleCRUD
            .SetDialect(SimpleCRUD.Dialect.SQLServer);
         _dbConnection = new SqlConnection
             (connection);
     }
     public void Dispose()
    {
         Dispose(true);
```

```
            GC.SuppressFinalize(this);
        }
        protected virtual void Dispose(bool disposing)
        {
            if (isDisposed)
                return;
            if (disposing)
                _dbConnection.Dispose();
            isDisposed = true;
        }
    }
```

We implement the `IDisposable` pattern in this class and set the SQL dialect to the SQL Server.

2. Add the following non-query method:

```
        public int ExecuteNonQuery(string sql)
        {
            try
            {
                _dbConnection.Open();
                return _dbConnection.Execute(sql);
            }
            finally
            {
                _dbConnection.Close();
            }
        }
```

This method executes raw SQL and returns the number of records affected by the SQL statement.

3. Add the following method to execute a non-query:

```
    public void ExecuteNonQuery(SqlCommandModel model)
        {
            try
            {
                _dbConnection.Open();
```

```
            var parameters = new DynamicParameters();
                foreach (
                SqlCommandParameterModel parameter in
                model.CommandParameters
            )
                parameters.Add(
                    parameter.ParameterName,
                    parameter.Value
                );
                _dbConnection.Query(
                model.CommandText,
                parameters,
                commandType: CommandType.StoredProcedure
            );
        }
        finally
    {
            _dbConnection.Close();
        }
}
```

This method takes a `SqlCommandModel` instance and builds up a `DynamicParameter` bag. Then, it executes a stored procedure defined by the model's `CommandText`.

4. Add the following generic scalar method:

```
public T ExecuteScalar<T>(string sql)
{
    try
    {
        _dbConnection.Open();
        return _dbConnection.ExecuteScalar
            <T>(sql);
}
    finally
    {
        if (_dbConnection != null
```

```
            && _dbConnection.State
                == ConnectionState.Open)
            _dbConnection.Close();
    }
}
```

This method takes a SQL statement and executes it, returning a single value of the required type.

5. Add the following method, which executes a stored procedure and returns a string:

```
public string ExecuteScalarSP(SqlCommandModel model)
{
        try
        {
            _dbConnection.Open();
            var parameters = new DynamicParameters();
                parameters.Add(
                model.CommandParameters[0]
                  .ParameterName,
             model.CommandParameters[0].Value
         );
            return _dbConnection.Query<Product>(
             model.CommandText,
             parameters,
             commandType: CommandType.StoredProcedure
         ).First().ProductName;
    }
        finally
        {
            if (
             _dbConnection != null
             && _dbConnection.State
                 == ConnectionState.Open)
             _dbConnection.Close();
    }
}
```

This method takes a `SqlCommandModel` instance and uses it to execute a stored procedure. Remember to add the missing `using` statements for `SqlCommandModel` to the class. The stored procedure execution returns a type of `IEnumerable<Product>`. So, we obtain the first product in the list and return its `ProductName`.

6. Add the following method, which executes raw SQL and returns a type of `IEnumerator<T>`:

```
public IEnumerator<T> ExecuteReader<T>(string sql)
    where T : class
{
        try
        {
            _dbConnection.Open();
            return _dbConnection.Query<T>(sql)
             .GetEnumerator();
    }
        finally
        {
            if (_dbConnection != null
             && _dbConnection.State
                 == ConnectionState.Open)
             _dbConnection.Close();
    }
}
```

This method executes a raw SQL string and returns a type of `IEnumerable<T>`.

7. Add the following method, which executes a stored procedure and returns a type of `IEnumerator<Product>`:

```
    public IEnumerator<Product> ExecuteReaderSP
        <Product>(
    SqlCommandModel model
)
    {
        try
```

```
        {
            _dbConnection.Open();
            var parameters = new DynamicParameters();
            foreach (SqlCommandParameterModel
                parameter in model.CommandParameters)
             parameters.Add(
                 parameter.ParameterName,
                 parameter.Value
             );
         return _dbConnection.Query<Product>(
             model.CommandText,
             parameters,
             commandType: CommandType.StoredProcedure
         ).GetEnumerator();
      }
      finally
      {
          if (_dbConnection != null
           && _dbConnection.State
           == ConnectionState.Open)
          _dbConnection.Close();
      }
  }
```

This method takes a `SqlCommandModel` instance and builds up a parameterized stored procedure that is executed. A type of `IEnumerator<Product>` is returned.

8. Add our final dapper method, which will obtain the first product name that matches the `productName` parameter:

```
public string GetProductNameSP(string productName)
    {
        try
        {
            _dbConnection.Open();
            var parameters = new DynamicParameters();
```

```
            parameters.Add("@ProductName",
                productName);
            return _dbConnection.Query<Product>(
             $"GetProductName", parameters,
             commandType: CommandType.StoredProcedure
             ).First().ProductName;
        }
            finally
            {
                if (_dbConnection != null
                 && _dbConnection.State
                 == ConnectionState.Open)
                  _dbConnection.Close();
            }
    }
```

This method takes a product name and executes the `GetProductName` stored procedure. The stored procedure matches all the products in the database whose product names are like the product name argument. Then, it gets the first product in the returned list and returns its product name.

That concludes our database and data access project setup in preparation for the benchmarking work we will be doing in the next chapter. Let's review what we have accomplished in this chapter.

Summary

In this chapter, we downloaded the **Northwind** SQL Server database script. Then, we added some stored procedures to insert, update, select, and delete data from the `Products` table.

After making sure that we have our database in place with the required stored procedures, we started a .NET 6.0 console application. We added our model class and data access classes for performing data access operations in Entity Framework, Dapper, and ADO.NET.

In the next chapter, we will be benchmarking the data access methods for each of these frameworks. In the *Further reading* section, you can further your knowledge of Entity Framework, Dapper, and ADO.NET using the links provided.

Further reading

To learn more about the topics that were covered in this chapter, take a look at the following resources:

- Entity Framework Core: `https://docs.microsoft.com/ef/core/`
- Dapper: `https://dapper-tutorial.net/dapper`
- ADO.NET: `https://dotnettutorials.net/course/ado-net-tutorial-for-beginners-and-professionals/`

11 Benchmarking Relational Data Access Frameworks

Data is extensively used in all aspects of our daily lives. In today's world of big data, the volumes of data being collected and stored for all kinds of analysis are phenomenal. When working with data, performance can slow down exponentially as the size of your data grows. Depending upon how much data you have to process, the time factor is often critical.

In a professional development environment, computer programmers don't always have access to the database server. Database server access is usually restricted for use by database developers and database administrators. With that in mind, this chapter is about benchmarking what code performs a database insert, update, read, and delete in the shortest possible time. In the *Further reading* section, there are links to documentation on database server performance that will help you to further improve the performance that you gain from working through this chapter.

In this chapter, we will be benchmarking three different ways of manipulating SQL Server database data. We will be performing a side-by-side comparison of Entity Framework, ADO.NET, and Dapper. After running the benchmarks for each of these data access and object mappers, you will be able to make an educated judgment call on the best form of data access and object mapping for your projects.

In this chapter, we will be covering the following topics:

- **Benchmarking data insertion methods**: In this section, we write the benchmarks for inserting data with ADO.NET, Entity Framework Core, and Dapper.NET with and without using stored procedures.
- **Benchmarking data selection methods**: In this section, we write the benchmarks for selecting data with ADO.NET, Entity Framework Core, and Dapper.NET with and without using stored procedures.
- **Benchmarking data editing methods**: In this section, we write the benchmarks for applying updates to data with ADO.NET, Entity Framework Core, and Dapper.NET with and without using stored procedures.
- **Benchmarking data deletion methods**: In this section, we write the benchmarks for deleting data with ADO.NET, Entity Framework Core, and Dapper.NET with and without using stored procedures.
- **The benchmarking results and their analysis**: In this section, we run the benchmarks that we wrote in the previous sections. We then analyze the results of our benchmark results to conclude the best way to perform various efficient data access and manipulation tasks.

After working through this chapter, you will have the skills needed to access and manipulate data with ADO.NET, Entity Framework, and Dapper.NET. You'll also be able to form your own judgment on which method of data access to use for your own projects.

> **Note**
> This chapter mainly involves you following along with writing a lot of code in preparation for running our data access benchmark methods in the last section. If you don't want to bother writing the code and just want to see the results, then jump to the last section of this chapter on the benchmarking results and their analysis. You can then jump to the areas of this chapter that are of most interest to you in helping you form your own opinions on the best data access methods for your needs. The source code is also available on GitHub to study for yourself.

Technical requirements

To master the skills presented in this chapter, it will be useful to have access to the following:

- Visual Studio 2022 or higher
- SQL Server 2019 or higher
- SQL Server Management Student 2019 or higher
- The book's source code: `https://github.com/PacktPublishing/High-Performance-Programming-in-CSharp-and-.NET/tree/master/CH10`

Benchmarking data insertion methods

In this section, we will be following on from the work we did in *Chapter 10, Setting Up Our Database Project*, by writing methods that will benchmark the performance of insert methods using ADO.NET, Entity Framework Core, and Dapper.NET. So, if you have not read *Chapter 10,* or looked at the source code, now would be a good time to do that.

The benchmarks written in this chapter will be run and the results will be analyzed in the last section. To save space due to chapter and page constraints, I will be leaving out references to `using` statements. Therefore, you will need to use Visual Studio's quick tips for adding missing `using` statements. Follow these steps to write our insertion method benchmarks:

1. Add the `BenchmarkDotNet` NuGet package.
2. Open the `BenchmarkTests` class and modify it as follows:

```
[MemoryDiagnoser]
[Orderer(SummaryOrderPolicy.Declared)]
[RankColumn]
public class BenchmarkTests
{
    [GlobalSetup]
    public void GlobalSetup()
    {
        InsertProductADNSP();
        InsertProductEFSP();
        InsertProductDDN();
```

```
    }
}
```

We have set our class up to execute benchmarks and summarize them in the order that they are declared, as well as diagnosing the memory usage and providing a performance ranking of the benchmarking methods. Then, we provided `GlobalSetup`, which is run before the benchmarks. This is to provide our benchmarks with data to select, update, and delete.

3. Add the `InsertProductADN` method:

```
[Benchmark]
public void InsertProductADN()
{
    string connectionString = SecretsManager
        .GetSecrets<DatabaseSettings>
            ("ConnectionString");
     AdoDotNetData adnData = new(connectionString);
    adnData.ExecuteNonQuery("INSERT INTO Products
      (ProductName, CategoryID, SupplierId,
         Discontinued) VALUES('ADO.NET Product', 1, 1,
           0)");
    adnData.Dispose();
}
```

This method obtains the connection string from the secrets file and creates a new `AdoDotNetData` instance by passing the connection string into its constructor. It then calls the `ExecuteNonQuery` method, passing into the method a raw SQL insert method. Once the query is run, the instance is disposed of.

4. Add the `InsertProductADNSP` method:

```
[Benchmark]
public void InsertProductADNSP()
{
    string connectionString = SecretsManager
        .GetSecrets<DatabaseSettings>
            ("ConnectionString");
```

```
        AdoDotNetData aaa = new(connectionString);
        SqlCommandModel model = new()
        {
            CommandText = "InsertProduct",
            CommandType = CommandType.StoredProcedure,
            CommandParameters =
            new SqlCommandParameterModel[] {
            new SqlCommandParameterModel() {
                ParameterName = "@ProductName",
                    DataType = DbType.String,
                    Value = "Dapper Product Edited"
                },
                new SqlCommandParameterModel() {
                    ParameterName = "@CategoryID",
                    DataType = DbType.Int32,
                    Value = 1
                }
                , new SqlCommandParameterModel() {
                    ParameterName = "@SupplierID",
                    DataType = DbType.Int32,
                    Value = 1
                }, new SqlCommandParameterModel() {
                    ParameterName = "@Discontinued",
                    DataType = DbType.Boolean,
                    Value = false
                }
            }
        };
        aaa.ExecuteNonQuery(model);
        aaa.Dispose();
    }
```

This method obtains the connection string from the secrets file and passes the string into the constructor of the `AdoDotNetData` class. It then creates a new `SqlCommandModel` that builds the properties for a stored procedure insert operation on the products table. It then calls the `ExecuteNonQuery` method, passing in the model that will be used to generate and execute the stored procedure call. The `AdoDotNetData` class is then disposed of.

5. Add the `InsertProductEF` method:
6. Add the `InsertProductEF` method:

```
[Benchmark]
public void InsertProductEF()
{
string connectionString = SecretsManager
      .GetSecrets<DatabaseSettings>
      ("ConnectionString");
EntityFrameworkCoreData efData
      = new(connectionString);
     Product product = new() {
          ProductName = "EF Product",
          CategoryID = 1,
          SupplierID = 1,
          Discontinued = false,
            QuantityPerUnit = "1"
      };
     efData.Products.Add(product);
     efData.SaveChanges();
     efData.Dispose();
}
```

This method obtains the connection string from the secrets file and passes it into the constructor of the `EntityFrameworkCoreData` class. It then creates a new product and adds that product to the `Products` collection. The changes are then saved, and the `EntityFrameworkCoreData` class is disposed of.

7. Now, add the `InsertProductEFSP` method:

```
[Benchmark]
public void InsertProductEFSP()
{
```

```
string connectionString = SecretsManager.
    GetSecrets<DatabaseSettings>
        ("ConnectionString");
 EntityFrameworkCoreData efData
    = new(connectionString);
 SqlCommandModel model = new()
 {
    CommandText = "EXEC InsertProduct
      @ProductName = {0}, @CategoryID = {1},
         @SupplierID = {2}, @Discontinued = {3}",
    CommandType = CommandType.StoredProcedure,
   CommandParameters
       = new SqlCommandParameterModel[] {
           new SqlCommandParameterModel() {
           ParameterName = "@ProductName",
           DataType = DbType.String,
            Value = "EF Product Edited"
           }
          , new SqlCommandParameterModel() {
           ParameterName = "@CategoryID",
           DataType = DbType.Int32,
           Value = 1
           }
          , new SqlCommandParameterModel() {
           ParameterName = "@SupplierID",
           DataType = DbType.Int32,
           Value = 1
           }
           , new SqlCommandParameterModel() {
           ParameterName = "@Discontinued",
           DataType = DbType.Boolean,
           Value = false
           }
       }
   };
 efData.ExecuteNonQuerySP(model);
```

```
        efData.Dispose();
    }
```

This method obtains the connection string from the secrets file and creates a new instance of the `EntityFrameworkCoreData` class. It then builds up the properties needed for the stored procedure insert via `SqlCommandModel`. Then, it executes the `ExecuteNonQuerySP` model, passing in the model that executes the insert stored procedure, and then disposes of the `EntityFrameworkCoreData` class.

8. Add the `InsertProductDDN` method:

```
[Benchmark]
public void InsertProductDDN()
{
    string connectionString = SecretsManager
        .GetSecrets<DatabaseSettings>
            ("ConnectionString");
    DapperDotNet ddnData = new(connectionString);
    int recordsAffected = ddnData
        .ExecuteNonQuery("INSERT INTO Products
            (ProductName, CategoryID, SupplierId,
                Discontinued) VALUES('Dapper.NET
                    Product', 1, 1, 0)");
    ddnData.Dispose();
}
```

This method obtains the connection string from the secrets file, creates a new instance of the `DapperDotNet` class, and executes a raw SQL insert statement by calling the `ExecuteNonQuery` method. It then disposes of the `DapperDotNet` class.

9. Add the `InsertProductDDNSP` method:

```
[Benchmark]
public void InsertProductDDNSP()
{
    string connectionString = SecretsManager
        .GetSecrets<DatabaseSettings>
            ("ConnectionString");
```

```
        DapperDotNet ddnData = new(connectionString);
        SqlCommandModel model = new() {
            CommandText = "InsertProduct",
           CommandType = CommandType.StoredProcedure,
           CommandParameters
               = new SqlCommandParameterModel[] {
                  new SqlCommandParameterModel() {
                   ParameterName = "@ProductName",
                   DataType = DbType.String,
                   Value = "Dapper Product" }
                 , new SqlCommandParameterModel() {
                   ParameterName = "@CategoryID",
                   DataType = DbType.Int32,
                   Value = 1 }
                   , new SqlCommandParameterModel() {
                   ParameterName = "@SupplierID",
                   DataType = DbType.Int32,
                   Value = 1 }
                   , new SqlCommandParameterModel() {
                   ParameterName = "@Discontinued",
                   DataType = DbType.Boolean,
                   Value = false }
               }
        };
        ddnData.ExecuteNonQuery(model);
        ddnData.Dispose();
    }
```

This method gets the connection string from the secrets file and creates a new `DapperDotNet` class. It then builds the `SqlCommandModel` properties required to execute the product insert stored procedure. Then, it calls the `ExecuteNonQuery` procedure, passing in the model that will execute the stored procedure. It then disposes of the `DapperDotNet` class.

That concludes our look at insert benchmarking methods. Now, we will start writing our selection benchmarking methods.

Benchmarking data selection methods

In this section, we will be writing our benchmarking methods that will test the performance of various data selection methods. These benchmarks will be run and analyzed in the last section of this chapter:

1. Add the `ReadScalarProductADN` method:

```
[Benchmark]
public void ReadScalarProductADN()
{
    string connectionString = SecretsManager
        .GetSecrets<DatabaseSettings>
            ("ConnectionString");
    AdoDotNetData adnData = new(connectionString);
    string productName = adnData
        .ExecuteScalar<string>("SELECT TOP 1
            ProductName FROM Products  WHERE Product
              Name LIKE 'ADO.NET Product%'");
   adnData.Dispose();
}
```

   This method obtains the connection from the `secrets` file, creates a new `AdoDotNetData` class, and executes the `ExecuteScalar` method, passing in a raw SQL statement that returns a string. It then disposes of the `AdoDotNet` class.

2. Add the `ReadScalarADNSP` method:

```
[Benchmark]
public void ReadScalarProductADNSP()
{
   string connectionString = SecretsManager
       .GetSecrets<DatabaseSettings>
           ("ConnectionString");
   AdoDotNetData aaa = new(connectionString);
   SqlCommandModel model = new SqlCommandModel() {
       CommandText = "GetProductName",
       CommandType = CommandType.StoredProcedure,
       CommandParameters
           = new SqlCommandParameterModel[] {
```

```
                new SqlCommandParameterModel() {
                ParameterName = "@ProductName",
                DataType = DbType.String,
                Value = "ADO.NET Product" }
            }
        };
        string productName
            = aaa.ExecuteScalar<string>(model);
        aaa.Dispose();
    }
```

This method obtains the connection string from the secrets file and creates a new instance of the `AdoDotNetData` class. It then builds `SqlCommandModel` up that contains the necessary properties to execute the scalar stored procedure. Then, it calls the `ExecuteScalar` method, passing in the model that executes the stored procedure, and returns the product name. It then disposes of the `AdoDotNetData` class.

3. Add the `ReadFilteredProductADN` method:

```
[Benchmark]
public void ReadFilteredProductADN()
{
    string connectionString = SecretsManager
        .GetSecrets<DatabaseSettings>
            ("ConnectionString");
    AdoDotNetData adnData = new(connectionString);
    IEnumerator<Product> data
        = adnData.ExecuteReader<Product>("SELECT *
            FROM Products  WHERE ProductName LIKE
                'ADO.NET Product'");
    adnData.Dispose();
}
```

This method gets the connection string from the secrets file and creates a new instance of the `AdoDotNetData` class. It then executes the `ExecuteReader` method, which takes a raw SQL statement and returns an enumerator of the `Product` type, and then disposes of the `AdoDotNetData` class.

4. Add the `ReadFilteredProductADNSP` method:

```
[Benchmark]
public void ReadFilteredProductADNSP()
{
    string connectionString = SecretsManager
        .GetSecrets<DatabaseSettings>
            ("ConnectionString");
    AdoDotNetData aaa = new(connectionString);
    SqlCommandModel model = new SqlCommandModel() {
        CommandText = "FilterProducts",
        CommandType = CommandType.StoredProcedure,
        CommandParameters
            = new SqlCommandParameterModel[] {
                new SqlCommandParameterModel() {
                ParameterName = "@ProductName",
                DataType = DbType.String,
                Value = "ADO.NET Product" }
            }
     };
    var data = aaa.ExecuteReader<dynamic>(model);
    aaa.Dispose();
}
```

This method gets the connection string from the secrets file and creates a new instance of the `AdoDotNetData` class. It then builds up `SqlCommandModel` that contains the properties that are required to execute the read stored procedure. Then, it executes the `ExecuteReader` method, which returns an enumerator, and then disposes of the `AdoDotNetData` class.

5. Add the `ReadScalarProductEF` method:

```
[Benchmark]
public void ReadScalarProductEF()
{
    string connectionString = SecretsManager
        .GetSecrets<DatabaseSettings>
            ("ConnectionString");
    EntityFrameworkCoreData efData
```

```
        = new(connectionString);
    string productName
        = efData.Products.FirstOrDefault(
            p => p.ProductName
             .Contains("EF Product")
        ).ProductName;
    efData.Dispose();
}
```

This method gets the connection string from the secrets file and creates a new instance of the `EntityFrameworkCore` method. It then gets the first item in the `Product` collection that matches the filter and assigns `ProductName`. Then, it disposes of the `EntityFrameworkCore` class.

6. Add `ReadScalarProductEFSP`:

```
[Benchmark]
public void ReadScalarProductEFSP()
{
    string connectionString = SecretsManager
        .GetSecrets<DatabaseSettings>
            ("ConnectionString");
    EntityFrameworkCoreData efData
        = new(connectionString);
    string productName = efData
        .ExecuteScalarSP("EF Product");
    efData.Dispose();
}
```

This method gets the connection string from the secrets file and then creates a new instance of the `EntityFrameworkCoreData` class. Then, it calls the `ExecuteScalarSP` method, passing in the name of the filter, returning the first `ProductName` that matches the filter, and then disposes of the `EntityFrameworkCoreData` class.

7. Add the `ReadFilteredProductsEF` method:

```
[Benchmark]
public void ReadFilteredProductsEF()
{
```

```
    string connectionString = SecretsManager
        .GetSecrets<DatabaseSettings>
            ("ConnectionString");
    EntityFrameworkCoreData efData
        = new(connectionString);
    IEnumerator<Product> products = efData.Products
        .Where(p => p.ProductName
        .Contains("EF Product")).GetEnumerator();
    efData.Dispose();
        products.Dispose();
}
```

This method gets the connection string from the secrets file and then creates an instance of the `EntityFrameworkCoreData` class. It then filters the products and returns an enumerator of products. Then, the method disposes of the `EntityFrameworkCoreData` class and the enumerator.

8. Add the `ReadFilteredProductsEFSP` method:

```
[Benchmark]
public void ReadFilteredProductsEFSP()
{
    string connectionString = SecretsManager
        .GetSecrets<DatabaseSettings>
            ("ConnectionString");
    EntityFrameworkCoreData efData
        = new(connectionString);
    IEnumerator<Product> products = efData
        .ExecuteReaderSP("EF Product");
    efData.Dispose();
        products.Dispose();
}
```

This method gets the secret from the secrets file and creates a new instance of the `EntityFrameworkCoreData` class. It then calls the `ExecuteReaderSP` method, which executes a stored procedure that returns an enumerator of the `Products` type. Then, the method disposes of the `EntityFrameworkCoreData` class and the enumerator.

9. Add the `ReadScalarProductDDN` method:

```
[Benchmark]
public void ReadScalarProductDDN()
{
    string connectionString = SecretsManager
        .GetSecrets<DatabaseSettings>
            ("ConnectionString");
    DapperDotNet ddnData = new(connectionString);
    string productName = ddnData
        .ExecuteScalar<string>("SELECT TOP 1
             ProductName FROM Products  WHERE Product
                 Name LIKE 'Dapper.NET Product%'");
    ddnData.Dispose();
}
```

This method gets the connection string from the secrets file and creates a new instance of the `DapperDotNet` class. It then executes the `ExecuteScalar` method, passing in a raw SQL statement that returns the top `ProductName` that matches the filter. Then, it disposes of the `DapperDotNet` class.

10. Add the `ReadScalarProductDDNSP` method:

```
[Benchmark]
public void ReadScalarProductDDNSP()
{
    string connectionString = SecretsManager
        .GetSecrets<DatabaseSettings>
            ("ConnectionString");
    DapperDotNet ddnData = new(connectionString);
    SqlCommandModel model = new() {
        CommandText = "GetProductName",
        CommandType = CommandType.StoredProcedure,
        CommandParameters
            = new SqlCommandParameterModel[] {
                new SqlCommandParameterModel() {
                ParameterName = "@ProductName",
                DataType = DbType.String,
                Value = "Dapper Product" }
```

```
            }
        };
        string productName
            = ddnData.ExecuteScalarSP(model);
        ddnData.Dispose();
    }
```

This method gets the connection string from the secrets file and creates a new instance of the `DapperDotNet` class. Then, the method builds `SqlCommandModel` that contains the properties necessary to execute a stored procedure. It then calls the `ExecuteScalarSP` method, passing in the model. The `ProductName` of the first matching product is returned. The method then disposes of the `DapperDotNet` class.

11. Add the `ReadFilteredProductsDDN` class:

```
    [Benchmark]
    public void ReadFilteredProductsDDN()
    {
        string connectionString = SecretsManager
            .GetSecrets<DatabaseSettings>
                ("ConnectionString");
        DapperDotNet ddnData = new(connectionString);
        IEnumerator<Product> data
            = ddnData.ExecuteReader<Product>("SELECT *
            FROM Products WHERE ProductName LIKE
                'Dapper.NET Product%'");
        ddnData.Dispose();
        data.Dispose();
    }
```

This method gets the connection string from the `secrets` file and then creates a new instance of the `DapperDotNet` class. It then calls the `ExecuteReader` method, passing in a raw SQL statement. An enumerator of the `Product` type is returned. `DapperDotNet` and the enumerator are then disposed of.

12. Add the `ReadFilteredProductsDDNSP` method:

```
    [Benchmark]
    public void ReadFilteredProductsDDNSP()
    {
```

```
    string connectionString = SecretsManager
        .GetSecrets<DatabaseSettings>
            ("ConnectionString");
    DapperDotNet ddnData = new(connectionString);
    SqlCommandModel model = new() {
        CommandText = "GetProductName",
        CommandType = CommandType.StoredProcedure,
        CommandParameters
            = new SqlCommandParameterModel[] {
                new SqlCommandParameterModel() {
                ParameterName = "@ProductName",
                DataType = DbType.String,
                Value = "Dapper.NET Product" }
            }
    };
    IEnumerator<Product> products
        = ddnData.ExecuteReaderSP<Product>(model);
    ddnData.Dispose();
}
```

This method gets the connection string from the secrets file and then creates an instance of the `DapperDotNet` class. It then builds up a `SqlCommandModel` that has the properties needed to execute a stored procedure. It then calls `ExcuteReaderSP`, passing in the model that returns an enumerator of the `Product` type.

We have now finished writing our selection benchmarks. Now, we'll move on to writing our update benchmarks.

Benchmarking data editing methods

In this section, we will be writing our benchmarks that test the performance of various update statements. These benchmarks will be run and analyzed in the final section of this chapter:

1. Add the `UpdateProductADN` method:

```
[Benchmark]
public void UpdateProductADN()
```

```
{
     string connectionString = SecretsManager
          .GetSecrets<DatabaseSettings>
               ("ConnectionString");
     AdoDotNetData adnData = new(connectionString);
    int recordsAffected
         = adnData.ExecuteNonQuery("UPDATE Products
              SET ProductName = 'ADO.NET Product -
                   Edited' WHERE ProductName =
                        'ADO.NET Product'");
     adnData.Dispose();
}
```

This method obtains the connection string from the secrets file and then creates a new instance of the AdoDotNetData class. It then calls the ExecuteNonQuery product, passing in a raw SQL statement, then returns the number of records affected and disposes of the AdoDotNetData class.

2. Add the UpdateProductADNSP method:

```
[Benchmark]
public void UpdateProductADNSP()
{
    string connectionString = SecretsManager
         .GetSecrets<DatabaseSettings>
              ("ConnectionString");
    AdoDotNetData aaa = new(connectionString);
    SqlCommandModel model = new() {
         CommandText = "UpdateProductName",
         CommandType = CommandType.StoredProcedure,
         CommandParameters
              = new SqlCommandParameterModel[] {
                   new SqlCommandParameterModel() {
                   ParameterName = "@OldProductName",
                   DataType = DbType.String,
                   Value = "ADO.NET Product" }
```

```
                , new SqlCommandParameterModel() {
                ParameterName = "@NewProductName",
                DataType = DbType.String,
                Value = "ADO.NET Product - Edited"}
                }
        };
        aaa.ExecuteNonQuery(model);
       aaa.Dispose();
   }
```

This method gets the connection string from the secrets file and creates a new instance of the `AdoDotNetData` class. `SqlCommandModel` is then built up with the properties needed to execute the update stored procedure. `ExecuteNonQuery` is then called with the model being passed in, and the stored procedure that performs the update is executed. The `AdoDotNetData` class is then disposed of.

3. Add the `UpdateProductEF` method:

```
[Benchmark]
public void UpdateProductEF()
{
    string connectionString = SecretsManager
        .GetSecrets<DatabaseSettings>
            ("ConnectionString");
    EntityFrameworkCoreData efData
    = new EntityFrameworkCoreData(connectionString);
    IQueryable<Product> products = efData.Products
    .Where(p => p.ProductName.Contains("EF
        Product"));
    foreach (Product product in products)
        product.ProductName = "EF Product Edited";
    efData.Products.UpdateRange(products);
     int recordsAffected = efData.SaveChanges();
    efData.Dispose();
}
```

This method gets the connection string from the secrets file and creates a new instance of the `EntityFrameworkCoreData` class. It then declares and assigns a queryable collection of products. This collection is then iterated with the name of each product updated. The `UpdateRange` method is then called on the `Products` collection, and the updated collection is passed in. The modifications are then saved, and the `EntityFrameworkCoreData` class is disposed of.

4. Add the `UpdateProductEFSP` method:

```
[Benchmark]
public void UpdateProductEFSP()
{
    string connectionString = SecretsManager
        .GetSecrets<DatabaseSettings>
            ("ConnectionString");
    EntityFrameworkCoreData efData =
        new(connectionString);
    SqlCommandModel model = new() {
        CommandText = "EXEC UpdateProductName
            @OldProductName = {0}, @NewProductName =
                {1}",
        CommandType = CommandType.StoredProcedure,
        CommandParameters
            = new SqlCommandParameterModel[] {
                new SqlCommandParameterModel() {
                ParameterName = "@OldProductName",
                DataType = DbType.String,
                Value = "EF Product" }
                , new SqlCommandParameterModel() {
                ParameterName = "@NewProductName",
                DataType = DbType.String,
                Value = "EF Product - Edited" }
                }
      };
    efData.ExecuteNonQuerySP(model);
    efData.Dispose();
}
```

This method gets the connection string from the `secrets` file and creates an instance of the `EntityFrameworkCoreData` class. It then builds up the `SqlCommandModel` that contains the properties needed to generate the call to the update stored procedure. The method then calls the `ExecuteNonQuerySP` procedure, which executes the stored procedure, passing in the model, and then disposes of the `EntityFrameworkCoreData` method.

5. Add the `UpdateProductDDN` method:

```
[Benchmark]
public void UpdateProductDDN()
{
    string connectionString = SecretsManager
        .GetSecrets<DatabaseSettings>
            ("ConnectionString");
    DapperDotNet ddnData = new(connectionString);
    int recordsAffected
        = ddnData.ExecuteNonQuery("UPDATE Products
            SET ProductName = 'Dapper.NET Product -
              Edited' WHERE ProductName = 'Dapper.NET
                Product'");
    ddnData.Dispose();
}
```

This method gets the connection string from the secrets file and creates a new instance of the `DapperDotNet` class. It then calls the `ExecuteNonQuery` method, passing in a raw SQL update statement. The number of records affected is returned, and the `DapperDotNet` class is disposed of.

6. Add the `UpdateProductDDNSP` method:

```
[Benchmark]
public void UpdateProductDDNSP()
{
string connectionString = SecretsManager
    .GetSecrets<DatabaseSettings>("ConnectionString");
DapperDotNet ddnData = new(connectionString);
SqlCommandModel model = new()
{
    CommandText = "UpdateProductName",
```

```
        CommandType = CommandType.StoredProcedure,
       CommandParameters = new SqlCommand
           ParameterModel[]{
           new SqlCommandParameterModel() {
               ParameterName = "@OldProductName",
               DataType = DbType.String,
               Value = "Dapper.NET Product - Edited" }
           , new SqlCommandParameterModel() {
                 ParameterName = "@NewProductName",
               DataType = DbType.String,
               Value = "Dapper.NET Product" }
       }
   };
   ddnData.ExecuteNonQuery(model);
   ddnData.Dispose();
   }
```

This method gets the connection string from the secrets file and creates a new instance of the `DapperDotNet` class. It then builds an `SQLCommandModel` in preparation for executing a stored procedure. It calls the `ExecuteNonQuery` method, passing in the model. The stored procedure is executed, and the method disposes of the `DapperDotNet` class.

This is the end of our look at the update benchmarks. Now for our final set of benchmark methods. In the next section, we will write our deletion benchmarks.

Benchmarking data deletion methods

In this section, we write our benchmarks for measuring the performance of our deletion methods. These benchmarks will be run and analyzed in the next section:

1. Add the `DeleteProductADN` method:

```
[Benchmark]
public void DeleteProductADN()
{
    string connectionString = SecretsManager
        .GetSecrets<DatabaseSettings>
            ("ConnectionString");
```

```
        AdoDotNetData adnData = new(connectionString);
        int recordsAffected
            = adnData.ExecuteNonQuery("DELETE FROM
                Products WHERE ProductName LIKE 'ADO.NET
                    Product%'");
        adnData.Dispose();
    }
```

This method gets the connection string from the secrets file. It then creates an instance of the `AdoDotNetData` class. Then, the method calls the `ExecuteNonQuery` method, passing into it a raw SQL delete statement. It then disposes of the `AdoDotNetData` class.

2. Add the `DeleteProductADNSP` method:

```
    [Benchmark]
    public void DeleteProductADNSP()
    {
        string connectionString = SecretsManager
            .GetSecrets<DatabaseSettings>
                ("ConnectionString");
        AdoDotNetData aaa = new(connectionString);
        SqlCommandModel model = new()
        {
            CommandText = "DeleteProduct",
            CommandType = CommandType.StoredProcedure,
            CommandParameters
                = new SqlCommandParameterModel[] {
                    new SqlCommandParameterModel() {
                    ParameterName = "@ProductName",
                    DataType = DbType.String,
                    Value = "ADO.NET Product - Edited"}
                }
        };
        aaa.ExecuteNonQuery(model);
        aaa.Dispose();
    }
```

This method gets the connection string from the secrets file, and then it creates an instance of the `AdoDotNetData` class. `SqlCommandModel` is built up with the properties required for the delete stored procedure execution. The model is then passed into the `ExecuteNonQuery` model, which executes the stored procedure, and the `AdoDotNetData` class is then disposed of.

3. Add the `DeleteProductEF` method:

```
[Benchmark]
public void DeleteProductEF()
{
    string connectionString = SecretsManager
        .GetSecrets<DatabaseSettings>
            ("ConnectionString");
    EntityFrameworkCoreData efData
    = new EntityFrameworkCoreData(connectionString);
    IQueryable<Product> products = efData.Products
    .Where(p => p.ProductName.Contains("EF Product"));
    efData.Products.RemoveRange(products);
    int recordsAffected = efData.SaveChanges();
    efData.Dispose();
}
```

This method gets the connection string from the secrets file and then creates an instance of the `EntityFrameworkCoreData` class. A queryable collection of products is then returned, matching the deletion criteria. This collection is then passed into the `RemoveRange` method of the `Products` collection, and the modification is saved with those items removed from the database. The method then disposes of the `EntityFrameworkCoreData` class.

4. Add the `DeleteProductEFSP` method:

```
[Benchmark]
public void DeleteProductEFSP()
{
    string connectionString = SecretsManager
        .GetSecrets<DatabaseSettings>
            ("ConnectionString");
    EntityFrameworkCoreData efData
        = new(connectionString);
```

```
    SqlCommandModel model = new() {
        CommandText = "EXEC DeleteProduct @ProductName
            = {0}",
        CommandType = CommandType.StoredProcedure,
        CommandParameters
            = new SqlCommandParameterModel[] {
                new SqlCommandParameterModel() {
                ParameterName = "@NewProductName",
                DataType = DbType.String,
                Value = "EF Product - Edited" }
                }
    };
    efData.ExecuteNonQuerySP(model);
    efData.Dispose();
}
```

This method gets the connection string from the secrets file and creates an instance of the `EntityFrameworkCoreData` class. It then builds up a `SqlCommandModel` that contains the properties of the deletion stored procedure. The `ExecuteNonQuerySP` method is called with the model that is passed in, the deletion stored procedure is executed, and the `EntityFrameworkCoreData` class is disposed of.

5. Add the `DeleteProductDDN` method:

```
[Benchmark]
public void DeleteProductDDN()
{
    string connectionString = SecretsManager
        .GetSecrets<DatabaseSettings>
            ("ConnectionString");
    DapperDotNet ddnData = new(connectionString);
    int recordsAffected
        = ddnData.ExecuteNonQuery("DELETE FROM
            Products WHERE ProductName LIKE
                'Dapper.NET Product%'");
    ddnData.Dispose();
}
```

This method gets the connection string from the secrets file and creates an instance of the `DapperDotNet` class. It then calls the `ExecuteNonQuery` method, passing into that method a raw SQL delete statement. The deletion is carried out and the number of records affected is returned. The `DapperDotNet` class is then disposed of.

6. Add the `DeleteProductDDNSP` method:

```
[Benchmark]
public void DeleteProductDDNSP()
{
    string connectionString = SecretsManager
        .GetSecrets<DatabaseSettings>
            ("ConnectionString");
    DapperDotNet ddnData = new(connectionString);
    SqlCommandModel model = new() {
        CommandText = "DeleteProduct",
        CommandType = CommandType.StoredProcedure,
        CommandParameters
            = new SqlCommandParameterModel[] {
              new SqlCommandParameterModel() {
              ParameterName = "@ProductName",
              DataType = DbType.String,
             Value = "Dapper.NET Product - Edited" }
            }
    };
    ddnData.ExecuteNonQuery(model);
    ddnData.Dispose();
}
```

This method gets the connection string from the secrets file and creates an instance of the DapperDotNet class. It then builds up the SqlCommandModel, containing stored procedure properties. This model is then passed into the ExecuteNonQuery method, which executes the stored procedure, and the DapperDotNet class is disposed of.

That was the last of our benchmarking methods. There is just one more job to do before we are able to run our benchmarks. Update the Program class as follows:

```
using BenchmarkDotNet.Running;
class Program
{
static void Main(string[] args)
    {
        BenchmarkRunner.Run<BenchmarkTests>();
}
}
```

The Main method executes the BenchmarkTests class. You can now do a release build to run the benchmarks. The program will take a while to execute, so you will need to be patient. In the next section, we will analyze the results of our various benchmarks to find out the most performant ways of performing inserts, selections, updates, and deletions.

The benchmarking results and their analysis

Before we analyze the results, it is worth noting some big data statistics from 2020. Google gets more than 40,000 queries per second. This equates to 3,456,000,000 queries per day. There are 65,000,000,000 WhatsApp business app messages sent per day. In the course of 24 hours, there are 1,440 minutes, which is 86,400 seconds, which is 86,400,000 milliseconds.

Here is our benchmark summary report:

Developer PowerShell

+ Developer PowerShell

```
// * Summary *

BenchmarkDotNet=v0.12.1, OS=Windows 10.0.19042
Intel Core i5-6300U CPU 2.40GHz (Skylake), 1 CPU, 4 logical and 2 physical cores
.NET Core SDK=6.0.100-preview.3.21202.5
  [Host]     : .NET Core 6.0.0 (CoreCLR 6.0.21.20104, CoreFX 6.0.21.20104), X64 RyuJIT
  DefaultJob : .NET Core 6.0.0 (CoreCLR 6.0.21.20104, CoreFX 6.0.21.20104), X64 RyuJIT
```

Method	Mean	Error	StdDev	Median	Rank	Gen 0	Gen 1	Gen 2	Allocated
InsertProductADN	2.092 ms	0.0418 ms	0.1150 ms	2.102 ms	9	37.1094	15.6250	-	71.31 KB
InsertProductADNSP	1.894 ms	0.0470 ms	0.1350 ms	1.866 ms	8	23.4375	5.8594	-	72.53 KB
ReadScalarProductADN	1.407 ms	0.0370 ms	0.1068 ms	1.390 ms	4	37.1094	13.6719	-	70.03 KB
ReadScalarProductADNSP	1.433 ms	0.0390 ms	0.1107 ms	1.416 ms	4	27.3438	7.8125	-	70.58 KB
ReadFilteredProductADN	1.084 ms	0.0400 ms	0.1153 ms	1.068 ms	1	35.1563	13.6719	-	68.04 KB
ReadFilteredProductADNSP	1.078 ms	0.0431 ms	0.1264 ms	1.067 ms	1	35.1563	9.7656	-	68.15 KB
UpdateProductADN	3.583 ms	0.2150 ms	0.6306 ms	3.475 ms	13	-	-	-	70.01 KB
UpdateProductADNSP	1.562 ms	0.0496 ms	0.1367 ms	1.541 ms	5	23.4375	5.8594	-	69.96 KB
DeleteProductADN	6.263 ms	0.8640 ms	2.3943 ms	5.775 ms	14	-	-	-	70.01 KB
DeleteProductADNSP	1.760 ms	0.1261 ms	0.3516 ms	1.588 ms	6	-	-	-	70.88 KB
InsertProductEF	2.916 ms	0.1014 ms	0.2942 ms	2.840 ms	12	62.5000	15.6250	-	124.29 KB
InsertProductEFSP	2.251 ms	0.0540 ms	0.1592 ms	2.261 ms	10	54.6875	15.6250	-	104.65 KB
ReadScalarProductEF	396.509 ms	23.6081 ms	68.8658 ms	360.339 ms	17	-	-	-	109.63 KB
ReadScalarProductEFSP	53.235 ms	1.6418 ms	4.6037 ms	51.883 ms	15	-	-	-	113.3 KB
ReadFilteredProductsEF	1.305 ms	0.0566 ms	0.1596 ms	1.294 ms	3	37.1094	9.7656	-	99.48 KB
ReadFilteredProductsEFSP	1.187 ms	0.0457 ms	0.1319 ms	1.171 ms	2	37.1094	9.7656	-	97.29 KB
UpdateProductEF	5,304.279 ms	51.0259 ms	45.2331 ms	5,309.797 ms	18	110000.0000	23000.0000	4000.0000	424807.66 KB
UpdateProductEFSP	1.964 ms	0.0952 ms	0.2685 ms	1.850 ms	8	-	-	-	104.28 KB
DeleteProductEF	386.716 ms	7.4865 ms	8.0105 ms	384.664 ms	17	-	-	-	129.72 KB
DeleteProductEFSP	2.012 ms	0.1439 ms	0.4174 ms	1.804 ms	8	-	-	-	103.23 KB
InsertProductDDN	2.058 ms	0.0466 ms	0.1374 ms	2.076 ms	9	37.1094	11.7188	-	71.42 KB
InsertProductDDNSP	1.841 ms	0.0442 ms	0.1289 ms	1.851 ms	7	39.0625	13.6719	-	75.24 KB
ReadScalarProductDDN	1.403 ms	0.0431 ms	0.1251 ms	1.381 ms	4	37.1094	9.7656	-	70.16 KB
ReadScalarProductDDNSP	1.514 ms	0.0528 ms	0.1497 ms	1.489 ms	5	37.1094	13.6719	-	71.84 KB
ReadFilteredProductsDDN	199.910 ms	3.9677 ms	9.0365 ms	200.890 ms	16	9666.6667	3333.3333	1333.3333	18113.34 KB
ReadFilteredProductsDDNSP	1.529 ms	0.0515 ms	0.1427 ms	1.510 ms	5	23.4375	5.8594	-	71.89 KB
UpdateProductDDN	2.297 ms	0.2973 ms	0.8719 ms	1.903 ms	11	-	-	-	70.13 KB
UpdateProductDDNSP	1.891 ms	0.1623 ms	0.4577 ms	1.714 ms	7	-	-	-	72.95 KB
DeleteProductDDN	2.522 ms	0.4970 ms	1.4497 ms	1.807 ms	11	-	-	-	70.3 KB
DeleteProductDDNSP	1.863 ms	0.1538 ms	0.4362 ms	1.763 ms	7	-	-	-	72.34 KB

Figure 11.1 – Data access benchmark summary

Let's discuss the insert statements first. The results are as follows:

- `InsertProductDDNSP` = 1.841 ms
- `InsertProductADNSP` = 1.894 ms
- `InsertProductDDN` = 2.058 ms
- `InsertProductADN` = 2.092 ms
- `InsertProductEF` = 2.196 ms
- `InsertProductEFSP` = 396.509 ms

From the summary, we can see that the best-performing insert statement is the Dapper.NET stored procedure insert, taking approximately 1.841 ms to execute, followed by `InsertProductADNSP`, which takes approximately 1.894 ms to execute. By far the worst performer is the `InsertProductEFSP` method, which takes 396.509 ms to execute. As we can see from these methods, even though we have six different ways of inserting data, they all perform at different speeds. When performance is a serious issue, your best option is to use Dapper.NET stored procedure execution or ADO.NET stored procedure execution when inserting data.

We will now look at scalar operations, starting with an ordered list of method performance:

- `ReadScalarProductDDN` = 1.403 ms
- `ReadScalarProductADN` = 1.407 ms
- `ReadScalarProductADNSP` = 1.433 ms
- `ReadScalarProductDDNSP` = 1.514 ms
- `ReadScalarProductEFSP` = 53.235 ms
- `ReadScalarProductEF` = 396.509 ms

Looking at these results, the Dapper.NET raw SQL execution takes approximately 1.403 ms, followed by the ADO.NET raw SQL execution at 1.407 ms. Both the Entity Framework Core methods perform much more slowly. So, when performance matters, you are best off using Dapper.NET or ADO.NET raw SQL queries to obtain scalar values.

Next are the filtered list queries. Here is an ordered list of the results:

- `ReadFilteredProductsADNSP` = 1.078 ms
- `ReadFilteredProductsADN` = 1.084 ms
- `ReadFilteredProductsEFSP` = 1.187 ms
- `ReadFilteredProductsEF` = 1.305 ms
- `ReadFilteredProductsDDNSP` = 1.529 ms
- `ReadFilteredProductsDDN` = 199.910 ms

As we can see from these results, ADO.NET raw SQL and stored procedure access perform the best at 1.078 ms and 1.084 ms, respectively. Surprisingly, this time it is Dapper.NET that performs the worst when it comes to raw SQL and stored procedure access. So, when performance matters for performing queries that return multiple records, you are best off using ADO.NET.

Now, we turn our attention to performing updates. Here is an ordered list of our results:

- `UpdateProductADNSP` = 1.562 ms
- `UpdateProductEFSP` = 1.964 ms
- `UpdateProductDDNSP` = 1.891 ms
- `UpdateProductDDN` = 2.297 ms
- `UpdateProductADN` = 3.583 ms
- `UpdateProductEF` = 5,304.279 ms

From these results, **the clear winner is the ADO.NET stored procedure access at 1.562 ms**. The worst performer is the Entity Framework Core update method. When performance matters, use ADO.NET stored procedures to update database records.

Finally, we'll look at our deletion benchmarks. Here is an ordered list of our results:

- `DeleteProductADNSP` = 1.760 ms
- `DeleteProductDDNSP` = 1.863 ms
- `DeleteProductEFSP` = 2.012 ms
- `DeleteProductDDN` = 2.522 ms
- `DeleteProductADN` = 6.263 ms
- `DeleteProductEF` = 386.716 ms

It can be seen that the worst performer is the Entity Framework Core method, taking about 386.716 ms to execute. On the other hand, the best performer is the ADO.NET stored procedure method, which takes only 1.760 ms, with the Dapper.NET stored procedure next, taking 1.863 ms. So, when performance matters, your best deletion strategy is to use ADO.NET stored procedures.

What can we summarize from these results?

Dapper.NET and ADO.NET come out on top when performing insert, read, update, and delete operations. The performance varies between raw SQL and stored procedure execution. When performance is critical, it would seem that the best strategy is rather than choosing just one framework and using only that one for all your data operations, to use a hybrid approach.

With a hybrid approach to data access, you will use a combination of data access frameworks. From each framework, you will decide on the best performer and use that for your data operation. In the case of our benchmarks, we would use two frameworks. The frameworks chosen are ADO.NET and Dapper.NET. This way, we could find the best performance possible for each type of data operation.

But given that these times only have millisecond differences, why does such performance matter?

Well, remember at the beginning of this section how we mentioned the big data statistics for 2020? The following table shows the performance of these methods when put into the context of big data search queries and app message storage:

Operation	Data Access Method	Per-Operation Execution Time (ms)	Operation Time (Processing Days)
3,456,000,000 Google searches per day	`ReadFilteredProductsADNSP`	1.078	43.12
3,456,000,000 Google searches per day	`ReadFilteredProductsDDN`	199.910	7,996.4
65,000,000,000 WhatsApp business messages per day	`InsertProductDDNSP`	1.841	1,385.01157
65,000,000,000 WhatsApp business Messages per day	`InsertProductEFSP`	396.509	298,299.595

Table 10.1 – Big data operation durations if SQL Server were used to store and read data

These benchmarks were run on an HP laptop with an Intel Core i5-6300U CPU 2.40 GHz (Skylake) processor. This is one CPU with four logical cores and two physical cores. I have 8 GB of RAM and a 256 GB SSD.

If the SQL Server on my laptop were used and I had the space available (which I don't) to store the WhatsApp business app message data, depending upon which method I used to insert the data, it would take between 1,385.01157 and 298,299.595 processing days on my laptop. If my laptop were used to retrieve Google search results from my SQL Server, then it would take between 43.12 and 7,996.4 processing days to retrieve those results.

This real-life application of the benchmarks to actual big data volumes based on the big data statistics of 2020 shows the importance of computer infrastructure and the type of investment that would be needed to make these searches and message sending and receiving instantaneous. It was important to have peak performance when working with such large datasets.

Tweaking large datasets through code can only go so far. That is why server computers have many more processors and disks, along with more memory, than your normal day-to-day workstations and home computers.

The key thing to take away from this chapter is that whenever you are deciding on a way forward to maximize performance, experiment and benchmark. Along with that, take the time to choose your physical infrastructure carefully.

Another thing to bear in mind when using a cloud host is the cost per data execution and cost per hour when running virtual machines. Then, there is the cost of data throughput and data storage saving and retrieval. With figures of apps such as Google and WhatsApp being in the billions, if you were to be that successful, could you imagine the running costs involved? That is why performance in today's competitive market is also so important. The quicker a piece of code executes in the cloud, the cheaper the price. The longer a piece of code takes to run, the more expensive it becomes.

As an example, if you have an Azure function that performs your data operations that are located in the West US region on the Consumption tier using a memory size of 128 with an execution time of 1.078 ms and 65,000,000,000 executions per month, then your bill for the month would be US$13,133.54. But if your execution time was 396.509 ms, then your bill for the month would be US$64,539.57. So, doing the same code action can mean a difference of 64,539.57 – 12,133.54 = US$52,406.03 per month on cloud expenditure operations. I am sure you would not want to spend that much money on such outgoings, and that does not even include the cost of the SQL Server instances!

That concludes this rather long chapter, and so we will now summarize what we have learned.

Summary

In this chapter, we learned how to perform inserts, selections, updates, and deletes in SQL Server. We learned how to perform these operations in different ways using pure ADO.NET, Entity Framework Core, and Dapper.NET. The different data operations were performed using raw SQL and stored procedures.

To understand the performance of each of these data access methods of the different data access frameworks, in this chapter, we benchmarked their runtime performance using `BenchmarkDotNet`. We saw that both Dapper.NET and ADO.NET performed better than Entity Framework Core in most cases and that even with these two frameworks, the performance varied considerably.

We concluded that rather than just adopting a single data access technology, in some situations where performance really matters, it could be beneficial to employ a hybrid approach to data access. With a hybrid approach, you use the best framework and method within that framework for the data access task in question. That way, you maximize your overall performance. This can also be critical in terms of keeping your infrastructure expenses down, especially when the infrastructure you are employing is a third-party cloud provider with your monthly bill being in the thousands of dollars.

But apart from computer code performance enhancements, we also studied big data volumes and calculated the number of processing days it would take to perform query and data insert operations when the volumes involved are in the billions. So, apart from code performance, we also came to understand that it is necessary to choose the right kind of infrastructure, which also comes at a price when using cloud services.

> **Note**
>
> Whatever you are doing, whenever performance is a critical business requirement, you are strongly advised to experiment and provide your own benchmarks. Based on your results, you can then choose your own methods of data access that you feel are most beneficial for your needs.

In the next chapter, we will be looking at improving the performance of SQL Server and Cosmos DB. But before we do, have a go at the following questions to see how well you have retained the information contained in this chapter. Also, there are very useful articles in the *Further reading* section that expand upon what has been covered in this chapter. This chapter purely focused on identifying the best data access methods in code using three different frameworks. But in the *Further reading* section, you will find topics that are specific to improving database performance that are well worth reading about.

Questions

1. Which data access method was fastest when inserting data?
2. Which data access method was fastest when selecting a scalar value?
3. Which data access method was fastest when selecting multiple records?
4. Which data access method was fastest when updating data?
5. Which data access method was fastest when deleting data?
6. Should you use one framework for all data access operations and why?

Further reading

- *Dapper vs Entity Framework vs ADO.NET Performance Benchmarking*: `https://www.exceptionnotfound.net/dapper-vs-entity-framework-vs-ado-net-performance-benchmarking/`
- Dapper tutorial: `https://dapper-tutorial.net/dapper`
- *ADO.NET Tutorial for Beginners and Professionals*: `https://dotnettutorials.net/course/ado-net-tutorial-for-beginners-and-professionals/`
- *SQL Server Database Performance Tuning*: `https://www.brentozar.com/sql/sql-server-performance-tuning/`
- Book – *High Performance SQL Server: Consistent Response for Mission-Critical Applications* by Benjamin Nevarez: `https://amzn.to/3gnUbe7`
- *Performance tips for Azure Cosmos DB and .NET*: `https://docs.microsoft.com/azure/cosmos-db/performance-tips-dotnet-sdk-v3-sql`
- *A technique for building high-performance databases with EF Core*: `https://www.thereformedprogrammer.net/a-technique-for-building-high-performance-databases-with-ef-core/`
- *How to improve SQL Server query performance in .NET*: `https://www.red-gate.com/products/dotnet-development/ants-performance-profiler/resources/how-to-improve-sql-server-query-performance-in-net`
- *Using Dapper and SQLKata in .NET Core for High-Performance Application*: `https://medium.com/geekculture/using-dapper-and-sqlkata-in-net-core-for-high-performance-application-716d5fd43210`
- *What are the best databases for a small .NET application?*: `https://www.slant.co/topics/274/~best-databases-for-a-small-net-application`

> **Point to Remember**
>
> Reading about performance in a book is all very good. But you should always do your own experimentation and benchmarking if performance is very important to you. Different hardware architecture and different programming styles will yield very different results, and this point is well worth remembering. Network usage, security software, and data volumes, along with file input and output, can all have an effect on the performance of your application.

12
Responsive User Interfaces

In this chapter, you will learn to write responsive user interfaces. You will write responsive **Windows Forms (WinForms)**, **Windows Presentation Foundation (WPF)**, ASP.NET, .NET MAUI, and WinUI applications. Using background worker threads, you will see how you can update and work with the **User Interface (UI)** in real time by running long-running tasks in the background.

In this chapter, we will be working through the following topics:

- **Building a responsive UI with WinForms**: In this section, you will write a simple WinForms application that remains responsive to user interaction while performing multiple tasks.
- **Building a responsive UI with WPF**: In this section, you will be writing a simple WPF application that remains responsive to user interaction while performing multiple tasks.
- **Building a responsive UI with ASP.NET**: In this section, you will be writing a simple ASP.NET application that remains responsive to user interaction while performing multiple tasks.

- **Building a responsive UI with .NET MAUI**: In this section, you will be writing a simple Xamarin.Forms application that remains responsive to user interaction while performing multiple tasks. You will then migrate the projects from Xamarin.Forms to .NET MAUI by updating the library references.
- **Building a responsive UI with WinUI**: In this section, you will be writing a simple WinUI application that remains responsive to user interaction while performing multiple tasks.

By working through this chapter, you will gain the skills to do the following:

- Use background worker threads to keep UIs responsive
- Use wait screens to provide updates when users are required to wait
- Use AJAX, WebSockets, SignalR, and gRPC/gRPC-Web to send and receive data and transfer assets
- Write responsive desktop, web, and mobile UIs

> **Note**
>
> For clarification, when speaking about responsive UIs in this chapter, we are not talking about the layout of the UI adapting to the device size or screen real estate. Instead, we are focused on making busy UIs responsive to user input instead of blocking the user from working during task execution.

Technical requirements

- Visual Studio 2022 or later.
- This chapter source code is available at `https://github.com/PacktPublishing/High-Performance-Programming-in-CSharp-and-.NET/tree/master/CH12`.

Building a responsive UI with WinForms

In this section, we will be building a very simple WinForms application that is **Dots Per Inch** (**DPI**)-aware and enables the user to continue working during long-running operations. The application has a splash screen with a progress bar and an updated label that provides visual feedback to the user that the application is busy loading. Once the loading progress has been completed, the splash screen closes, and the main window is displayed.

On the main window, there is a label that gets updated every time you click on the increment count button, a paged table that you can navigate through using the buttons provided, and a progress indicator for a long-running task that also has a cancel button.

While the long-running task is executing, you can move the window around, increment the label by clicking the increment count button, and you can page through the data. If you choose to, you can also cancel the long-running task.

When the long-running task is completed, canceled, or encounters an error, the task progress panel is hidden.

Enabling DPI awareness and long file path awareness

In this section, we will configure a WinForms application so that it looks good on high-DPI screens and normal-DPI large screens. We also configure it to be aware of long file paths. Follow these steps:

1. Start a new .NET 6 WinForms application and call it `CH12_ResponsiveWinForms`.
2. Add a new *application manifest* file.
3. Open the `app.manifest` file and update the `compatibility` section as follows:

```
<compatibility xmlns="urn:schemas-microsoft-
    com:compatibility.v1">
    <application>
      <supportedOS
        Id="{e2011457-1546-43c5-a5fe-008deee3d3f0}" />
      <supportedOS
        Id="{35138b9a-5d96-4fbd-8e2d-a2440225f93a}" />
      <supportedOS
        Id="{4a2f28e3-53b9-4441-ba9c-d69d4a4a6e38}" />
      <supportedOS
        Id="{1f676c76-80e1-4239-95bb-83d0f6d0da78}" />
      <supportedOS
        Id="{8e0f7a12-bfb3-4fe8-b9a5-48fd50a15a9a}" />
    </application>
  </compatibility>
```

This XML code enables DPI awareness in WinForms applications from Windows Vista upward.

4. Uncomment the following `application` section:

```
<application xmlns="urn:schemas-microsoft-com:asm.v3">
    <windowsSettings>
        <dpiAware xmlns="">
            True
        </dpiAware>
        <longPathAware xmlns="">
            True
        </longPathAware>
    </windowsSettings>
</application>
```

   This code informs the compiler that the application is aware of long paths and DPI settings. With these settings in place, the application will now scale for different screen DPI settings and will be able to handle long paths that are 256 characters long.

In the next section, we will add a splash screen with loading progress feedback.

Adding a splash screen that updates with loading progress

Applications can load very quickly, or they can load quite slowly. When they are loading, the user is unaware of what the application is doing. You may choose to display a splash screen as part of your application branding. If your application loads fast, then you may need to add a delay for a short period such as 3 seconds to enable the user to see the splash screen. Otherwise, all the user may see is a quick screen flicker.

If the application has some heavy loading operations that take time to process, the user can think there is an issue and that the program has crashed. So, it is good practice to provide a splash screen that provides visual feedback to the user. This way, the user knows that the application is busy processing and has not crashed. When users see such feedback, they are more patient and will wait until the application has loaded.

In this section, we add a splash screen with visual feedback. The main window simulates several loading operations with a delay to the UI. Then, the splash screen is closed and the main window is displayed. We will now start adding the necessary code:

1. Add a new form called `SplashScreenForm`, and change its **FormBorderStyle** property to **None** and its **StartPosition** property to **CentreScreen**. Change the **BackColor** property to **ActiveCaptionText**.

2. Add a **ProgressBar** component called `LoadingProgressBar` to the form and dock it to the bottom of the form.
3. Add a label to **SplashScreenForm** called `LoadingProgressLabel` and dock it to the bottom of the form so that it appears just above the progress bar. Set the **Text** property to **Loading. Please wait...** and **Font | Size** to **12**. Change the **ForeColor** property to **HighlightWhite**. Set **Margin | All** and **Padding | All** to **8**.
4. Add another label to **SplashScreenForm** called `TitleLabel` with the **Text** property set to **Responsive WinForms Example**, **ForeColor** set to **HighlightText**, **Font | Size** set to **32**, and **Location** set to **29, 126**.
5. Rename **Form1** `MainForm` and open the form. Double-click on WindowsForm. This will open the code window.
6. Add the following `using` statements to the `MainForm` class:

```
using System;
using System.Collections.Generic;
using System.ComponentModel;
using System.Threading;
using System.Windows.Forms;
```

These `using` statements provide all that we need for our splash screen's code to function.

7. Add the following member variables to the `MainForm` class:

```
private int _clickCounter;
    private int _operationNumber;
        private int _offset = 0;
    private int _pageSize = 10;
    private int _currentPage = 1;
```

These member variables will be referenced by the various methods in our `MainForm` class to provide paging, in-memory data storage, and store the click count and operation number of the operation being processed.

8. Update the `MainForm_Load` method as follows:

```
private void MainForm_Load(object sender, EventArgs e)
{
    SplashScreenForm splashScreen = new
        SplashScreenForm();
     splashScreen.Show(this);
```

```
        for (int x = 1; x <= 100; x++)
        {
            Thread.Sleep(500);
            splashScreen.UpdateProgress(x, $"Progress
               Update: Performing load operation {x}
                   of 100...");
            Application.DoEvents();
        }
        splashScreen.Close();
    }
```

This code creates our splash screen and then iterates 100 times, simulating many loading operations. Each iteration causes the UI thread to sleep for half a second, updates the splash screen progress, and releases the thread so that other threads can do their work by calling `Application.DoEvents()`.

9. Open **SplashScreenForm** and view its code. Add the following method:

```
public void UpdateProgress(int value, string message)
{
        LoadingProgressBar.Value = value;
        LoadingProgressLabel.Text = message;
        Invalidate();
}
```

This code takes input from the `MainForm` class and updates the splash screen's label and progress bar, providing feedback to the user that the application is loading and making progress.

We have now completed the progress bar. If you run the code, you will see the following splash screen:

Figure 12.1 – The WinForms splash screen

Now that our splash screen is working, let's add our label and button that displays an incremental count of button clicks.

Adding the increment count button and label

To demonstrate non-blocking of the UI when a long operation is executing, we will have a label that is updated with text every time a user clicks a button. We will need to perform the following tasks in our code:

1. Add a label called `ClickCounterLabel` to `MainForm` and dock it to the top. Set its text to an empty string and its text properties to **Segoe UI** and **36pt**, with **TextAlign** set to **MiddleCenter**.
2. Add a button called `IncrementCountButton` to the form and dock it to the top of the form. Set its text to **&Increment Text**.
3. *Double-click* on the button to generate its click event. Update the code of the click event with the following code:

```
private void IncrementCountButton_Click(object sender,
    EventArgs e)
    {
        _clickCounter++;
        ClickCounterLabel.Text = $"You have clicked
            the button {_clickCounter} times.";
}
```

   Each time the user clicks the button, the `_clickCounter` variable is incremented by one. The `ClickCounterLabel` text is then updated, informing the user of how many times they have clicked the button.

The next thing we will be doing is adding a table with paged navigation. We shall be doing that in the following section.

Adding a table with paged data

In this section, we will be adding a table with paged navigation. This will demonstrate that the user can still interact with the page through data in a WinForms application, even when long operations are running in the background. Let's begin:

1. Add **DataGridView** to the **MainForm** design window, call it `DataTable`, and set its **Dock** property to **Fill**.

2. Add **FlowLayoutPanel** underneath **DataGridView** called `DataPagingPanel`, with its **Dock** property set to **Bottom**.
3. Add a button to **FlowLayoutPanel** called `FirstButton`, with the text set to |<<. *Double-click* the button to generate the click event. Then, return to the design window.
4. Add a button to **FlowLayoutPanel** called `PreviousButton` with the text set to <<. *Double-click* the button to generate the click event. Then, return to the design window.
5. Add a textbox called `PageTextBox` to **FlowLayoutPanel**.
6. Add a button called `NextButton` to **FlowLayoutPanel**, with the text set to >>. *Double-click* the button to generate its click event. Then, return to the design window.
7. Add a button called `LastButton` to **FlowLayoutPanel**, with the text set to >>|. *Double-click* the button to generate its click event. This time, stay in the code view, as we have completed what we need to do on the UI for this section.
8. Add the `BuildCollection` method:

```
private void BuildCollection()
{
    _products = new();
    for (int x = 1; x <= 100; x++)
    {
        _products.Add(new Product { Id = x, Name =
            $"Product {x}" });
    }
}
```

This method builds a collection of `100` products.

9. Add the call to the `BuildCollection` method to the `MainForm_Load` method before the `SplashScreenForm` instantiation line.
10. After the line that closes the splash screen, add the following two lines of code:

```
DataTable.DataSource = PagedProducts();
PageTextBox.Text = $"Page {_currentPage} of
    {PageCount()}";
```

This code sets the data source for our **DataGridView** control to a page of the dataset via the call to the `PagedProducts` method.

11. Add the `PagedProducts` method:

```
private List<Product> PagedProducts()
    {
        return _products.GetRange(_offset, _pageSize);
}
```

This method returns a range from the `_products` collection. The `_offset` variable stores the index value that forms the starting point of the returned collection, and the `_pageSize` variable stores the number of records to be returned for a page.

12. Add the `PageCount` method:

```
private int PageCount()
    {
        return _products.Count / _pageSize;
}
```

This method obtains the number of products contained within the `_products` collection, divides that number by the `_pageSize` variable, and then returns the result. The result is the number of data pages that we can navigate through.

13. Update the `FirstButton_Click` method as follows:

```
private void FirstButton_Click(object sender,
    EventArgs e)
    {
        if (_currentPage > 1)
        {
            _offset = 0;
            _currentPage = 1;
            PageTextBox.Text = $"Page {_currentPage}
                of {PageCount()}";
            DataTable.DataSource = PagedProducts();
        }
}
```

This code moves to the first page in the dataset and updates the UI accordingly.

14. Update the `PreviousButton_Click` method with the following code:

```
private void PreviousButton_Click(object sender,
    EventArgs e)
{
    if (_currentPage > 1)
    {
        _offset -= _pageSize;
        _currentPage--;
        PageTextBox.Text = $"Page {_currentPage}
            of {PageCount()}";
        DataTable.DataSource = PagedProducts();
    }
}
```

This code moves to the previous page in the dataset and updates the UI accordingly.

15. Add the `NextButton_Click` method code:

```
private void NextButton_Click(object sender,
    EventArgs e)
    {
        if (_currentPage < PageCount())
        {
            _offset += _pageSize;
            _currentPage++;
            PageTextBox.Text = $"Page {_currentPage}
                of {PageCount()}";
            DataTable.DataSource = PagedProducts();
        }
}
```

This code moves to the next page of the dataset and updates the UI accordingly.

16. Add the `LastButton_Click` method code:

```
private void LastButton_Click(object sender,
    EventArgs e)
```

```
    {
        if (_currentPage < PageCount())
        {
            _offset = _products.Count - _pageSize;
            _currentPage = PageCount();
            PageTextBox.Text = $"Page {_currentPage}
                of {PageCount()}";
            DataTable.DataSource = PagedProducts();
        }
    }
```

This method moves to the last page of the dataset and updates the UI accordingly.

17. Finally, add the `Product` class:

```
internal class Product
{
    public int Id { get; set; }
    public string Name { get; set; }
    public string Description { get; set; } = "It is a
      long established fact that a reader will be
        distracted by the readable content of a page
          when looking at its layout.";
    public float Price { get; set; } = 9.99F;
    public int Units { get; set; } = 100;
}
```

This class is the `Product` class that our `MainForm` uses to build its list of products within its `BuildCollection` method.

We have now built our paged data table, and we have our increment button and label in place. The final thing to do with our form is to add our long-running task, to show that user interactions are still possible without being blocked by long-running tasks. This will be the topic of our next section.

Running long-running tasks in the background

In this section, we are going to upgrade our UI to show the progress of a long-running task that is running in the background. The user will be able to cancel the long-running task at any time. When the task is completed, whatever state it is in, the long-running task update progress controls will be hidden from the user. Let's start adding the code:

1. Add a **CommandButton** component called `LongRunningOperationCancelButton` and set its text to `&Cancel long running operation`.
2. Add a **StatusStrip** component and call it `StatusBar`.
3. Add a **ToolStripProgressBar** component called `TaskProgressBar`.
4. Add a **ToolStripLabel** component called `StatusLabel` and make sure its text property is empty.
5. Add a **BackgroundWorker** component called `CollectionBuilderBackgroundWorker`.
6. Add a **BackgroundWorker** component called `LongRunningProcessBackgroundWorker`.
7. In the `MainForm` class, add the following three lines to the constructor:

```
LongRunningProcessBackgroundWorker.DoWork +=
    LongRunningProcessBackgroundWorker_DoWork;
LongRunningProcessBackgroundWorker.ProgressChanged +=
    LongRunningProcessBackgroundWorker_ProgressChanged;
LongRunningProcessBackgroundWorker
    .RunWorkerCompleted += LongRunning
        ProcessBackgroundWorker_RunWorkerCompleted;
```

   This code adds the handlers for our `BackgroundWorker`, which will be responsible for executing the long-running task.

8. Add the following method call to the last line of the `MainForm_Load` method before the closing brace: `LongRunningProcess();`.
9. Add the following `LongRunningProcess` method:

```
private void LongRunningProcess()
    {
        if (LongRunningProcessBackgroundWorker.IsBusy
```

```
                != true)
            {
                LongRunningProcessBackgroundWorker
                    .RunWorkerAsync();
            }
        }
```

If `LongRunningProcessBackgroundWorker` is not busy, then the `RunWorkerAsync` method called `LongRunningProcessBackground Worker_DoWork` is run.

10. Add `LongRunningProcessBackgroundWorker_DoWork` to the `MainForm` class:

```
private void LongRunningProcessBackgroundWorker_DoWork
    (object sender, DoWorkEventArgs e)
    {
        BackgroundWorker worker = sender as
            BackgroundWorker;

    for (int i = 1; i <= 100; i++)
        {
            if (worker.CancellationPending == true)
            {
                e.Cancel = true;
                break;
            }
            else
            {
                _operationNumber = i;
                System.Threading.Thread.Sleep(100);
                worker.ReportProgress((i / 100)
                    * 100);
           }
       }
}
```

We are casting the sender as `BackgroundWorker` and assigning it to our local worker variable. Then, we iterate 100 times. Each time we iterate, we set the `_operationNumber` variable to the loop count variable value, sleep for 100 milliseconds, and then call the `ReportProgress` method of the worker passing in the percentage of work done.

11. Add the `LongRunningProcessBackgroundWorker_ProgressChanged` method to the `MainForm` class:

```
private void LongRunningProcessBackgroundWorker
    _ProgressChanged(object sender, ProgressChanged
        EventArgs e)
    {
        StatusLabel.Text
        = ($"Progress: {_operationNumber}%");
    TaskProgressBar.Value = _operationNumber;
        if (_operationNumber == 100)
    {
            Thread.Sleep(100);
            LongRunningOperationCancelButton
            .Visible = false;
        StatusBar.Visible = false;
    }
}
```

This code updates the UI with the progress of the long-running task. If all the operations have been completed, the task cancel button and status bar are hidden from the user.

12. Add the `LongRunningProcessBackgroundWorker_RunWorkerCompleted` method to the `MainForm` class:

```
private void LongRunningProcessBackgroundWorker
    _RunWorkerCompleted(object sender,
        RunWorkerCompletedEventArgs e)
```

```
    {
        if (e.Cancelled == true)
            StatusLabel.Text = "Canceled!";
        else if (e.Error != null)
            StatusLabel.Text = "Error: " +
                e.Error.Message;
        else
            StatusLabel.Text = "Done!";
    }
```

When the long-running task is completed, this method executes `StatusLabel.Text` to the outcome of the method, with the outcomes being either `Cancelled`, `Error`, or `Done`.

13. Our final piece of code to write before we complete and run our WinForms application is to add code to the `LongRunningOperationButton_Click` method to `MainClass`, as follows:

```
private void LongRunningOperationCancelButton
    _Click(object sender, EventArgs e)
{
    if (LongRunningProcessBackgroundWorker
        .WorkerSupportsCancellation == true)
    {
        LongRunningProcessBackgroundWorker
            .CancelAsync();
        LongRunningOperationCancelButton.Visible =
            false;
        StatusBar.Visible = false;
    }
}
```

This code checks to see whether the task supports cancellation. If it does, then the task is canceled, and the cancel button and status bar are hidden from the user.

14. Run the code. You should see the splash screen shown in *Figure 12.1*. Then, you should see the main window resembling what is shown in *Figure 12.2*. Move the window about and click on the increment count button. Also, click the paging buttons to move between data pages of the dataset, and cancel the task. You should see that the window is completely responsive to your input, as follows:

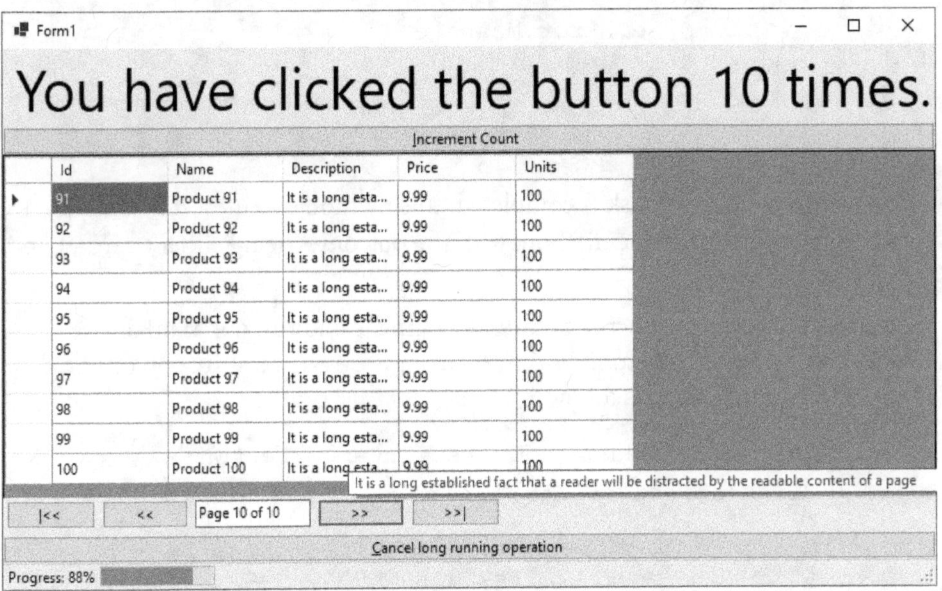

Figure 12.2 – The Windows Forms main application window

As you can see, we have written a WinForms application that has a lot going on. We have a splash screen that provides visual feedback to the user so that they do not think that the application has crashed in any way, and we have a UI that remains responsive to user input during a long-running task.

Now that we have a working WinForms application, let's turn our attention to WPF. In the next section, we will apply what we have learned with our WinForms application to a WPF application.

Building a responsive UI with WPF

In this section, we are going to build the same kind of interface as we did for the WinForms application, but this time, it will be using WPF. We will now start writing our code:

1. Create a new WPF application called `CH12_ResponsiveWPF` and make sure to select **.NET 6.0** as the target framework.
2. Add the `Product` class to the project. It is the same code that we used in our WinForms application.
3. Add a new Window called `SplashWindow`.
4. Modify the **SplashWindow** XAML as follows:

```
<Window x:Class="CH12_ResponsiveWPF.SplashWindow"
       xmlns=""
       xmlns:x=""
       xmlns:d=""
       xmlns:mc=""
       xmlns:local="clr-namespace:CH12_ResponsiveWPF"
       mc:Ignorable="d"
       Background="White"
       Foreground="White"
       WindowStyle="None"
       WindowStartupLocation="CenterScreen"
       Title="SplashWindow" Height="450" Width="800">
    <StackPanel HorizontalAlignment="Center"
     VerticalAlignment="Center">
     <Label TextBlock.FontSize="32"
         Content="Responsive WPF Example" />
     <Label x:Name="LoadingProgressLabel"
         TextBlock.FontSize="12"
             Content="Loading..." />
     <ProgressBar x:Name="LoadingProgressBar"
         Minimum="0" Maximum="100" />
    </StackPanel>
</Window>
```

The XAML we have just updated declares a stack panel with two labels and a progress bar. The first label displays the title, and the second label displays loading progress along with the progress bar.

5. Add the following method to the `SplashWindow` class:

```
public void UpdateProgress(int value, string message)
{
    LoadingProgressBar.Value = value;
    LoadingProgressLabel.Content = message;
    InvalidateVisual();
}
```

This code will be called by the `MainWindow` class and is responsible for updating the progress indicators on **SplashWindow**.

6. Open the `MainWindow.xaml` file and replace the existing XAML with the following:

```
<StackPanel HorizontalAlignment="Stretch"
    VerticalAlignment="Stretch" Background="Red">
    <Label x:Name="CounterLabel" FontSize="32"
        Foreground="Yellow" Margin="8" Padding="8" />
        <Button x:Name="IncrementCounterButton"
        Content="Increment Counter"
        Click="IncrementCounterButton_Click"
        HorizontalAlignment="Center" Padding="8"
        Margin="0, 0, 0 , 8" />
        <DataGrid x:Name="DataTable" />
        <StackPanel Orientation="Horizontal"
            HorizontalAlignment="Center" Margin="0, 4,
                0, 4">
            <Button x:Name="FirstButton"
            Content="|&lt;&lt;"
          Click="FirstButton_Click" Margin="4"
          Padding="8" />
            <Button x:Name="PreviousButton"
            Content="&lt;&lt;"
            Click="PreviousButton_Click" Margin="4"
            Padding="8" />
```

```
            <Label x:Name="PageLabel"
            Background="White" Foreground="Black"
            Width="110" Height="32"
            VerticalContentAlignment="Center" />
            <Button x:Name="NextButton"
            Content="&gt;&gt;"
            Click="NextButton_Click" Margin="4"
            Padding="8" />
            <Button x:Name="LastButton"
            Content="&gt;&gt;|"
            Click="LastButton_Click" Margin="4"
            Padding="8" />
        </StackPanel>

        <StackPanel x:Name="StatusPanel"
        VerticalAlignment="Bottom"
        Orientation="Horizontal" Background="Yellow">
            <Label x:Name="StatusLabel"
            Content="Progress Update: ..." />
            <ProgressBar x:Name="TaskProgressBar"
            Minimum="0" Maximum="100" Width="500" />
            <Button x:Name="CancelTaskButton"
            Content="Cancel Task"
            Click="CancelTaskButton_Click" />
        </StackPanel>
    </StackPanel>
```

This XAML provides a status panel that will show the progress of any background tasks, an increment label and an increment button, a data grid, and a navigation panel for paging through different pages of data.

7. Add the following `using` statements to the `MainWindow.xaml.cs` file:

```
using System;
using System.Collections;
using System.Collections.Generic;
using System.ComponentModel;
using System.Threading;
```

```
using System.Windows;
using System.Windows.Threading;
```

These `using` statements are needed for our WPF window to function without error.

8. Add the following member variables to the `MainWindow` class:

```
private int _clickCounter;
private int _operationNumber;
private List<Product> _products;
private int _offset = 0;
private int _pageSize = 10;
private int _currentPage = 1;
BackgroundWorker _worker;
```

Here, we have the same variables that we had with our WinForms application, except we also declare a background worker.

9. Update the `MainWindow` constructor with the following code:

```
public MainWindow()
{
    InitializeComponent();
    BuildCollection();
    SplashWindow splashWindow = new SplashWindow();
    splashWindow.Show();
    for (int x = 1; x <= 100; x++)
    {
        Thread.Sleep(100);
        splashWindow.UpdateProgress(x, $"Progress
            Update: Performing load operation {x} of
                100...");
        DoEvents();
}
    splashWindow.Close();
    DataTable.ItemsSource = PagedData();
    PageLabel.Content = $"Page {_currentPage} of
        {PageCount()}";
```

```
     _worker = new BackgroundWorker();
    _worker.WorkerReportsProgress = true;
    _worker.WorkerSupportsCancellation = true;
    _worker.DoWork += Worker_DoWork;
    _worker.ProgressChanged += Worker_ProgressChanged;
    _worker.RunWorkerCompleted +=
        Worker_RunWorkerCompleted;
    _worker.RunWorkerAsync();
}
```

This code is pretty much the same as our WinForms load method. The only real difference is that all our initialization code is in the constructor.

10. Add the `Worker_DoWork` method:

```
private void Worker_DoWork(object sender,
    DoWorkEventArgs e)
{
    BackgroundWorker worker = sender as
        BackgroundWorker;
for (int i = 1; i <= 100; i++)
    {
        if (worker.CancellationPending == true)
        {
            e.Cancel = true;
            break;
    }
        else
        {
            _operationNumber = i;
            System.Threading.Thread.Sleep(100);
             worker.ReportProgress((i / 100) * 100);
        }
    }
}
```

This code simulates the work of 100 operations with a small delay for each operation.

11. Add the `Worker_ProgressChanged` method code:

```
private void Worker_ProgressChanged(object sender,
    ProgressChangedEventArgs e)
{
    StatusLabel.Content = ($"Progress:
        {_operationNumber}%");
    TaskProgressBar.Value = _operationNumber;
}
```

This code updates the progress indicators for the long-running task.

12. Add the `Worker_RunWorkerCompleted` method:

```
private void Worker_RunWorkerCompleted(object sender,
    RunWorkerCompletedEventArgs e)
{
    if (e.Cancelled == true)
        StatusLabel.Content = "Cancelled!";
    else if (e.Error != null)
        StatusLabel.Content = "Error: " + e.Error.
          Message;
    else
        StatusLabel.Content = "Done!";
Thread.Sleep(1500);
    StatusPanel.Visibility = Visibility.Collapsed;
}
```

This method reports the result of the long-running task and then hides the status panel from the end user.

13. Add the `PagedData` method:

```
private IEnumerable PagedData()
{
    return _products.GetRange(_offset, _pageSize);
}
```

This method returns a page of data whose index starts at `_offset`, with the number of returned rows defined by `_pageSize`.

14. Add the `DoEvents` method:

```
public static void DoEvents()
{
    Application.Current.Dispatcher
        .Invoke(DispatcherPriority.Render,
            new Action(delegate {
                // Your operation goes here.
            }));
}
```

This code performs like the WinForms `Application.DoEvents()` code. You can place your non-UI blocking code here, and update the UI.

15. Add the `BuildCollection` method:

```
private void BuildCollection()
{
    _products = new();
    for (int x = 1; x <= 100; x++)
    {
        _products.Add(new Product { Id = x, Name =
            $"Product {x}" });
    }
}
```

The `BuildCollection` method builds our dataset of 100 products.

16. Add the `PageCount` method:

```
private int PageCount()
{
    return _products.Count / _pageSize;
}
```

The `PageCount` method works out how many pages of data there are based on the dataset size, divided by the page size, and then returns the result.

17. Add the `FirstButton_Click` method:

```
private void FirstButton_Click(object sender,
    RoutedEventArgs e)
```

```
{
    if (_currentPage > 1)
    {
        _offset = 0;
        _currentPage = 1;
        PageLabel.Content = $"Page {_currentPage} of
            {PageCount()}";
        DataTable.ItemsSource = PagedData();
    }
}
```

When executed, this method navigates to the first record in our dataset and upgrades the UI accordingly.

18. Add the `PreviousButton_Click` method:

```
private void PreviousButton_Click(object sender,
    RoutedEventArgs e)
{
    if (_currentPage > 1)
    {
        _offset -= _pageSize;
        _currentPage--;
        PageLabel.Content = $"Page {_currentPage} of
            {PageCount()}";
        DataTable.ItemsSource = PagedData();
}
}
```

This method will move to the previous page of the dataset and update the UI accordingly.

19. Add the `NextButton_Click` code:

```
private void NextButton_Click(object sender,
    RoutedEventArgs e)
{
    if (_currentPage < PageCount())
    {
```

```
            _offset += _pageSize;
            _currentPage++;
            PageLabel.Content = $"Page {_currentPage} of
                {PageCount()}";
            DataTable.ItemsSource = PagedData();
        }
    }
```

This method moves to the next page of the dataset and updates the UI accordingly.

20. Add the `LastButton_Click` method:

```
private void LastButton_Click(object sender,
    RoutedEventArgs e)
{
    if (_currentPage < PageCount())
    {
        _offset = _products.Count - _pageSize;
        _currentPage = PageCount();
        PageLabel.Content = $"Page {_currentPage} of
            {PageCount()}";
        DataTable.ItemsSource = PagedData();
    }
}
```

This method moves to the last dataset page and updates the UI accordingly.

21. Add the `IncrementCounterButton_Click` method:

```
private void IncrementCounterButton_Click(object
    sender, RoutedEventArgs e)
{
    _clickCounter++;
    CounterLabel.Content = $"You have clicked the
        button {_clickCounter} times.";
}
```

Each time you click `IncrementCounterButton`, this method will increment the `_clickCounter` variable and report on the screen how many times you have clicked the button.

22. Add the final WPF method called `CancelTaskButton_Click`:

```
private void CancelTaskButton_Click(object sender,
    RoutedEventArgs e)
{
    if (_worker.WorkerSupportsCancellation == true)
        _worker.CancelAsync();
}
```

This method cancels the long-running task if it supports cancellation.

23. Run the WPF application. You will find that you are presented with the splash screen showing the loading progress, as displayed here:

Responsive WPF Example

Progress Update: Performing load operation 96 of 100...

Figure 12.3 – The WPF application's splash screen

When the loading completes, the splash screen closes and you are presented with the main window. While a long-running task is in progress, you can move the window about, click on the increment counter button, navigate through the paged data, and cancel the long-running task.

As you can see from the following screenshot, we have everything in place that provides visual feedback of progress to end users and a UI that remains responsive to user input during a long-running task:

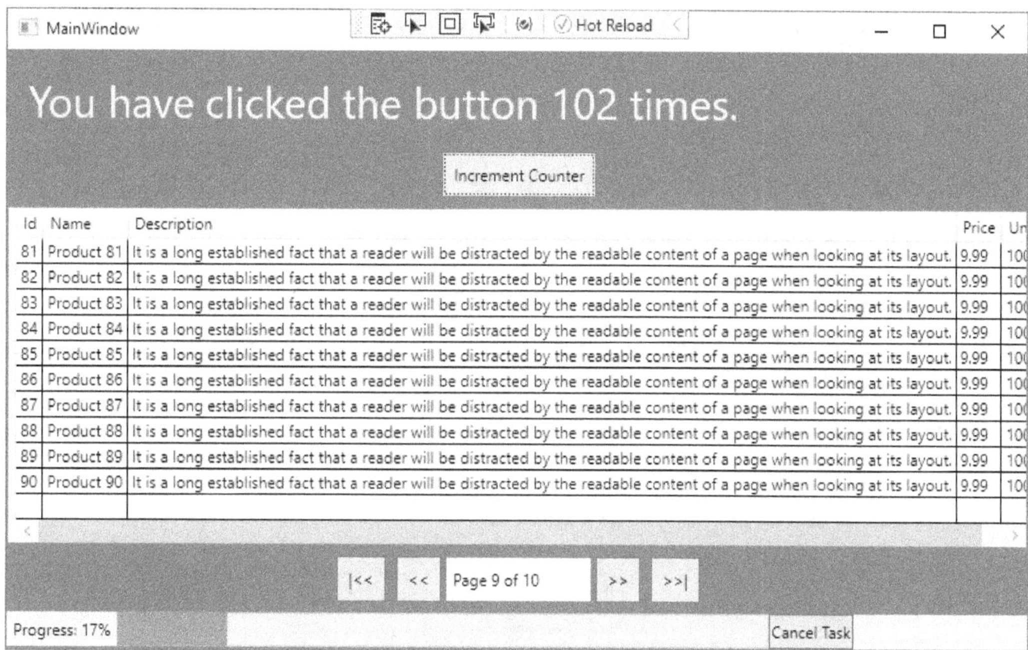

Figure 12.4 – The WPF application's main window

In the next section, we will look at how to keep ASP.NET UI responsive to user input.

Building a responsive UI with ASP.NET

In this section, we will be looking at ways to assist ASP.NET applications in being quick and responsive. We will start by looking at memory and distributed caching. Then, we will look at how you can update a section of a page using AJAX. Next, we will move on to write a real-time chat application with SignalR. We will then take a look at using WebSockets in our ASP.NET applications.

> **Note**
>
> We will not be covering gRPC-Web in this chapter, as we have already covered that topic with example code in *Chapter 9*, *Enhancing the Performance of Networked Applications*, in which we looked at gRPC for non-web applications and gRPC-Web for web applications. In this chapter, we also implemented a simple Blazor web application using gRPC-Web, so you can refer to this chapter for gRPC/gRPC-Web.

Let's begin looking at a responsive ASP.NET application by focusing on caching. There are two kinds of caching we will be looking at. These are **memory caching** and **distributed caching**. In the next section, we will implement memory caching.

Implementing memory caching

Web applications load resources over the network we all know as the internet. Accessing, downloading, and rendering resources from the internet takes varying degrees of time. Time can vary due to network traffic, the quality of the network, and computer system resources. Is there a way in which we can speed this process up? Well, yes. We can implement caching. But what exactly is caching?

Caching is the local storage of frequently accessed resources for faster access and processing.

In this section, you will see how we can easily implement in-memory caching in ASP.NET. To implement in-memory caching, follow these steps:

1. Start a new ASP.NET Core web app (`Model-View-Controller`) project and call it `CH12_ResponsiveASPNET`.
2. Add the `Microsoft.Extensions.Caching.Memory` NuGet package. If Visual Studio cannot install it, run the following command in the Package Manager:

```
Install-Package Microsoft.Extensions.Caching.Memory -
    Version 6.0.0-preview.7.21377.19
```

3. In the `HomeController` class, add the statement `using Microsoft.Extensions.Caching.Memory`.
4. Add the following member variables:

```
private readonly ILogger<HomeController> _logger;
private IMemoryCache _memoryCache;
```

This code declares the variables that will store our logger and memory cache objects.

5. Update the `HomeController` constructor, as shown next:

```
public HomeController(ILogger<HomeController> logger,
    IMemoryCache memoryCache)
{
  _logger = logger;
```

```
        _memoryCache = memoryCache;
    }
```

In this code, the logger and memory cache objects that we will be using are injected into our class, and we pass in variables to set our member variables.

6. Add the `GetMemoryCacheTime` method:

```
    private DateTime GetMemoryCacheTime()
    {
        DateTime currentTime;
        bool alreadyExists = _memoryCache.TryGetValue
            ("CachedTime", out currentTime);
        if (!alreadyExists)
        {
        currentTime = DateTime.UtcNow.ToLocalTime();
        _memoryCache.Set(
        "CachedTime",
        currentTime,  MemoryCacheEntryExtensions
            .SetSlidingExpiration(
            new MemoryCacheEntryOptions() {
                SlidingExpiration
                    = TimeSpan.FromMinutes(5) },
                TimeSpan.FromMinutes(5)
         ));
        }
        return currentTime;
    }
```

Here, we are checking whether our `CachedTime` variable exists in the memory cache. If it does exist, then the out variable called `currentTime` is set and the cached time is returned. Otherwise, we get the current time and store it in the memory cache with a sliding expiration value, and then we return the cached time.

7. Update the `Index` method with this code:

```
    [HttpGet]
    public string Index()
```

```
{
DateTime memoryCacheTime = GetMemoryCacheTime();
return $"Current Time: {DateTime.UtcNow.ToLocalTime()}
    \nMemory Cache Time: {memoryCacheTime}";
}
```

The `Index` controller method returns a string. This string that is returned is the cached time.

8. Run the project and navigate to `https://localhost:5001/Home`. You should see something like the following output:

   Current Time: 12/07/221 20:18:25

   Memory Cache Time: 12/07/2021 20:18:25

   As you can see, the time did not exist in the cache, and so was added to the cache before it was returned.

> **Note**
>
> The setting of port numbers is dependent on the availability of ports. Whatever port you choose, it will not work if it is in use by another program.

9. Now, refresh the page, and you should see different values for the current time and the memory cache time:

   Current Time: 12/07/2021 20:21:21

   Memory Cache Time: 12/07/2021 20:18:25

   You can clearly see that the memory cache time is older than the current time. This shows that we have stored the time in the in-memory cache and retrieved it successfully.

Implementing in-memory caching is really easy in ASP.NET, and you can enhance the page load and render time by storing and retrieving data from the in-memory cache. Now that we have looked at the in-memory cache, we will turn our attention to distributed cache.

Implementing distributed caching

In this section, we will be using the same ASP.NET web project and controller to implement distributed caching. What do we mean by distributed caching? Distributed caching extends the concept of local caching to include caching over several computers. Such caching enables the scaling of transactional data. You would mainly use distributed caching to store application data that resides in a database, and data related to web sessions. In this section, we use Redis for our caching. Redis is an in-memory data structure store, used as a distributed, in-memory key-value database, cache, and message broker, with optional durability. To implement distributed caching, perform the following:

1. Add the `Microsoft.Extensions.Caching.Redis` NuGet package to the web package. You can use the following command:

```
Install-Package Microsoft.Extensions.Caching.Redis -
    Version 2.2.0
```

2. In the `HomeController` class, add the `using Microsoft.Extensions.Caching.Distributed` statement.
3. Add the following member variable:

```
private IDistributedCache _distributedCache;
```

   This variable will hold our distributed cache object that gets injected via the constructor.

4. Now, update the constructor code:

```
public HomeController(ILogger<HomeController> logger,
    IMemoryCache memoryCache, IDistributedCache
        distributedCache)
{
    _logger = logger;
    _memoryCache = memoryCache;
    _distributedCache = distributedCache;
}
```

   We are injecting the distributed cache object and setting our member variable.

5. To use our distributed cache, we will need to encode and decode Base64 strings. Add the following two methods:

```
private static string Base64Encode(string text)
{
    byte[] bytes = Encoding.UTF8.GetBytes(text);
    return Convert.ToBase64String(bytes);
}
public static string Base64Decode(string text)
{
    byte[] bytes = Convert.FromBase64String(text);
    return Encoding.UTF8.GetString(bytes);
}
```

In these two methods, we are encoding a string as a Base64 encoded string, and we are also decoding strings from Base64 to UTF8.

6. Add the `GetDistriutedCacheString` method:

```
private string GetDistributedCacheString()
{
  string data = _distributedCache.GetString
      ("StringValue");
  if (data == null)
  {
     data = Base64Encode($"Hello, World!
          {DateTime.UtcNow.ToLocalTime()}");
     _distributedCache.Set("StringValue",
     Convert.FromBase64String(data),
    new DistributedCacheEntryOptions()
    {
            AbsoluteExpiration
            = DateTime.UtcNow.AddMinutes(10),
    });
    data = Base64Decode(data);
  }
return data;
}
```

In this code, we obtain string data from the cache. If it exists, then we return it. If it does not exist, then we save the Base64 encoded version of the string to the cache with an absolute expiry being set and then return the Base64 decoded version of the string as a UTF encoded string.

7. Update the `HomeController.Index` method, as shown here:

```
[HttpGet]
public string Index()
{
DateTime memoryCacheTime = GetMemoryCacheTime();
string data = GetDistributedCacheString();
return $"Current Time: {DateTime.UtcNow.ToLocalTime()}
    \nMemory Cache Time: {memoryCacheTime}
        \nDistributed Cache String: {data}";
}
```

This code obtains in-memory cache and distributed cache stored data and outputs it to the user, showing the current time, the in-memory cached time, and the data stored in the distributed cache.

8. Run the program and navigate to `https://localhost:5001`. You should see the following output:

```
Current Time: 12/07/2021 21:05:59
Memory Cache Time: 12/07/2021 21:05:59
Distributed Cache String: Hello, World! 12/07/2021
21:05:59
```

We can see that the memory cached time and distributed cache string have both just been added to the cache because they are the same as the current time. Now, refresh your browser. You should see that both cached values are older than the current time as shown:

```
Current Time: 12/07/2021 21:08:13
Memory Cache Time: 12/07/2021 21:05:59
Distributed Cache String: Hello, World! 12/07/2021
21:05:59
It is plain to see that both cached values already
existed in the cache, since they are older than the
current time.
```

In this and the previous section, you have seen how easy it is to add in-memory and distributed caching to our application. Both forms of caching can be really useful in improving the performance of your ASP.NET web applications. In the next section, we will look at how to update a small section of the currently displayed page using AJAX.

Using AJAX to update part of the currently displayed page

In this section, we will use AJAX to update a part of a page that is currently being displayed. This saves us from having to load the whole page. Let's start writing our AJAX example:

1. Right-click on the `Controllers` folder. From the context menu, select **Add | Controller…**. Then, select **MVC Controller – Empty**.
2. Call the new controller `AjaxController` and open the class.
3. Update the controller by adding the following method:

```
[Route("Ajax/Demo")]
public IActionResult AjaxDemo()
{
    return new JsonResult("Ajax Demo Result");
}
```

   This method when called will return a JSON result, which in our case is a simple string.

4. *Right-click* on the `Index` method and select **Add View**. This will create a view for the Ajax controller called `index.cshtml`.
5. Update the `Views/Ajax/index.cshtml` file with the following HTML and JavaScript code:

```
<!DOCTYPE html>
<html>
    <head>
        <meta name="viewport" content="width=device-
            width" />
        <title>Ajax Example</title>
    </head>
    <body>
        <fieldset>
```

```
            <legend>Ajax Demonstration</legend>
            <form>
                <input type="button" value="Ajax
                Demonstration" id="ajaxDemonstration
                Button" />
                <br />
                <span id="ajaxDemoResult"></span>
            </form>
        </fieldset>

        <script
           src="https://code.jquery.com/jquery-
               3.6.0.slim.min.js"
            integrity="sha256-u7e5khyithlIdTpu22P
                HhENmPcRdFiHRjhAuHcs05RI="
            crossorigin="anonymous"
        >

        </script>
        <script>
            $(document).ready(function( ) {
                $('#ajaxDemonstrationButton')
                    .click(function() {
                    $.ajax({
                        type: 'GET',
                        url: '/Ajax/Demo',
                        success: function (result) {
                          $('#ajaxDemoResult')
                              .html(result);
                        }
                    });
                });
            });
        </script>
    </body>
</html>
```

We have an HTML form. That form has a button that, when pressed, will execute JavaScript that will retrieve AJAX data by executing our `AjaxDemo` action method. This will result in our JSON string being displayed on the page.

6. Run the project and navigate to `http://localhost:5001/Ajax`. You should see the following:

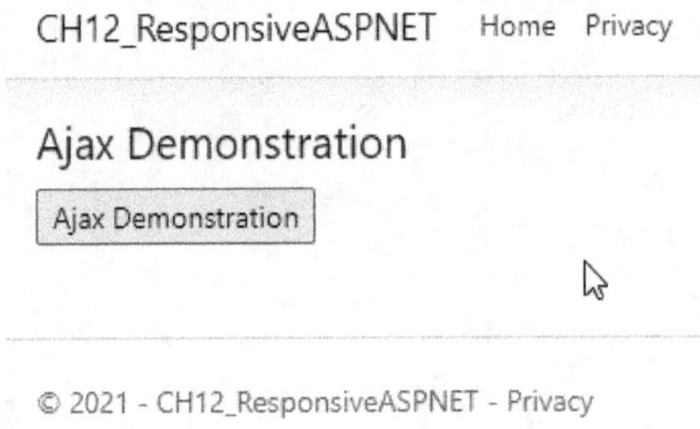

Figure 12.5 – The AJAX demo before AJAX is retrieved

As you can see, our page is loaded without our JSON string. Now, click the **Ajax Demonstration** button. You now see the following:

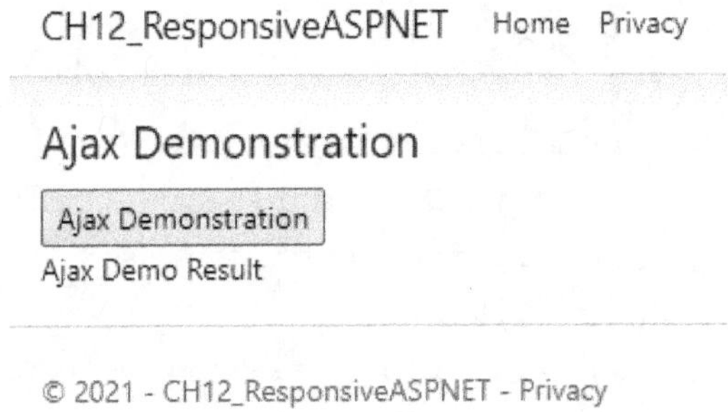

Figure 12.6 – The AJAX demo displaying the JSON string retrieved using AJAX

After clicking the button, we can see that the AJAX action retrieved our JSON string and displayed it on the page without a complete page load.

We have seen how to update a portion of a page using AJAX, and before that, we saw how to implement in-memory and distributed caching. In the next section, we will look at how to implement WebSockets.

Implementing WebSockets

In this section, we will be implementing **WebSockets**. You may have heard of WebSockets, but what are they? A WebSocket is a full-duplex communication protocol for communication over a single TCP connection. To find out more about the WebSocket specification, you can look up the IETF RFC 6455 from 2011 (`https://www.rfc-editor.org/rfc/rfc6455.txt`).

What do we use WebSockets for? Well, we can use them to open a single two-way interactive session between browsers and servers. That way, we can negate server polling, send messages to a server, and receive responses via events. Thus, making our applications event-driven.

In our WebSockets demonstration, we will click a button. It will open a WebSocket, send a message, receive a response, and then close the connection. The communication between our browser and the server will be output to our web page. So, let's get started with writing our WebSocket example:

1. Add a new controller called `WebSocketsController`.
2. Right-click the `Index` method and select **Add View**.
3. Update the `Views/WebSockets/Index.cshtml` file as follows:

```
<script type = "text/javascript">
    function WebSocketExample (){
        var socket = new WebSocket("wss://
            javascript.info/article/websocket/
                demo/hello");
        var messages = document.getElementById
            ('messages')
        var innerHTML = messages.innerHTML;
        socket.onopen = function(e) {
            innerHTML += '<p>[open] Connection
                established</p>';
            messages.innerHTML += innerHTML;
            innerHTML += '<p>Sending to server</p>';
            messages.innerHTML += innerHTML;
            socket.send('WebSocket message!');
        };
```

```
        socket.onmessage = function(event) {
            innerHTML += `<p>[message] Data received
                from server: ${event.data}</p>`;
        };
        socket.onclose = function(event) {
            if (event.wasClean) {
                innerHTML += `<p>[close] Connection
                closed cleanly, code=${event.code}
                reason=${event.reason}</p>`;
            messages.innerHTML = innerHTML;
            } else {
         // e.g. server process killed or network down
         // event.code is usually 1006 in this case
        innerHTML += '<p>[close] Connection died</p>';
                messages.innerHTML = innerHTML;
            }
        };
        socket.onerror = function(error) {
            innerHTML += `<p>[error]
                ${error.message}</p>`;
            messages.innerHTML = innerHTML;
        };
    }
</script>
<p>Click the following button to see the function in
    action</p>
<input type = "button" onclick = "WebSocketExample()"
    value = "Display">
<p id="messages" onload="WebSocketExample()"></p>
```

When a WebSocket is opened via the button click, the `messages` paragraph is updated with messages, and then a message is sent to the server. When the server responds, the messages paragraph is then updated to inform the user that the server has responded. If an error occurs, then a message is displayed to the user. The WebSocket is then closed and a message is displayed on the page.

4. Run the code and navigate to `http://localhost:5001/WebSockets`. Click on the button, and you should end up with the following:

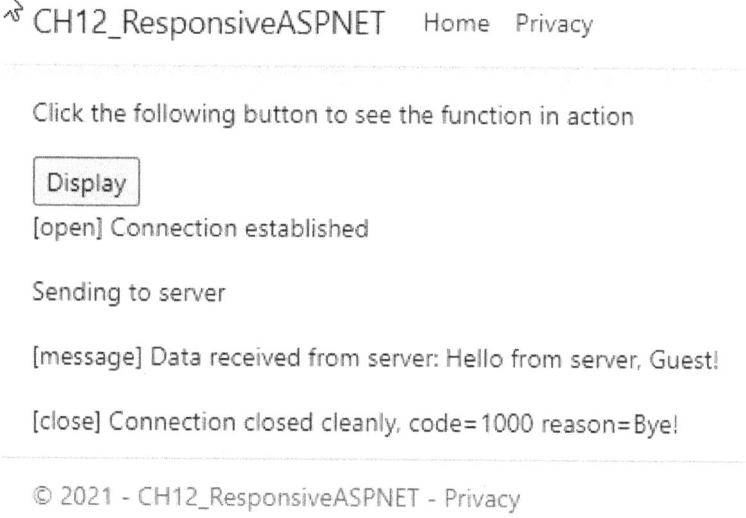

Figure 12.7 – The end result of clicking on the button and executing our WebSocket example

There is not that much code to WebSockets. In this example, we have sent a simple message and received a response. All our code to do this exists in the `CSHTML` file of our view. In the next section, we will look at writing a real-time chat program using SignalR.

Implementing a real-time chat application using SignalR

In this section, we will learn how to write real-time functionality in an ASP.NET web application using SignalR. We will demonstrate SignalR in action by writing a simple chat application. We will now begin writing the application:

1. *Right-click* the project and select **Add** | **Client-Side Library** from the context menu, and fill in the details as shown in *Figure 12.8*. Then, click the **Install** button:

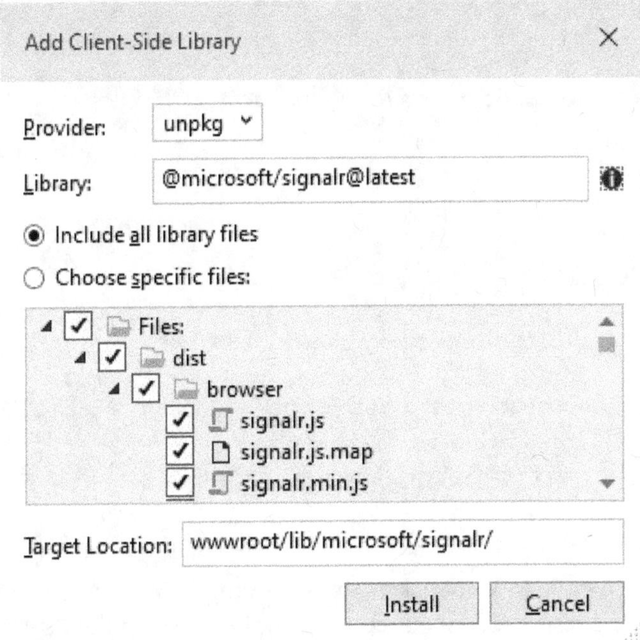

Figure 12.8 – The Add Client-Side Library configured to install SignalR

2. Copy the `wwwroot/lib/microsoft/signalr` library and paste it into the `wwwroot/js` folder.
3. Add a new controller called `SignalRController`.
4. Add a folder called `Hubs` under the main project root.
5. Add a class to the `Hubs` folder called `ChatHub`. Then, update the `ChatHub` class, as shown here:

```
public class ChatHub : Hub
{
    public async Task SendMessage(
```

```
            string user, string message
        )
        {
            await Clients.All
            .SendAsync(
                "ReceiveMessage", user, message
            );
        }
    }
```

We have our SignalR hub class in place, and our `SendMessage` method sends a message to the specified user asynchronously.

6. *Right-click* on the `Index` method in the `SignalRController` class and select **Add View** from the context menu.
7. In the `Views/SignalR/Index.cshtml` file, replace the existing contents with the following code:

```
@page
<div class="container">
<div class="row"> </div>
    <div class="row">
        <div class="col-2">User</div>
          <div class="col-4">
            <input type="text"
                id="userInput" />
        </div>
    </div>
    <div class="row">
        <div class="col-2">Message</div>
          <div class="col-4">
            <input type="text"
                id="messageInput" />
        </div>
    </div>
    <div class="row"> </div>
    <div class="row">
        <div class="col-6">
```

```
                <input type="button"
               id="sendButton" value="Send Message" />
            </div>
        </div>
    </div>
    <div class="row">
    <div class="col-12">
               <hr />
        </div>
    </div>
    <div class="row">
    <div class="col-6">
               <ul id="messagesList"></ul>
        </div>
    </div>
    <script src="~/js/signalr/dist/browser/signalr.js">
       </script>
    <script src="~/js/chat.js"></script>
```

We have put together a chat UI. The script uses SignalR. All we need to do now is add our JavaScript that makes our UI interactive.

8. In the `wwwroot/js` folder, add a file called `chat.js` with the following code:

```
"use strict";
var connection = new signalR.HubConnectionBuilder()
    .withUrl("/chatHub").build();
document.getElementById("sendButton").disabled = true;
connection.on("ReceiveMessage", function (user,
    message) {
        var li = document.createElement("li");
     document.getElementById("messagesList")
         .appendChild(li);
     li.textContent = `${user} says ${message}`;
});
connection.start().then(function () {
        document.getElementById("sendButton")
         .disabled = false;
```

```
}).catch(function (err) {
    return console.error(err.toString());
});
document.getElementById("sendButton")
    .addEventListener("click", function (event) {
        var user = document
            .getElementById("userInput").value;
            var message = document
            .getElementById("messageInput").value;
        connection.invoke(
            "SendMessage", user, message
        ).catch(function (err) {
                return console.error(err.toString());
            });
            event.preventDefault();
    });
```

We have added JavaScript that makes our UI interactive. This code manages the sending of chat messages between users.

9. In the `Program` class, add the following services:

```
services.AddRazorPages();
services.AddSignalR();
```

This code adds SignalR to our available services so that we can pass SignalR requests to SignalR.

> **Note**
>
> If using the new minimal template, the code is `builder.Services.AddRazorPages(); builder.Services.AddSignalR();`.

10. Update the `Program` class to include the mapped route to our `ChatHub`:

```
app.MapHub<ChatHub>("/chatHub");
```

We have included the route to our `ChatHub` so that our chat application knows how to handle incoming requests.

11. Run the code and navigate to `https://localhost:5001/SignalR`. You will need two browser instances side by side. Enter usernames and messages in each browser and click on the **Send Message** button. Each time you enter text, it will appear on the receiver's chat page, as you can see here:

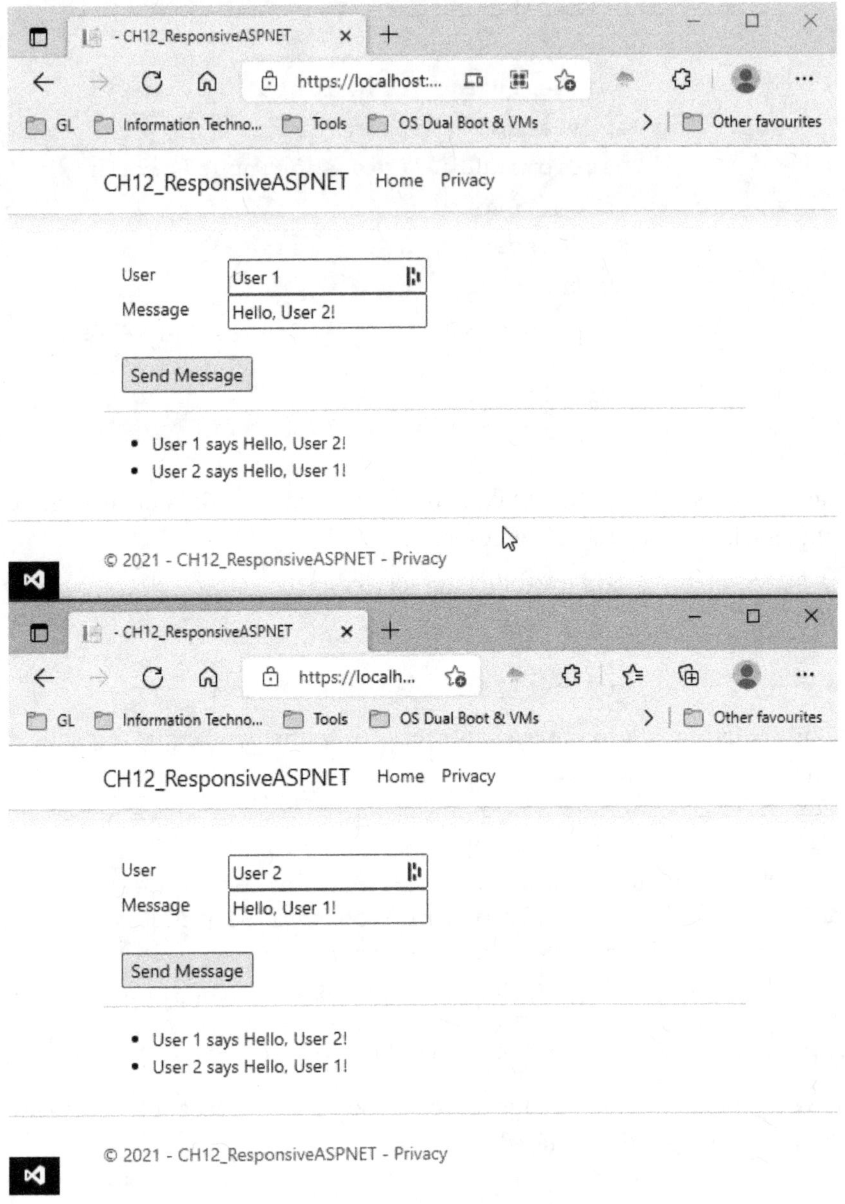

Figure 12.9 – Our SignalR application in action

It was fairly straightforward setting up and running our SignalR. As you can see, SignalR is an excellent choice for real-time communication, and I am sure you will be able to take this knowledge further in the web applications you write. That concludes our work on ASP.NET in this chapter. So, let's now move on to look at .NET MAUI in the next section.

Building responsive UIs with .NET MAUI

Microsoft .NET MAUI is the new version of Xamarin.Forms. There have been some significant changes between Xamarin.Forms version 5.0 and .NET MAUI (Xamarin. Forms version 6.0). The biggest change in MAUI has been to combine the Android, iOS, and macOS projects into a main project. While the code specific to Windows still resides in its own project, Microsoft is working to include the Windows code in the main project. This will lead to us having one single project for writing cross-platform applications using C# and XAML. Let's have a look at some of the other improvements to building cross-platform applications using .NET MAUI.

> **Note**
>
> If you are using an early version of MAUI, to run the Windows project, you will need to set the Windows project as the startup project and deploy the project. Once the project is deployed, you can run the application from the Windows start menu.

Layouts

Another significant change made in .NET MAUI is that the original layouts used by Xamarin.Forms projects have been moved to `Microsoft.Maui.Controls.Compatibility` namespace. By default, MAUI will use new layouts. These layouts are based on a new `LayoutManager` that has been written for performance, consistency, and maintainability. The new layouts are `Grid`, `FlexLayout`, and `StackLayout` (`HorizontalStackLayout` and `VerticalStackLayout`). Microsoft encourages you to select the stack layout that best suits your needs. You are also encouraged to replace legacy layouts with new layouts.

The default spacing values for the new layouts have been standardized to the value of 0. Having these values as 0 sets the expectation that you will set your own preferred values to meet your design requirements. It is best to set these values in your global styles as follows:

```
<ResourceDictionary>
    <Style TargetType="StackLayout">
```

```
        <Setter Property="Spacing" Value="8"/>
    </Style>

    <Style TargetType="Grid">
        <Setter Property="ColumnSpacing" Value="8"/>
        <Setter Property="RowSpacing" Value="8"/>
    </Style>
</ResourceDictionary>
```

Let's move on to look at accessibility improvements.

Accessibility

Microsoft regularly meets with developers who are heavily invested in making applications that meet the highest accessibility rating. This led Microsoft to remove the `TabIndex` and `IsTabStop` properties, as they ended up being confusing and not meeting accessibility needs. For better accessibility, you can improve a screen reader's ability to know the order of reading a UI by implementing a thoughtful design. If you need to take control over the order of UI components, Microsoft recommends that you use the `SemanticOrderView` component.

SetSemanticFocus and Announce

Screen readers are an essential part of an application that is accessible and friendly. To aid these applications' performance in being able to read the correct components, there is a new `SemanticExtensions` class. As part of this class, there is a new method called `SetSematicFocus`. This method enables the setting of a screen reader's focus to a specific element.

> **Note**
>
> At the time of writing, `SetSemanticFocus` and `Announce` are only available for iOS, Android, and Mac Catalyst.

Here is a XAML example of setting semantic focus:

```
<VerticalStackLayout>
    <Label
        Text="SemanticExtensions:"
        TextColor="Black"
        FontAttributes="Bold"
```

```
        FontSize="14"
        Margin="0,8"/>
    <Button
        Text="Semantic focus is applied to the label that
            follows upon the button being pressed."
        FontSize="12"
        Clicked="LabelFocusButton_Clicked"/>
    <Label
        x:Name="SomeLabel"
        Text="Hello, I am able to receive semantic focus!"
        FontSize="12"/>
</VerticalStackLayout>
```

In this XAML, we have an instruction label and a button for the user to press. When the button is pressed, the click event will set the semantic focus to `semanticFocusLabel`. Here is the click event code:

```
private void LabelFocusButton_Clicked(object sender,
    EventArgs e)
{
    SomeLabel.SetSemanticFocus();
}
```

The following code enables the screen reader to make an announcement:

```
SemanticScreenReader.Announce(
    "Make your applications accessible to MAUI users!"
);
```

Another accessibility addition is automatic font scaling.

Font scaling

By default, all components now have automatic font scaling, and it is enabled by default. That means that when your users change their text scaling on the various platforms, your application's text will scale to their chosen settings automatically. You can turn automatic font scaling off for control with the following markup: `FontAutoScalingEnabled="False"`. Changing the attribute to `True` or removing it will turn font auto-scaling back on.

BlazorWebView

Using BlazorWebView, you can host Blazor websites in your Microsoft MAUI applications. This enables your Blazor website to make use of native platform functionality and various user controls. You can add `BlazorWebView` to a XAML page and point it to the root of your Blazor application:

```
<BlazorWebView HostPage="wwwroot/index.html"
               Services="{StaticResource Services}">
     <BlazorWebView.RootComponent>
          <RootComponent Selector="#app"
                ComponentType="{x:Type local:Main}" />
     </BlazorWebView.RootComponent>
</BlazorWebView>
```

As you can see from the XAML, the root of our Blazor application is `wwwroot/index.html`. In the next section, we will take a look at WinUI 3.

> **Note**
>
> As of June 20, 2022, MAUI is generally available, but to develop MAUI applications, you will need to install a .NET 2022 preview.

Building a responsive UI with MAUI

In this section, we will build a simple responsive UI using MAUI. Until MAUI is included with Visual Studio 2022, you will need to ensure you use Visual Studio 2022 Preview:

1. Start a new .NET MAUI app and call it `CH12_ResponsiveMAUI`.
2. Add a new folder called `Api`.
3. In the `Api` folder, add a class called `PropertyChangedNotifier` and replace its contents with the following code:

```
namespace CH12_ResponsiveMAUI.Api
{
    using System.ComponentModel;
    using System.Runtime.CompilerServices;

    public class PropertyChangeNotifier :
        INotifyPropertyChanged
```

```
    {
        public event PropertyChangedEventHandler
            PropertyChanged;

        protected void OnPropertyChanged
            ([CallerMemberName] string propertyName =
               null)
        {
            PropertyChanged?.Invoke(this, new
                PropertyChangedEventArgs
                   (propertyName));
        }
    }
}
```

This code is a base class that implements the `INotifyPropertyChanged` interface.

4. Add a new folder called `Data`.
5. Add a new class to the `Data` folder called `BaseEntity` with the following properties:

```
public int Id { get; set; }
public DateTime CreatedDate { get; set; }
public DateTime ModifiedDate { get; set; }
```

These are base properties for our entities that will inherit this class.

6. Add a new interface to the `Data` folder called `IRepository` and replace the class with the following code:

```
public interface IRepository<T> where T : BaseEntity
{
        T GetById(int id);
        T FirstOrDefault(Func<T, bool> query);
        void Add(T entity);
        void Update(T entity);
        void Remove(T entity);
        List<T> GetAll();
        List<T> Filter(Func<T, bool> query);
```

```
        int Count();
        int FilteredCount(Func<T, bool> query);
}
```

This interface will be implemented by all our repositories.

7. Add a class called `BaseRepository` to the `Data` folder and update the class with the following code:

```
public class BaseRepository<T> : IRepository<T> where
    T : BaseEntity
{
        protected ICollection<T> Context;
        public BaseRepository(ICollection<T> context)
        {
            if (context == null)
                throw new ArgumentNullException
                    ("context");
            Context = context;
        }
}
```

This class is a generic base repository that implements the `IRepository` interface. The context for storing data is of type `ICollection`, and we set `Context` to the collection passed in as a parameter.

8. Add the `Add` method:

```
public void Add(T entity)
{
Context.Add(entity);
}
```

This code adds an entity to our collection.

9. Add the `Count` method:

```
public int Count()
{
if (Context != null)
     return Context.Count;
return 0;
```

```
}
```

This code returns the count of all the entities in our collection.

10. Add the `Filter` method:

```
public List<T> Filter(Func<T, bool> query)
{
return Context.Where(query).ToList();
}
```

This code takes a query and returns a filtered list of items.

11. Add the `FilteredCount` method:

```
public int FilteredCount(Func<T, bool> query)
{
    return Context.Where(query).Count();
}
```

This code returns the items in our filtered list.

12. Add the `FirstOrDefault` method:

```
public T FirstOrDefault(Func<T, bool> query)
{
    return Context.Where(query).FirstOrDefault();
}
```

This method returns the first record to match our query. If there is no match, then the default value is returned instead.

13. Add the `GetAll` method:

```
public List<T> GetAll()
{
return Context.ToList();
}
```

The method returns all the items in our list.

14. Add the `GetById` method:

```
public T GetById(int id)
{
return Context.Where(t => t.Id == id)
```

```
        .FirstOrDefault();
}
```

This method gets an item from the list, as identified by its ID number.

15. Add the `Remove` method:

```
public void Remove(T entity)
{
Context.Remove(entity);
}
```

This method removes an entity from the collection.

16. Add the `Update` method:

```
public void Update(T entity)
{
T item = Context.FirstOrDefault(t => t.Id ==
    entity.Id);
int index = Context.ToList().IndexOf(item);
if (index != -1)
    Context.ToList()[index] = entity;
}
```

This method updates an entity in the collection.

17. Add a new class to the `Data` folder and call it `PeopleRepository`. Then, update the class definition as follows:

```
internal class PeopleRepository : BaseRepository
    <Person>
{
public PeopleRepository(ICollection<Person> context) :
    base(context)
{
}
}
```

This class creates a new repository of type `Person`.

18. Add a new folder with a class called `Person`. Then, update the class as follows:

```
public class Person : BaseEntity
{
    public string FirstName { get; set; }
            public string LastName { get; set; }
    }
```

This class inherits our `BaseEntity` class and adds the properties `FirstName` and `LastName`.

19. Add a new folder called `ViewModels` and a new class called `ViewModelBase`. Update the class definition as shown:

```
public class ViewModelBase<T> : PropertyChangeNotifier
{
bool _isRefreshing;
public ObservableCollection<T> Entities { get; private
   set; } = new ObservableCollection<T>();
public bool IsRefreshing
{
         get { return _isRefreshing; }
         set
                  {
                        _isRefreshing = value;
                        OnPropertyChanged();
                  }
            }
}
```

This class is the base view model class for all our view models. It can be cast to any type, and it implements `PropertyChangeNotifer`.

20. Add `PeopleViewModel`:

```
    public class PeopleViewModel :
        ViewModelBase<Person>
    {
        public PeopleViewModel()
```

```
        {
            SeedPeopleRepository();
        }

        private void SeedPeopleRepository()
        {
            Entities.Add(new Person { Id = 1,
            FirstName = "Person", LastName = "One",
            CreatedDate = DateTime.Now, ModifiedDate =
            DateTime.Now });
            Entities.Add(new Person { Id = 2,
            FirstName = "Person", LastName = "Two",
            CreatedDate = DateTime.Now, ModifiedDate =
            DateTime.Now });
            Entities.Add(new Person { Id = 3,
            FirstName = "Person", LastName = "Three",
            CreatedDate = DateTime.Now, ModifiedDate =
            DateTime.Now });
            Entities.Add(new Person { Id = 4,
            FirstName = "Person", LastName = "Four",
            CreatedDate = DateTime.Now, ModifiedDate =
            DateTime.Now });
            Entities.Add(new Person { Id = 5,
            FirstName = "Person", LastName = "Five",
            CreatedDate = DateTime.Now, ModifiedDate =
            DateTime.Now });
            Entities.Add(new Person { Id = 6,
            FirstName = "Person", LastName = "Six",
            CreatedDate = DateTime.Now, ModifiedDate =
            DateTime.Now });
            Entities.Add(new Person { Id = 7,
            FirstName = "Person", LastName = "Seven",
            CreatedDate = DateTime.Now, ModifiedDate =
            DateTime.Now });
            Entities.Add(new Person { Id = 8,
            FirstName = "Person", LastName = "Eight",
```

```
CreatedDate = DateTime.Now, ModifiedDate =
DateTime.Now });
Entities.Add(new Person { Id = 9,
FirstName = "Person", LastName = "Nine",
CreatedDate = DateTime.Now, ModifiedDate =
DateTime.Now });
Entities.Add(new Person { Id = 10,
FirstName = "Person", LastName = "Ten",
CreatedDate = DateTime.Now, ModifiedDate =
DateTime.Now });
Entities.Add(new Person { Id = 11,
FirstName = "Person", LastName = "Eleven",
CreatedDate = DateTime.Now, ModifiedDate =
DateTime.Now });
Entities.Add(new Person { Id = 12,
FirstName = "Person", LastName = "Twelve",
CreatedDate = DateTime.Now, ModifiedDate =
DateTime.Now });
Entities.Add(new Person { Id = 13,
FirstName = "Person", LastName =
"Thirteen", CreatedDate = DateTime.Now,
ModifiedDate = DateTime.Now });
Entities.Add(new Person { Id = 14,
FirstName = "Person", LastName =
"Fourteen", CreatedDate = DateTime.Now,
ModifiedDate = DateTime.Now });
Entities.Add(new Person { Id = 15,
FirstName = "Person", LastName =
"Fifteen", CreatedDate = DateTime.Now,
ModifiedDate = DateTime.Now });
Entities.Add(new Person { Id = 16,
FirstName = "Person", LastName =
"Sixteen", CreatedDate = DateTime.Now,
ModifiedDate = DateTime.Now });
Entities.Add(new Person { Id = 17,
FirstName = "Person", LastName =
```

```
                "Seventeen", CreatedDate = DateTime.Now,
                ModifiedDate = DateTime.Now });
                Entities.Add(new Person { Id = 18,
                FirstName = "Person", LastName =
                "Eighteen", CreatedDate = DateTime.Now,
                ModifiedDate = DateTime.Now });
                Entities.Add(new Person { Id = 19,
                FirstName = "Person", LastName =
                "Ninetenn", CreatedDate = DateTime.Now,
                ModifiedDate = DateTime.Now });
                Entities.Add(new Person { Id = 20,
                FirstName = "Person", LastName = "Twenty",
                CreatedDate = DateTime.Now, ModifiedDate =
                DateTime.Now });
            }
        }
```

This code seeds our collection with people.

21. Add a new page to the root of the project called `SplashPage`:

```
public partial class SplashPage : ContentPage,
    INotifyPropertyChanged
{
    Timer _timer;
    double _progress;
    public event PropertyChangedEventHandler
        PropertyChanged;
    public SplashPage()
    {
        InitializeComponent();
        _timer = new Timer(new TimerCallback((s) =>
            ReportProgress()), null, TimeSpan.Zero,
                TimeSpan.FromSeconds(3));
    }
    ~SplashPage() => _timer.Dispose();
}
```

Our SplashPage is a loading page that will display progress to the user in the form of a progress bar and label. The class inherits from the Content page and implements the INotifyPropertyChanged event. We have a timer whose callback is a method for reporting loading progress.

22. Add the ReportProgress method:

```
private void ReportProgress()
{
    _timer.Dispose();

    Task.Run(() =>
    {
        // Run code here

        for (int i = 0; i <= 100; i++)
        {
            Thread.Sleep(250);
            _progress = (double)i / 100;
            SafeInvokeInMainThread
                (UpdateProgress);
        }
        SafeInvokeInMainThread(LoadMainPage);
    });
}
```

This method stops the timer and runs the code to update the application loading progress status. It uses a safe invoke method that will update the splash screen.

23. Add the LoadMainPage method:

```
private void LoadMainPage()
{
Application.Current.MainPage = new AppShell(new
    BaseEntity() { Id = 1, CreatedDate = DateTime.Now,
        ModifiedDate = DateTime.Now });
Shell.Current.GoToAsync("//main");
}
```

This method sets the application's `MainPage` to `AppShell` and passes in a parameter of type `BaseEntity`.

24. Add the `SaveInvokeInMaInThread` method:

```
private void SafeInvokeInMainThread(Action action)
{
        if (DeviceInfo.Platform ==
            DevicePlatform.WinUI)
        {
            Application.Current.Dispatcher
                .Dispatch(action);
        }
        else
        {
            MainThread.BeginInvokeOnMainThread
                (action);
        }
}
```

This code performs a safe invocation on the main thread to update the UI. The method checks the device the application is running on before calling the correct method for the device.

25. Add the `UpdateProgress` method:

```
private void UpdateProgress()
    {
        LoadingProgressBar.ProgressTo(_progress, 500,
            Easing.Linear);
        LoadingProgressLabel.Text = $"Progress Update:
             Performing load operation {(int)
                 (_progress * 100)} of 100...";
}
```

This method updates the progress bar and the label.

26. Update the `SplashPage` XAML, as shown here:

```
<?xml version="1.0" encoding="utf-8" ?>
<ContentPage xmlns="
      http://schemas.microsoft.com/dotnet/2021/maui"
```

```
    xmlns:x="http://schemas.microsoft.com/winfx/
        2009/xaml"
    x:Class="CH12_ResponsiveMAUI.SplashPage"
    Title="SplashPage">
  <VerticalStackLayout VerticalOptions="Center">
      <StackLayout HorizontalOptions="Center"
          VerticalOptions="Center">
          <Label FontSize="32" Text="Responsive
              MAUI Example" />
          <Label x:Name="LoadingProgressLabel"
              FontSize="12" Text="Loading..." />
          <ProgressBar x:Name="LoadingProgressBar"
              Progress="0" />
      </StackLayout>
  </VerticalStackLayout>
</ContentPage>
```

This markup contains our UI definition that will be updated by the code when it runs.

27. Update `MainPage` by replacing the current XAML with the following XAML:

```
<?xml version="1.0" encoding="utf-8" ?>
<ContentPage xmlns=
    "http://schemas.microsoft.com/dotnet/2021/maui"
    xmlns:x="http://schemas.microsoft.com/winfx/
        2009/xaml"
    x:Class="CH12_ResponsiveMAUI.MainPage">

  <ScrollView>
      <HorizontalStackLayout
          Spacing="25"
          Padding="30,0"
          VerticalOptions="Center">

          <StackLayout Margin="20"
              HorizontalOptions="Start">
              <CollectionView x:Name=
```

```
                "collectionView" ItemsSource="{Binding
                  Entities}">
                    <CollectionView.ItemTemplate>
                        <DataTemplate>
                            <Grid Padding="10">
                                <Grid.RowDefinitions>
                                    <RowDefinition
                                    Height="Auto" />
                                </Grid.RowDefinitions>
                                <Grid.ColumnDefinitions>
                                    <ColumnDefinition
                                     Width="Auto" />
                                      <ColumnDefinition
                                       Width="Auto" />
                                      <ColumnDefinition
                                       Width="Auto" />
                                      <ColumnDefinition
                                       Width="Auto" />
                                      <ColumnDefinition
                                       Width="Auto" />
                               </Grid.ColumnDefinitions>
                               <Label Grid.Column="1"
                               Text="{Binding Id}"
                               FontAttributes="Bold" />
                               <Label Grid.Column="2"
                               Text="{Binding
                               FirstName}"
                               FontAttributes="Bold" />
                               <Label Grid.Column="3"
                               Text="{Binding LastName}"
                               FontAttributes="Bold" />
                               <Label Grid.Column="4"
                               Text="{Binding
                               CreatedDate}"
                               FontAttributes="Bold" />
                               <Label Grid.Column="5"
```

```
                Text="{Binding
                ModifiedDate}"
                FontAttributes="Bold" />
            </Grid>
        </DataTemplate>
    </CollectionView.ItemTemplate>
</CollectionView>
</StackLayout>

<StackLayout HorizontalOptions="End">
    <Image
    Source="dotnet_bot.png"
    SemanticProperties.Description="Cute
        dot net bot waving hi to you!"
    HeightRequest="200"
    HorizontalOptions="Center" />

    <Label
    Text="Hello, World!"
    SemanticProperties.HeadingLevel=
        "Level1"
    FontSize="32"
    HorizontalOptions="Center" />

    <Label
    Text="Welcome to .NET Multi-platform
        App UI"
    SemanticProperties.HeadingLevel=
        "Level2"
    SemanticProperties.Description=
       "Welcome to dot net Multi platform
          App U I"
    FontSize="18"
    HorizontalOptions="Center" />

    <Button
```

```
                x:Name="CounterBtn"
                Text="Click me"
                SemanticProperties.Hint="Counts the
                    number of times you click"
                Clicked="OnCounterClicked"
                HorizontalOptions="Center" />
            </StackLayout>

        </HorizontalStackLayout>
    </ScrollView>

</ContentPage>
```

This code updates the original source by adding a table of people.

28. Add a `PeopleRepository` class variable and update the constructor of the `MainPage` class, as shown here:

```
    PeopleRepository _peopleRepository;

    public MainPage()
    {
        InitializeComponent();
        BindingContext = new PeopleViewModel();
}
```

This code modifies our `MainPage` by setting its `BindingContext` to `PeopleViewModel`.

29. Run the code, and you should see the following screen:

Figure 12.10 – The splash page

The following screen is what you'll see next:

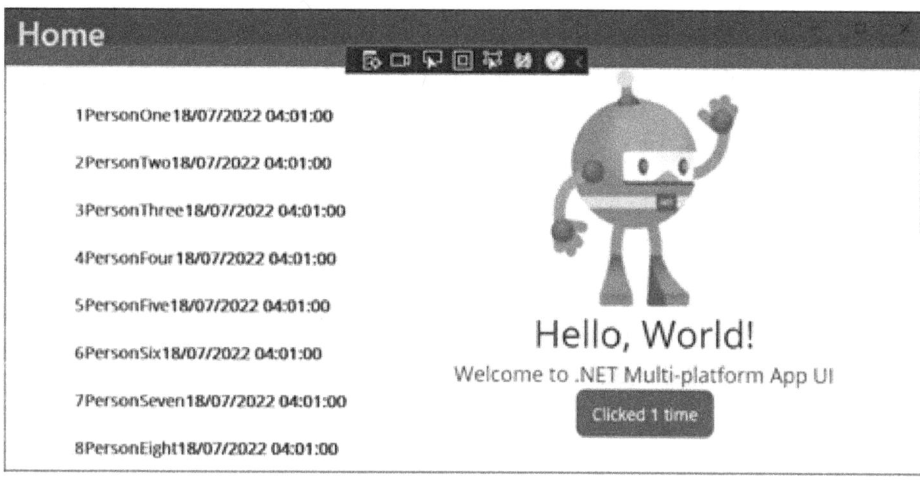

Figure 12.11 – The main form with a table in a scroll view with a button that responds to clicks

We have managed to build a responsive splash screen that also populates a table and responds to button clicks. That concludes our look at MAUI. We will now move on to WinUI 3.

Building a responsive UI with WinUI 3

In this section, we will look at how to provide user feedback using the `ProgressRing` component while performing a long-running operation in WinUI 3 applications. When your users trigger a long-running operation that holds up the UI, it is a good idea to provide user feedback until the operation completes. Let's write a simple application that simulates a long-running operation using the following steps:

1. Start a new WinUI3 application and call it `CH12_ResponsiveWinUI3`.
2. Open `MainWindow.xaml` and replace the existing XAML between the Window tags with the following XAML:

```
<StackPanel VerticalAlignment="Center"
    HorizontalAlignment="Center">
    <ProgressRing x:Name="ProgressRingIndicator1"
        IsActive="{x:Bind IsWorking, Mode=OneWay}"
          Visibility="{x:Bind IsWorking, Mode=OneWay}"
              />
    <Button x:Name="DoWorkButton" Content="Do Work"
       Click="DoWorkButton_Click" />
```

```
        <TextBlock x:Name="MessageTextBlock" />
    </StackPanel>
```

We have used OneWay binding to bind our the ProgressRing class' IsActive and Visibility properties to the IsWorking property.

3. In the code behind the class, implement the INotifyPropertyChanged interface.
4. Add the following members to the class:

```
private DispatcherTimer _dispatcherTimer;
public event PropertyChangedEventHandler
    PropertyChanged;
private bool _isWorking;
```

_dispatcherTimer will be used to simulate a long-running operation. The PropertyChanged event will be used to notify ProgressRing that the IsWorking property has changed, and the _isWorking variable will be updated to let ProgressRing know to either show or hide itself.

5. Add a method to raise the PropertyChanged event if it is not null:

```
private void NotifyPropertyChanged(string property)
{
if (PropertyChanged != null)
{
PropertyChanged(this,
    new PropertyChangedEventArgs(property));
}
}
```

When we set the IsWorking property, we call this method so that the PropertyChanged event is raised.

6. Add the following three lines to the constructor:

```
_dispatcherTimer = new DispatcherTimer();
_dispatcherTimer.Interval = TimeSpan.FromSeconds(10);
_dispatcherTimer.Tick += DispatcherTimer_Tick;
```

These three lines instantiate our DispatcherTimer, set its interval to 10 seconds, and add the Tick event handler.

7. We will now add the `DispatcherTimer_Tick` event handler:

```
private void DispatcherTimer_Tick(object sender,
    object e)
{
_dispatcherTimer.Stop();
_dispatcherTimer.Tick -= DispatcherTimer_Tick;
IsWorking = false;
MessageTextBlock.Text = "Work completed.";
}
```

   We stop the timer and remove the event handler to stop it from firing again and being held in memory. Then, we set the `IsWorking` property to `false`, which results in `ProgressRing` being hidden and made inactive. Then, we add a message to `MessageTextBlock`.

8. Now, add the `IsWorking` property:

```
public bool IsWorking
{
get { return _isWorking; }
set
    {
            _isWorking = value;
        NotifyPropertyChanged("IsWorking");
}
}
```

9. When setting our property, we call the `NotifyPropertyChanged` method that raises the `PropertyChanged` event to let `ProgressRing` know that the property has changed.

10. Now, add the code for the button click:

```
private void DoWorkButton_Click(object sender,
    RoutedEventArgs e)
{
DoWorkButton.Visibility = Visibility.Collapsed;
```

```
    IsWorking = true;
    _dispatcherTimer.Start();
}
```

We collapse our button, as it is no longer needed. Set the `IsWorking` property to `true`, and start our `DispatcherTimer`.

11. Run the code. You should see a single button that says **Do Work**. Click on the button. The button should disappear and be replaced by `ProgressRing` for 10 seconds. Then, `ProgressRing` should disappear and be replaced with the text **Work completed**.

Now that we have concluded our look at responsive UIs, let's summarize what we have learned.

Summary

In this chapter, you learned how to work with various UI frameworks to make UIs responsive. First, we looked at WinForms. With WinForms, we enabled DPI and long file path awareness. We also ensured that despite running long background tasks, we could page through data in a table and perform other UI operations, and we also added a splash screen that updates with the loading progress.

With WPF, we managed to produce a window that has a long-running task that can be canceled with progress indication. It also has a paged data table and button that, when clicked, updates the click count label.

Then, we looked at memory caching and distributed caching in ASP.NET. We also used AJAX to update part of the currently displayed page and looked at WebSockets and SignalR. We implemented a real-time ASP.NET chat application using SignalR.

We then went on to look at MAUI. In particular, we looked at layouts, accessibility, and `BlazorWebView`. Finally, we looked at WinUI 3 and how to provide user feedback when a long-running process is taking place.

In the next chapter, we will be looking at distributed systems. But first, try answering the questions in the next section, and then do some further reading to enhance your knowledge of responsive UIs.

Questions

1. How can you make a WinForms application scale properly on high-DPI screens or normal-DPI large screens?
2. How do you cope with long file paths on Windows?
3. How can you keep users engaged when your application takes a long time to start?
4. How can you keep an application responsive to user input when you have a long-running process in operation?
5. What caching methods can you use to speed up access to resources?
6. How can you load only part of a web page?
7. Name two frameworks for performing network data transfer and real-time networked communication?
8. Name three accessibility methods available in MAUI.
9. How do you include an existing Blazor web application in an MAUI project?
10. When your application is already loaded, and a user kicks off a long-running operation, what controls can you use to provide user feedback so that users don't think your WinUI 3 application has crashed?

Further reading

- *Which is best? WebSockets or SignalR*: `https://dotnetplaybook.com/which-is-best-websockets-or-signalr/`
- *Why is SignalR/messagepack 2 times faster than gRPC/protobuf?*: `https://github.com/grpc/grpc-dotnet/issues/812`
- *Tutorial: Get started with ASP.NET Core SignalR*: `https://docs.microsoft.com/aspnet/core/tutorials/signalr?view=aspnetcore-5.0&tabs=visual-studio`
- *WebSocket*: `https://javascript.info/websocket`
- *Migrate your app from Xamarin.Forms*: `https://docs.microsoft.com/dotnet/maui/get-started/migrate`
- *Xamarin.Forms Made Easy*: `https://winstongubantes.blogspot.com/2018/09/backgrounding-with-xamarinforms-easy-way.html`

- *Xamarin – Working with threads*: `https://lukealderton.com/blog/posts/2016/october/xamarin-forms-working-with-threads/`
- Creating Android emulators on Windows: `https://docs.microsoft.com/xamarin/android/get-started/installation/android-emulator/device-manager?tabs=windows&pivots=windows`
- Installing the Microsoft OpenJDK: `https://docs.microsoft.com/xamarin/android/get-started/installation/openjdk`
- *Single-project MSIX Packaging Tools for VS 2022*: `https://marketplace.visualstudio.com/items?itemName=ProjectReunion.MicrosoftSingleProjectMSIXPackagingToolsDev17`
- *Improving rendering performance with Blazor component virtualization*: `https://www.daveabrock.com/2020/10/20/blazor-component-virtualization/#:~:text=Improve%20rendering%20performance%20with%20Blazor%20component%20virtualization%20Use,the%20entire%20HTML%20tree%20loads%20from%20the%20server.`
- *How to Reuse Xamarin.Forms Custom Renderers in .NET MAUI*: `https://www.syncfusion.com/blogs/post/how-to-reuse-xamarin-forms-custom-renderers-in-net-maui.aspx`
- *Announcing .NET MAUI Preview 7*: `https://devblogs.microsoft.com/dotnet/announcing-net-maui-preview-7/`
- *.NET Multi-platform App UI*: `https://dotnet.microsoft.com/en-us/apps/maui`

13
Distributed Systems

In this chapter, you will learn about distributed applications and how you can improve their performance. You will understand how to build performant applications using the **Command Query Responsibility Separation** (**CQRS**) software design pattern, event sourcing, and microservices. You will learn how to use cloud providers such as Microsoft Azure to build scalable distributed solutions using Cosmos DB, Azure Functions, and the open source Pulumi infrastructure tool.

In this chapter, we will cover the following topics:

- **Implementing the CQRS design pattern**: In this section, we will implement the CQRS design pattern with a sample project that demonstrates the separation of commands and queries.
- **Implementing event sourcing**: Many resources always show event sourcing with CQRS. But in this section, we will write a sample project that demonstrates pure event sourcing without CQRS. By doing this, you will know how to implement CQRS and event sourcing individually and be able to combine them to work together.
- **Using Microsoft Azure for distributed systems**: In this section, we will provide a high-level overview of Azure Functions – specifically Durable Azure Functions – for providing robust, secure, and scalable serverless code that performs well in a distributed environment. We will also look at the difference between containers and serverless, and when to use one over the other.

- **Managing your cloud infrastructure with Pulumi**: Managing Azure resources can become unwieldy, especially when the number of microservices you deploy increases. So, in this section, we will look at how Pulumi allows you to manage your cloud infrastructure and resources using pure C# that you can include in your build, test, and deploy pipelines.

By completing this chapter, you will gain the following skills:

- You will be able to separate commands and queries into different services.
- You will be able to persist state changes as sequences of state-changing events.
- You will be able to understand the difference between containers and serverless, and you will be able to know when to use one over the other.
- You will understand the different types of Durable Azure Function types and design patterns so that you can use them to build serverless functions.
- You will be able to manage your cloud using Pulumi.

Technical requirements

You'll need the following components to follow along with this chapter and perform the necessary programming tasks:

- Visual Studio 2022 or later
- This book's source code: `https://github.com/PacktPublishing/High-Performance-Programming-in-CSharp-and-.NET/tree/master/CH13`
- Optional: A Microsoft Azure account
- Optional: A Pulumi subscription

Implementing the CQRS design pattern

In this section, we will look at the **Command Query Responsibility Separation** (**CQRS**) design pattern. In simple terms, a command is a method that performs an action, while a query is a method that returns data. Commands do not perform queries, and queries do not perform commands. Commands can have separate models for queries. Now, let's write a simple console application that demonstrates how easy it is to implement this pattern, which is used extensively in microservice development:

1. Start a new console application called `CH13_CQRSPattern`.

2. Add a new class called `CQRSBasedClass`.
3. Add the `SleepCommand` method:

```
public void SleepCommand(int milliseconds)
{
    Thread.Sleep(milliseconds);
}
```

Our `SleepCommand` method is an example of a command. It takes in a parameter that is several milliseconds in length. A command is then executed that causes the current thread to sleep for the number of milliseconds specified by the caller. This particular command does not return a value.

4. Add the `DateTimeQuery` method:

```
public DateTime DateTimeQuery()
{
    return DateTime.Now;
}
```

Our `DateTimeQuery` method is an example of a query. It is a parameterless query, although queries can have parameters. The query does not perform any commands. It simply returns the current date and time to the caller.

5. In the `Program` class, add the `ExecuteCommand` method:

```
private static void ExecuteCommand()
{
    new CQRSBasedClass().SleepCommand(1000);
}
```

The `ExecuteCommand` method executes `SleepCommand` in our `CQRSBasedClass`, which causes the current thread to sleep for 1 second.

6. Add the `ExecuteQuery` method:

```
private static DateTime ExecuteQuery()
{
    return new CQRSBasedClass().DateTimeQuery();
}
```

The `ExecuteQuery` method executes `DateTimeQuery` in our `CQRSBasedClass`, which queries the current date and time and returns the value.

7. Update the `Program.cs` file, as follows:

```
Console.WriteLine("Hello, World! This is the most
    simple example of CQRS in action.");
ExecuteCommand();
Console.WriteLine($"The current date and time is:
  {ExecuteQuery()}.");
```

   We start our program by writing a message to the console. Then, we call `ExecuteCommand`. Finally, we write a message to the console that includes the current date and time that is returned by the call to `ExecuteQuery`.

As you can see, in its most basic form, the CQRS pattern is really simple. A command performs an action and nothing else, while a query performs a query and nothing else. We can shift commands into their own command class so that the only purpose of the class is to execute commands. We can do the same with queries by placing them in their own query class so that all the query class does is returns queries.

If you study this book's source code, you will see that we have done this. We have a `CommandClass` with a command called `Sleep`. We also have a `QueryClass` with a query called `Now`. CQRS is the enabling pattern that's used in microservice development. It is often used in conjunction with message brokers, message buses, message sending and receiving, domain modeling, domain events, event sourcing, eventual consistency, separate read and write models, and **domain-driven design** (**DDD**). This is where people tend to become lost. But even though the CQRS pattern is used alongside all these, the pattern itself is very simple, and it enables these other patterns and technologies to gel nicely together.

In terms of database operations, you can think of the `add`, `edit`, `delete`, and `update` operations as commands, and you can think of `select` operations as queries.

Now that we have a simple understanding of the CQRS pattern, in the next section, we will turn our attention to understanding and implementing event sourcing.

Implementing event sourcing

When you consider documents in a document store and records in a database, these are normally a business's point of truth. Their state is the source of truth.

Event sourcing record events become your source of truth rather than the state of data in tables, or the state of documents in document stores.

So, instead of using the state as a point of truth, we can use recorded events as a source of truth.

In the old days of programming, this was known as an audit trail. I remember working on a database several years ago. It had an audit table. In that table, there was a record of all the actions that were carried out on the database and by whom. We could tell when data operations took place, what those data operations were, and who or what process was carried out those data operations. Then, if anything went wrong with the database, we could analyze that table and know which operation caused the resulting problems. To store this information, we would use database triggers that fired on every `add`, `update`, `delete`, and `read` operation. These triggers were events that fired upon a data operation that recorded what data modifications had taken place, who made those modifications, why they made the changes, and at what date and time those modifications took place.

In this section, we will look at event sourcing, which records events that become your source of truth. Events allow you to understand how you arrived at a particular state at a particular point in time.

An easy way to understand the benefits of event sourcing is to have a look at your bank statement. When you receive your bank statement, you start with the balance that was carried over from the previous month. Then, you see a list of transactions that took place during the period covered by the statement, which consists of money entering your account and money leaving your account. Each of these transactions is an event. These events can be money transfer in, money transfer out, direct debit payment, interest payment, standing order payment, bank charge payment, payment of goods, salary/wages being paid in, and so forth.

When you consider this scenario, your bank statement shows how you came to have the money come in and leave your account. But from a database point of view, just by looking at the data, this is not so easy. When you look at the data, you usually have to write a query that joins multiple tables in a relational database together to reveal the facts of how the state of your account changed. But you don't necessarily know the context that led to those changes being made.

However, in the same scenario, when you store events, you are storing facts. These facts are based on true events that happened in the past, and that is why they can be trusted.

As for transactional logs, they inform you of what state changes took place. However, they don't necessarily tell you why those state changes were made. On the other hand, when you store events, they inform you of what state changes were made, and the reason why those state changes were made.

Events are stored as aggregates in an append-only form. An aggregate is a consistency guard. You can see the state changes and the context that led to those changes. This means that you can revert the state to the last known consistent state at a particular point in time by replaying events forward or backward. You can use the event log to provide an audit trail. Information such as why and when can be very useful to various business functions, such as senior management, marketing, finance, and resource planning, since event logs are full of very valuable business information.

Going back to our example scenario, an event represents a fact that took place in our banking domain. Each event in our banking system is a source of truth from which our bank account's current state is derived. Such facts are immutable business facts.

Our banking events will follow the normal methodology of providing state information, metadata that provides contextual information, the date and time when it happened, and other information that is necessary and appropriate.

Let's look at an example of how we can aggregate events so that they arrive at a specific state for our bank account:

- Events:
  - A dividend of £39 was issued by the investment firm to the customer at 12:43 A.M. on June 12, 2021.
  - A dividend of £39 was paid into the customer's bank account by the investment firm at 12:45 A.M. on June 12, 2021.
- Events
  - A salary of £2,300 paid was into the customer's bank account by the employer at 12:00 A.M. on July 25, 2021, using BACS.
  - A standing order of £230 was transferred from the customer's bank account into their savings account to build up an emergency stash at 09:11 A.M. on July 26, 2021.
  - A direct debit of £432 was paid to the local authority for rent from the customer's bank account at 07:00 P.M. on July 25, 2021, using the relevant Android banking app.
  - A direct debit of £103 was paid to the local authority for council tax at 08:29 P.M. on July 26, 2021, using online banking.
  - £23.79 was paid for groceries to the merchant by the customer using contactless payment at 09:35 P.M. on July 27, 2021.

As you can see from our banking scenario, when we use events as our fact-based points of truth, we see the full context of where the money is coming from, where the money is going, and by which method, what the amount is, and the precise date and time when it takes place.

These events ensure the data is in a consistent state, that there is an audit trail, and that valuable information is provided that allows business decisions to be made based on trustworthy facts.

Continuing with our banking scenario, each bank account would have a stream and unique identifier. All events that occur against that bank account will be recorded via its stream. So, we end up with one stream per aggregate. In our banking scenario, our aggregate is the group of events that take place against a specific bank account.

Event sourcing example project

In this section, we will write a simple event sourcing application that also provides examples of usage. To implement the project, follow these steps:

1. Start a new .NET 6.0 console application and call it `CH13_EventSourcing`.
2. Add a public interface called `IEvent` with an empty method body. This is a convenient interface for marking any object an event.
3. Add a new public interface called `IRegisterable` and add the following method:

```
void RegisterWithEventAggregator(IEventAggregator
    eventAggregator);
```

This method allows registerable objects to register themselves with an event aggregator.

4. Add a new public interface called `IEventAggregator` and add the following methods:

```
void Register(IRegisterable registerable);
void Register<T>(EventHandler<T> eventhandler) where
    T : IEvent;
void RaiseEvent(IEvent evt);
```

The `Register` method is used for registering objects of the `IRegisterable` type with the event aggregator. The `Register<T>` method registers an event handler of the `T` type for the specified object type. Finally, the `RaiseEvent` executes the event that was passed in as a parameter.

5. Add a new class called `EventHandler` and replace its contents with the following code:

```
namespace CH13_EventSourcing;
public delegate void EventHandler<T>(T evt) where T :
    IEvent;
```

   This delegate defines our event handler, which is of the `T` type, for events of the `IEvent` type.

6. Add a new class called `SingleThreadedEventAggregator` that implements the `IEventAggregator` interface.

7. Add the following dictionary field to hold our event handlers:

```
IDictionary<Type, IList<EventHandler<IEvent>>>
    _eventHandlers;
```

   This dictionary defines a list of event handlers of the `IEvent` type for objects of a specified type.

8. Add the following constructor:

```
public SingleThreadedEventAggregator()
{
    _eventHandlers = new Dictionary<Type,
        IList<EventHandler<IEvent>>>();
}
```

   Here, we instantiate our dictionary of event handlers.

9. Update the `Register` method, as shown here:

```
public void Register(IRegisterable registerable)
{
    registerable.RegisterWithEventAggregator(this);
}
```

   This method registers our event aggregator with the registerable type that was passed in.

10. Update the `Register<T>` method, as shown here:

```
public void Register<T>(EventHandler<T> eventHandler)
    where T : IEvent
{
```

```
        if (!_eventHandlers.ContainsKey(typeof(T)))
        {
            _eventHandlers[typeof(T)] = new
                List<EventHandler<IEvent>>();
        }
        var eventHandlerList = _eventHandlers[typeof(T)];
        eventHandlerList.Add(evt => eventHandler
            ((T)evt));
    }
```

This method checks our dictionary to see if it contains a key of the specified type; if it doesn't, it adds one. Then, it creates a new event handler list of the specified type and adds the event handler.

11. Update the `RaiseEvent` method:

```
    public void RaiseEvent(IEvent evt)
    {
        IList<EventHandler<IEvent>> eventHandlerList;
        if (_eventHandlers.TryGetValue(evt.GetType(),
            out eventHandlerList))
        {
            foreach (EventHandler<IEvent> eventHandler in
                eventHandlerList)
            {
                eventHandler.Invoke(evt);
            }
        }
    }
```

This method gets a list of event handlers for the event that was passed in and loops through them, invoking them.

12. Add a new class called `MultiThreadedEventAggregator` that implements the `IEventAggregator` interface.
13. Add the following dictionary to the class:

```
    IDictionary<Type, IList<EventHandler<IEvent>>>
        _eventHandlers;
```

This dictionary will hold a list of event handlers and their events.

14. Add the following constructor:

```
public MultiThreadedEventAggregator()
{
        _eventHandlers = new ConcurrentDictionary<Type,
            IList<EventHandler<IEvent>>>();
}
```

Our constructor initializes our list of event handlers. Notice that we are using a concurrent dictionary to handle multi-threaded scenarios.

15. Add the following method:

```
public void Register(IRegisterable registerable)
{
        registerable.RegisterWithEventAggregator(this);
}
```

This method registers the event handlers of the registerable object with the multi-threaded event aggregator.

16. Add the following `Register` method:

```
public void Register<T>(EventHandler<T> eventHandler)
    where T : IEvent
{
        if (!_eventHandlers.ContainsKey(typeof(T)))
        {
            _eventHandlers[typeof(T)] = new
                List<EventHandler<IEvent>>();
        }
        var eventHandlerList = _eventHandlers[typeof(T)];
        eventHandlerList.Add(evt => eventHandler((T)evt));
}
```

This method checks our dictionary to see if it contains a key of the specified type; if it doesn't, it adds one. Then, it creates a new event handler list of the specified type and adds the event handler.

17. Add the `RaiseEvent` method:

```
public void RaiseEvent(IEvent evt)
{
    IList<EventHandler<IEvent>> eventHandlerList;
    if (_eventHandlers.TryGetValue(evt.GetType(), out
        eventHandlerList))
     {
         Parallel.ForEach(eventHandlerList,
             eventHandler =>
             {
                 eventHandler.Invoke(evt);
             });
     }
}
```

This method loops through all our event handlers stored in the event handler list and invokes them for the specified event that was passed in as an argument.

This is the completed base project. Now, let's look at an example of using our event sourcing code.

18. Add a folder called `BankApp`.
19. Add the following `DividendPayment` class to the `BankApp` folder:

```
internal class DividendPayment : IEvent
{
    public string From { get; set; }
    public string To { get; set; }
    public DateTime PaymentDate { get; set; }
    public Decimal Amount { get; set; }
}
```

This class defines our dividend payment event. This event provides information on a dividend payment regarding who sent the payment, to whom the payment was made, the date of the payment, and the amount of the payment.

20. Add the `InvalidDateException` class to the `BankApp` folder:

```
internal sealed class InvalidDateException : Exception
{
    public InvalidDateException() : base()
    {
    }
    public InvalidDateException(string? message)
        : base(message)
    {
    }
     public InvalidDateException(string? message,
         Exception? innerException) : base(message,
             innerException)
    {
    }
}
```

This class implements the `System.Exception` class and will be used to inform others that an exception occurred due to an incorrect date.

21. Add the `StandingOrderPayment` class to the `BankApp` folder:

```
internal class StandingOrderPayment : IEvent
{
    public string From { get; set; }
    public string To { get; set; }
    public DateOnly StartDate { get; set; }
    public decimal Amount { get; set; }
}
```

This class defines our standing order payment event, which informs us of who pays the standing order and to whom, the start date of the standing order, and the amount to be paid.

22. Add the `EventHandlers` class to the `BankApp` folder, and update it as follows:

```
internal class EventHandlers : IRegisterable
{
}
```

Our class implements the `IRegisterable` interface and will be used to register our events with the event aggregator that was used for those events.

23. Add the following property and constructor:

```
public string Name { get; }
public EventHandlers(string name)
{
    Name = name;
}
```

This property is set in the constructor to label the `EventHandlers` class for easy human reference.

24. Add the following registration code:

```
public void RegisterWithEventAggregator
    (IEventAggregator eventAggregator)
{
    eventAggregator.Register<DividendPayment>
        (OnDividendPayment);
    eventAggregator.Register<StandingOrderPayment>
        (OnStandingOrderPayment);
}
```

This method registers the events and the event handlers for the dividend payments and standing orders with the event aggregator.

25. Add the following handler method for dividend payments:

```
private void OnDividendPayment(DividendPayment evt)
{
    Console.WriteLine($"Dividend paid by {evt.From} to
        {evt.To} on {evt.PaymentDate} of
            £{evt.Amount}.");
}
```

Every time a dividend payment is made, this event handler is called, and the properties of the dividend payment event are logged to the console window.

26. Add the following handler method for standing order payments:

```
private void OnStandingOrderPayment
    (StandingOrderPayment evt)
```

```
{
    try
    {
        Console.WriteLine($"Standing order paid by
        {evt.From} to {evt.To} on {GetStanding
        OrderDate(evt.StartDate)} of
        £{evt.Amount}.");
    }
    catch (InvalidDateException idex)
    {
        Console.WriteLine(idex.Message);
    }
}
```

Every time a standing order payment is paid, this event handler is called. The properties of the standing order payment event are written out on the console. During this process, the payment date is checked to see if it is valid; if it's not, then an `InvalidDateException` is raised.

27. Add the `GetStandingOrderDate` method:

```
private static DateTime GetStandingOrderDate(DateOnly
    startDate)
{
    if (DateTime.UtcNow.Ticks < startDate.ToDateTime
        (TimeOnly.FromTimeSpan(TimeSpan.Zero)).Ticks)
        throw new InvalidDateException("Invalid
          Date: Payment date cannot be before
            standing order start date!");
    if (DateTime.Now.Day < startDate.Day)
        throw new InvalidDateException("InvalidDate:
          Payment cannot be made before the standing
            order month pay day.");
    return DateTime.Now;
}
```

This method takes the start date of the standing order and checks the date against the current date. An exception is thrown if the date is before the standing order start date or is not on or after the payment date for the month. Otherwise, the current date and time are returned.

28. Replace the text in the `Program.cs` class with the following:

```
using CH13_EventSourcing;
using CH13_EventSourcing.BankApp;
using EventHandlers = CH13_EventSourcing.BankApp
    .EventHandlers;
SingleThreadedEventAggregator eventAggregator = new();
EventHandlers eventHandlers = new("Payment Event
    Handlers");
DividendPayment dividendPayment = new DividendPayment
    { From = "Company Name", To = "Customer Name",
        PaymentDate = DateTime.Now, Amount = 23.45M };
StandingOrderPayment standingOrderPayment = new
    StandingOrderPayment { From = "Customer Name", To
        = "Company One", StartDate = DateOnly.Parse
            ("25/02/2022") };
eventAggregator.Register(eventHandlers);
eventAggregator.RaiseEvent(dividendPayment);
eventAggregator.RaiseEvent(standingOrderPayment);
```

This is our application entry point. We create an event aggregator that is single-threaded. Then, we create an instance of the `EventHandlers` class and pass it in the text that shows these event handlers are used to handle payment events. Next, we create two events – one for dividend payments and the other being for standing order payments. The instance of the `EventHandlers` class is then passed into the event aggregator so that the event handlers can be registered. Finally, the events for the dividend payment and standing order are raised.

29. Run the program. You should see something similar to the following output:

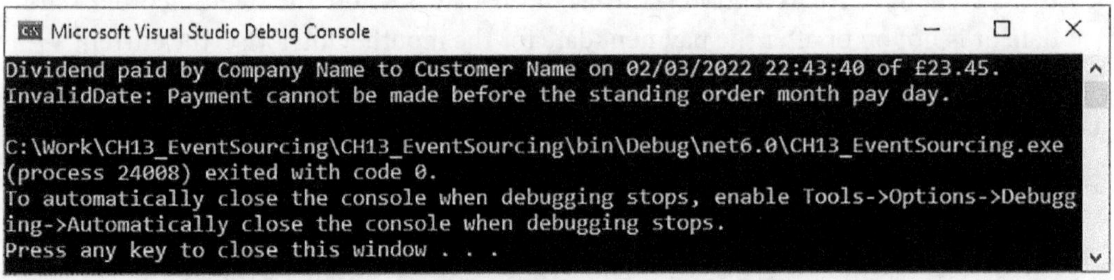

```
Microsoft Visual Studio Debug Console
Dividend paid by Company Name to Customer Name on 02/03/2022 22:43:40 of £23.45.
InvalidDate: Payment cannot be made before the standing order month pay day.

C:\Work\CH13_EventSourcing\CH13_EventSourcing\bin\Debug\net6.0\CH13_EventSourcing.exe
(process 24008) exited with code 0.
To automatically close the console when debugging stops, enable Tools->Options->Debugg
ing->Automatically close the console when debugging stops.
Press any key to close this window . . .
```

Figure 13.1 – The output of our event source application

With that, you have coded and run an event sourcing application. Before that, you did the same with a CQRS application. By writing these two applications, you have seen pure CQRS and pure event sourcing at work. With this knowledge, you can now write applications that use these patterns individually or that combine them so that they work together. In the next section, we will provide a high-level overview of Microsoft Azure in terms of writing distributed systems.

Using Microsoft Azure for distributed systems

In this section, we will learn how to use Azure to implement durable microservices using serverless features, namely Azure Functions.

What is Azure? As I am sure you are aware by now, Microsoft Azure is Microsoft's cloud offering for hosting your databases, APIs, and data resources. It also has many other forms of cloud offerings. Microsoft Azure consists of paid services, free for 1-year services, and always free services. You are advised to review their different cloud services and compare them with other providers to suit your needs. Pay particular attention to which services are free, along with their usage limits, and which services you will have to pay for.

Let's name some good reasons to host your applications and databases in the cloud instead of on-premises. Well, you don't have to pay for hardware or electricity costs for a start. Then, there is the aspect of scaling up and scaling out when your existing infrastructure meets the maximum capacity. Hardware can become obsolete very quickly as the complexity of the needs of software and its users grows. So, there are many reasons to use the cloud that you will need to carefully consider, and with those reasons, there will be both pros and cons. Therefore, when deciding to use the cloud, make sure you research, document, and price everything so that you start on the right footing. This will make system management, maintenance, and business growth much easier in the long run. If you get things right from the start of your endeavors, then you will save yourself potential headaches further down the line!

A microservice is normally a simple web service that receives a request and sends a response. Many kinds of microservices exist, such as film and music streaming services and document upload and retrieval services. In the DDD of microservices, the microservice will normally have a data source. On Azure, this could be a file held in blob storage, data stored in an Azure SQL Server relational database, or even data stored in an Azure Cosmos DB NoSQL database.

Modern microservice implementations are relying less on containerization that uses tools such as Docker and Kubernetes and more on pure serverless options such as Azure Functions. The beauty of an Azure Function is that it is only active for the period of the call. Once the function has done what it is required to do, it simply goes to sleep. An Azure Function also uses fewer computing resources and power compared to containerized solutions. The only downside is that you must manage many Azure Functions. And so, just like with containerization, you will need some way to orchestrate all your Azure Functions in an easy-to-maintain, extended, and useful way.

Azure Functions

An Azure Function is a unit of work. When you implement Azure Functions, you do not have to concern yourself with provisioning and managing infrastructure, since Azure Functions is one of Microsoft's serverless computing offerings.

Serverless computing is managed by the serverless provider. This means that the serverless computing provider is responsible for investing heavily into provisioning and managing the infrastructure that hosts your serverless computing services, such as Azure Functions. This means you get to save money on hardware and electricity costs, and can fully concentrate your efforts on developing, testing, deploying, and maintaining your serverless projects.

Microsoft's investment into serverless computing provides your Azure Functions with networking, service discovery, routing, and events to facilitate high-performance communication between your functions and other aspects of your software system architecture.

An Azure Function normally consists of one or more inputs that you can bind and trigger, and outputs that you can bind to, with your custom code sitting between the inputs and outputs, as shown in the following diagram:

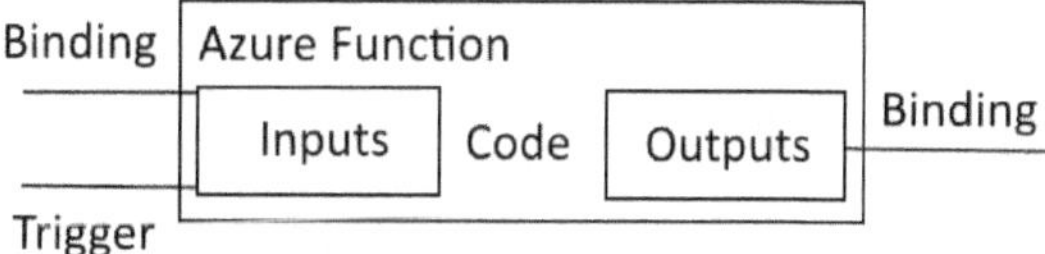

Figure 13.2 – High-level Microsoft Azure Functions concept diagram

Azure Functions are excellent tools to use when developing distributed systems. But the complexity of using Azure Functions begins to materialize when the number of Azure Functions in your projects starts to grow. Managing large numbers of Azure Functions requires a form of orchestration. Orchestration makes managing many Azure Functions more straightforward for the infrastructure team. The orchestration to employ for Azure Functions is Durable Azure Function.

Durable Azure Functions

You can execute Azure Functions with stateful orchestration using durable functions. Azure Functions provide an extension known as Durable Functions. Durable function applications consist of multiple Azure Functions. Each function in a durable function orchestration can perform a different role and/or function. The different types of durable functions are activity, orchestrator, entity, and client. Let's take a brief look at each type of durable function.

Durable function type – activity

A basic unit of work is defined as an activity function within the orchestration of a durable function. This means that when an orchestrated function performs multiple tasks, such as data validation, reading data, and updating data, each of these tasks will be executed by a durable activity function. Once a durable activity function has been completed, it may return data to the function that orchestrated the activity.

Activity functions are defined by activity triggers. `DurableActivityContext` is passed in as a parameter. Event triggers can be bound to JSON-serializable objects that can be used to pass input data into functions. Since an activity function can only have single values passed to them, you can overcome this limitation by using arrays, complex types, and tuples.

> **Note**
> Activity functions can only be triggered from an orchestrator function and are only guaranteed to run at least once by the Durable Task Framework. Because we don't know how many times an activity might be called, Microsoft recommends that you make durable activity functions idempotent whenever possible.

Durable function type – orchestrator

Use the orchestrator function type when you need to control what actions are executed, and the order that you need to execute them.

Durable function type – entity

A durable entity can be invoked by client and orchestrator functions and is triggered by an entity trigger. A durable entity function is used to read and update an object's state.

Durable function type – client

A durable client function is defined using a durable client output binding. Client functions are used to start orchestrator and entity functions since, on the Azure portal, these functions cannot be triggered by button clicks.

Durable function patterns

There are several patterns that you can use to manage your durable functions. These include the following:

- Aggregator (stateful entities)
- Async HTTP APIs
- Fan-out/fan-in
- Function chaining
- Human interaction
- Monitoring

The aggregator (stateful entities) pattern

In this pattern, a single addressable entity is used to aggregate event data that takes place over a certain period. The data that's passed into an aggregator can come from multiple sources. Data may be spread over time and can be delivered in batches. You can process data upon arrival and make the aggregated data available for querying by external clients.

In the aggregator pattern, the aggregator function should be run in a single process or VM. The main reason is due to the complexity of concurrency control when it's used with normal functions that are stateless.

Async HTTP APIs

Factors that affect the time it takes for an API call to complete include volume and latency, as well as other factors beyond your control. Durable functions have a built-in mechanism for working with the execution of long-running functions, and the durable function's runtime is also responsible for managing the state.

Fan-out/fan-in

Durable functions allow you to execute functions in parallel and on the results of tasks.

Function chaining

When using ordinary functions with service bus queues, you have more complexity when it comes to error handling, and it can be hard to visualize the relationship between a function and a queue.

However, when you use a durable function, you have one location where you can set the order of your functions, storage queues are automatically managed by the durable function, and if errors occur in any of the activities, they get propagated back to the orchestration function.

Human interaction

Durable functions can be used to escalate processes that have not received human interaction within an agreed timeline.

Monitoring (actors)

When you need to perform a recurring task, such as releasing system resources, durable functions provide a flexible way for you to manage recurrence intervals, use a single orchestration to manage multiple monitor processes, and manage the lifetime of a task.

Containers and serverless

Container and serverless technologies all have a valid place in the microservice ecosystem. The primary thought process is to know their strengths and weaknesses to help you choose the best option for your needs.

Containers

Containers are a good option for you if you have legacy code that you want to migrate to a more modern platform and code base. You do not have to rewrite your legacy code base, such as web services and batch processes, immediately. You can place them within a container and deploy them to the cloud. Then, when time, money, and resources become available, you can plan for and implement the rewriting of your legacy projects.

When you rely on third-party dependencies, cost and PaaS availability can be an issue. Sites such as Docker Hub provide access to many readily available containers for various third-party dependencies that you can pull and deploy.

Local development of multiple microservices can be simplified with Docker Compose files. You can add as many services as you need to a Docker Compose file and start them all up when they are required.

Using Kubernetes clusters, an ingress controller is used to expose only those services you want to be exposed to. This allows you to provide secure code with a limited footprint that makes life hard for hackers.

Some downsides to containers are that they can encourage the use of older development techniques that are more heavyweight and require more computing power. This can lead to an increase in computing costs. Containers also need a core number of cluster nodes that are always running, adding to your costs.

Serverless

External services can be integrated with serverless technology such as Azure Functions. Rapid application development is promoted by the simplified programming model of serverless computing.

When programming serverless code, you are encouraged to use an event-driven approach to your functions. Such code is easily scalable and can be easily rewritten or discarded as your business evolves.

Serverless code supports *scale to zero* as functions only ever run when they are needed and do not run when they are not needed. This helps reduce running costs as resource consumption is very minimal compared to services such as cluster nodes, which are always running.

Rapid scale-out of serverless code is another advantage of such technologies, as you only ever pay for the running time of the function.

Serverless functions can pose a security risk, so you must take steps to ensure your functions are safe and secure.

Now that you know about the strengths and weaknesses of containers and serverless functions, and you have reviewed the various types of durable functions available in Microsoft Azure, as well as some durable function patterns, let's look at managing our cloud infrastructures in C# with Pulumi.

Managing your cloud infrastructure with Pulumi

In this section, you will learn how to manage your cloud infrastructure using Pulumi. With cloud infrastructure, it is important to be consistent. One way to achieve this is to remove the human element, which is prone to error, and automate as much as you can. An important aspect of the cloud that can be readily automated is infrastructure provisioning tasks. And that's where Pulumi comes in.

With Pulumi, you can code **Infrastructure as Code** (**IaC**) solutions. Code and configuration files are used to manage and provision the infrastructure that your software will run on.

Pulumi projects can be written in various programming languages such as Python, VB.NET, F#, and C#. We are interested in using C# for our Pulumi projects. You can use Pulumi to do the following:

- Specify your infrastructure.
- Automate how cloud resources are created, updated, and deleted.
- Use IDEs and code editors such as Visual Studio and Visual Studio Code.
- Catch mistakes during compilation.
- Enforce security, compliance, and best practices.
- Use existing NuGet libraries as well as code your own libraries.
- Use Kubernetes, Docker containers, Azure Functions, and Cosmos DB to build applications that are easy to scale.

> **Note**
> To follow along, you will need to have Chocolatey installed since it will be used as the package manager for installing Pulumi. You will also need to have a Microsoft Azure account to deploy your IaC. On Windows, when using the command line, make sure you are using PowerShell and that you are running it as an Administrator.

Now, let's look at a very simple example of provisioning blob storage, adding files to blob storage, and destroying our provisioned resources. The following steps will provision, use, and delete Azure blob storage:

1. Install Pulumi with the following command:

```
> choco install pulumi
```

2. Ensure that you have .NET 6 SDK or higher installed.
3. Configure Pulumi's access to your Microsoft Azure account by typing the following command:

```
az login
```

> **Note**
>
> Your credentials will never be sent to pulumi.com, and they will only be used by Pulumi for authentication purposes when managing and provisioning resources.

4. At this point, you are ready to start using Pulumi. If the `az` term is not recognized, try the following command:

```
Invoke-WebRequest -Uri https://aka.ms/
installazurecliwindows -OutFile .\AzureCLI.msi; Start-
Process msiexec.exe -Wait -ArgumentList '/I
AzureCLI.msi /quiet'; rm .\AzureCLI.msi
```

5. Create a new project using the following commands:

```
> Mkdir CH13_Pulumi
> cd CH13_Pulumi
> pulumi new azure-csharp
```

You will be asked to enter your token, or you can simply press *Enter* to log into Pulumi and have Pulumi obtain your token for you. If you don't have one, you can create one quite easily at this stage. Once you are logged in, you will be asked a series of questions in PowerShell. You can simply accept all the defaults.

6. Open the project in Visual Studio. Let's review the project files:

A. `Pulumi.yaml` is used to define the project.

B. `Pulumi.dev.yaml` is used to store configuration values for your stack.

C. `Program.cs` is the entry point for your project.

D. `MyStack.cs` is used to define your stack resources.

This class creates an Azure resource group and a storage account. The primary key for the storage account is then exported. You will find the location for the resource group in the `Pulumi.dev.yaml` file with the `azure-native:location` property name.

7. Now, deploy your stack with the following command:

```
Pulumi up
```

When prompted, select **Yes** to deploy your stack to Azure.

8. At this stage, you should be able to log into your Azure account and see the newly created resource, and that it is a storage account.

9. Add an HTML file to your project called `index.html` and edit the file by adding some HTML content and saving it. Here is some sample content:

```
<html><head><title>Sample
  HTML</title></head><body><h1>Hello, World!</h1>
<hr /><p>This is a sample paragraph.</p></body></html>
```

10. Add the following code to the `MyStack.cs` class immediately after the code block that creates the Azure storage account resource:

```
// Enable static website support
var staticWebsite = new StorageAccountStaticWebsite(
     "staticWebsite",
     new StorageAccountStaticWebsiteArgs
{
     AccountName = storageAccount.Name,
     ResourceGroupName = resourceGroup.Name,
     IndexDocument = "index.html",
});
```

With that, we have created a new static website resource that utilizes the storage account we've just created.

11. Next, add the following code after the code shown in *Step 10*:

```
// Upload the file
var index_html = new Blob("index.html", new BlobArgs
{
ResourceGroupName = resourceGroup.Name,
AccountName = storageAccount.Name,
ContainerName = staticWebsite.ContainerName,
Source = new FileAsset("index.html"),
ContentType = "text/html",
});
```

Here, we used our cloud resources and a local `FileAsset` to upload our `index.html` file to blob storage.

12. At the end of the constructor, add the following code:

```
// Web endpoint to the website
this.StaticEndpoint = storageAccount
    .PrimaryEndpoints.Apply(
        primaryEndpoints => primaryEndpoints.Web
    );
```

This code configures the web endpoint to our static website.

13. Add the following property just above the constructor:

```
[Output]
public Output<string> StaticEndpoint { get; set; }
```

This property provides our static website endpoint.

14. Now, it's time to deploy our changes by typing the following command:

```
pulumi up
```

This will upload the `index.html` file to blob storage and make our static website available to the public. You should see a URL that you can use to view the web page that you created and uploaded. The file should be visible in your blob storage, which you can view via the Azure portal or Azure Storage Explorer.

15. Once you are satisfied that the preceding code has worked for you, it is time to destroy the resources. Type the following command:

```
pulumi destroy
```

If you want to destroy the entire stack, type the following command:

```
pulumi stack rm dev
```

With that, the stack has been completely removed from Pulumi.

In this section, you learned how to manage your Azure stack with Pulumi. By using Visual Studio and the PowerShell command line, you created an Azure resource account and assigned blob storage to it. Then, you created a static website resource and used the cloud resources and local `FileAsset` to upload the static website, which consisted of a single file called `index.html`. You were able to view the file in blob storage and view the web page in your browser.

In the next section, we will look at some performance considerations for distributed systems.

Performance considerations for distributed computing

We now know how to develop distributed systems. But what about their performance? What kinds of things should we be aware of in terms of the performance of distributed systems?

The first consideration is the network connection between clients and servers. TCP collisions can result in lost packets of information. This can corrupt communication between multiple devices and cause connections to time out. The most common reason for TCP collisions is when two or more computers share the same IP address.

No computer on the same network should have the same address as another computer on the same network. This results in unpredictable network behavior that is detrimental to the performance and stability of a networked application. If you experience this situation, simply change the IP address of one of the computers to a different IP address.

Another problem that can result in slow network communication is **Domain Name Resolution** (**DNS**). If DNS is not correctly set, then accessing a network resource such as a web page or web service may take longer than expected and cause a connection or request to time out. It is worth noting that there is usually more than one DNS on a distributed network. You have the DNS server of the external network and your router, which has a DNS for your local network. Either of these could be responsible for slow DNS resolution. Some steps you can take to resolve DNS issues are as follows:

1. Check your network connectivity.
2. Check that your DNS addresses are correct and in the right order.
3. Ping the computer name, IP address, or base URL, such as google.co.uk, that you are trying to access to see if it responds or times out.
4. Identify the nameservers in use using `nslookup`.
5. Check the DNS suffix.
6. Check that the DNS settings have been configured to pull the DNS IP address from the DHCP server.

7. Use `ipconfig` to release and renew the DHCP address and DNS information.
8. Check the DNS server to see if any services need to be restarted or if the server needs to be rebooted.
9. Sometimes, the information on the router becomes stale, so a quick solution is to reboot the router.
10. Every so often, an ISP will run into problems of their own that affect you. In these cases, you will need to communicate with them to understand the problem and get some indication of when things will be back to normal.

Distributed firewalls may be employed to protect business networks. Misconfiguration of firewalls can result in resource access being denied or invisible. If machines are unable to access distributed resources, then the distributed firewall is a good place to start. If the distributed firewall is configured correctly, then check client and server firewalls to see if they are enabled or disabled and whether they are correctly configured or not.

For example, I have dealt with a lot of SQL Server problems. Some have been DNS and DHCP issues, but the most common issues are SQL Server configuration and firewall configurations. SQL Server uses dynamic ports. But sometimes, these can clash, as can fixed ports. Also, I have found that for SQL Server to work on many networks, the Named Pipes and TCP protocols must be enabled. Once these protocols have been changed within SQL Server Configuration Manager, you need to restart the instance of SQL Server they apply to, followed by the SQL Server Browser service. If you have firewalls in place, then the SQL Server executable for the instance will need to be added to the firewall as an application exception. If you need to use specific ports, then you would need to add port exceptions. Standard port exceptions for SQL Server are `1433` for TCP and `1434` for UDP.

Sometimes, even after the aforementioned SQL Server troubleshooting has been completed, networked applications can still not see the SQL Server instance. When this happens, a workaround is to recreate the database connection string with the following format: `IP_ADDRESS,PORT_NUMBER\INSTANCE_NAME`.

Another problem that can affect SQL Server connectivity in a distributed setting is the SQL Server driver that is installed and used. If you use specific versions of a SQL Server native client, then you will need to ensure that that particular version of the native client is installed on all computers for them to be able to connect to SQL Server. The way around this is to realize that the SQL Server driver is installed by default on all Windows computers, both server and client. If you use this driver, then you do not have to worry about rolling out SQL Server Native Clients to various computers that are part of your distributed system.

Another area of performance is the database querying aspect. The same query to obtain a set of results can be written in so many different ways to obtain the required results. This is especially true with larger result sets that have more joins. Dynamic SQL can also perform slowly. Therefore, speeding up queries can improve a database-driven distributed application significantly. You can use SQL Server Profile and review SQL Server Execution Plans to identify bottlenecks and rewrite SQL so that it's more performant. You can also add missing indexes, correct incorrect indexing, and use pre-compiled stored procedures for performance enhancement purposes.

SQL Server can become corrupt and fail for many reasons, so it must be regularly updated with security patches. Here, you can use Always-On and failover clusters to keep connections alive and switch between SQL Servers when a server is down or needs to be taken offline for maintenance.

The number of connections to a resource can also overload a distributed system to the point that clients cannot connect. To overcome this, you can employ load balancing so that when a resource server reaches a certain peak, clients are sent to an alternative server for those resources.

Another common oversight when sharing networked resources is network permissions. Sometimes, a folder may not be shared that should be shared. A really nasty problem can be that of a permissions hierarchy that is enforced via group policy that overrides even a network domain administrator's ability to do their job.

It is important to fully document your group policy and permissions structure for current and future staff. With a clear document that diagrammatically shows the permission groups and hierarchies, as well as provides a list of resources and their permission sets, if someone or an app is having trouble accessing resources, such documents can ease the pain of troubleshooting such problems.

It is also worth noting that sometimes, System and TrustedInstaller take control of certain resources and prevent you from being able to access resources locally and across a network. This can result in you having to override the ownership of that network or local location and file resource.

Computer security software can also significantly slow down network traffic and even stop programs from working. The usual culprits are the firewall, as we mentioned previously, and antivirus software. If your software is not code-signed with an authority-approved code signing certificate, then DLLs and executables can be quarantined and identified as harmful software. This is what is known as being identified as a false positive. You can either sign your software, add your software as an application or folder exception, or pass your software to security firms to assess your software and update their software to prevent this from happening in the future.

Antivirus software can also slow applications down when all network traffic and even local files are real-time scanned. An example of this is educational software that pulls audio files across the network during assessments. A characteristic that identifies this being the case is when the audio files are backed up and fired together. To overcome this problem, you can update the antivirus software by adding the application, its folders, and its resources as folder and or application/file exceptions.

The size of resources also affects network performance. The larger the size, the longer the time to request and receive a resource. Here, you can reduce the size of resources such as images, video, and audio files using various compression techniques. You can also zip resources up and transmit them before they need to be accessed, such as at application startup. You can store resources in the local cache once they have been requested and received.

When the workload increases to the point that your current system cannot handle it, you have two options: scale up vertically or scale out horizontally. Scaling up involves increasing the physical computing capacity to cope with the increased workload. Scaling out is when you add more servers to cope with the increased workload. At the time of writing, the way forward for many companies is to use server VMs and containers and have containers running in container management software such as Docker and Kubernetes on cloud platforms such as Azure, AWS, Google Cloud, and others.

Large libraries and executables can be made smaller by moving code into microservices such as Azure Functions. Azure Functions is an event-driven, compute-on-demand experience that extends the existing Azure application platform with capabilities to implement code triggered by events occurring in Azure or third-party services, as well as on-premises systems. These online services can then scale up and down and run only when they are required to do so. This has the added advantage of providing cost savings, such as electricity and equipment costs.

You can also use tools such as the browser developer tools and Postman to monitor application and network performance.

Now, let's summarize what we have learned.

Summary

In this chapter, we started by looking at the implementation of the CQRS design pattern. Then, we looked at an implementation of event sourcing. You can use both these patterns by themselves, though they can also be combined to provide very powerful and functional microservices.

Then, we took a high-level look at using Microsoft Azure for writing distributed systems. The benefits and negative aspects of containers and serverless functions were covered to help you understand when to use each technology.

In terms of Microsoft Azure, we focused mainly on Azure Functions. Specifically, we looked at Durable Azure Functions. We identified the various types of durable functions and various durable function patterns.

Now, take some time to answer this chapter's questions to see how much you have retained from this chapter. Please review the *Further reading* section to build upon what you have learned in this chapter.

In the next chapter, we will be looking at multithreaded programming in C#.

Questions

Answer the following questions to test your knowledge of this chapter:

1. What does CQRS stand for?
2. Why do we use the CQRS pattern when developing microservices?
3. What is event sourcing?
4. Why do we use event sourcing?
5. What are containers?
6. Why would we use containers?
7. What are serverless functions?
8. Why should we use serverless functions?
9. What are durable functions?
10. What are the different types of durable functions?
11. What types of durable function patterns are there?
12. What is Pulumi?
13. Why would we use Pulumi?

Further reading

To learn more about the topics that were covered in this chapter, take a look at the following resources:

- *Getting started with Pulumi on Azure*: `https://www.pulumi.com/docs/get-started/azure/`
- *Building Modern Cloud Applications using Pulumi and .NET Core*: `https://devblogs.microsoft.com/dotnet/building-modern-cloud-applications-using-pulumi-and-net-core/`
- *Orchestration Using Durable Azure Functions*: `https://blog.kiprosh.com/orchestration-using-durable-azure-function/`
- *Durable Functions Orchestrations*: `https://docs.microsoft.com/en-us/azure/azure-functions/durable/durable-functions-orchestrations?tabs=csharp`
- *Best Practices for Durable Functions Patterns*: `https://www.serverless360.com/blog/azure-durable-functions-patterns-best-practices`
- *Chapters 9* and *10* of *Clean Code in C#* by Jason Alls: `https://www.amazon.co.uk/Clean-Code-application-performance-practices-ebook/dp/B08614MS6S`
- *10 Ways to Troubleshoot DNS Resolution Issues*: `https://techgenix.com/10-Ways-Troubleshoot-DNS-Resolution-Issues/`

Part 3: Threading and Concurrency

Part 3 covers threading, parallel processing, and asynchronous processing. We discuss various ways to process code synchronously, asynchronously, and in parallel. In doing so, we learn how to reduce the time it takes to process a series of tasks, and how we can utilize the number of CPUs and cores.

This part contains the following chapters:

14 Multi-Threaded Programming

In this chapter, you will learn about **multi-threaded programming**. You will learn what threads are and about background and foreground threads. Then, you will learn how to pass data into threads before you run them. You will also learn how to pause, interrupt, destroy, schedule, and cancel threads.

In this chapter, we will be covering the following topics:

- **Understanding threads and threading**: This section covers the life cycle of threads.
- **Creating threads with and without parameters**: This section provides examples of thread creation with and without parameters.
- **Pausing and interrupting threads**: This section covers how to pause and interrupt threads.
- **Destroying and canceling threads**: This section covers destroying and canceling threads.
- **Scheduling threads**: This section covers how to schedule threads.
- **Thread synchronization and locks**: This section covers how to synchronize threads, protect resources, and prevent deadlocks and race conditions.

By the end of this chapter, you will have gained the following skills:

- You will understand threads and threading.
- You will be able to create threads with and without parameters.
- You will be able to pause and interrupt threads.
- You will be able to destroy and cancel threads.
- You will be able to schedule threads.

Technical requirements

To ensure that you benefit from this chapter, you should have the following requirements:

- Visual Studio 2022
- The book's source code from the following link: `https://github.com/PacktPublishing/High-Performance-Programming-in-CSharp-and-.NET/tree/master/CH14`.

Understanding threads and threading

In this section, we will understand the life cycle of threads. Threads in C# have a life cycle as follows:

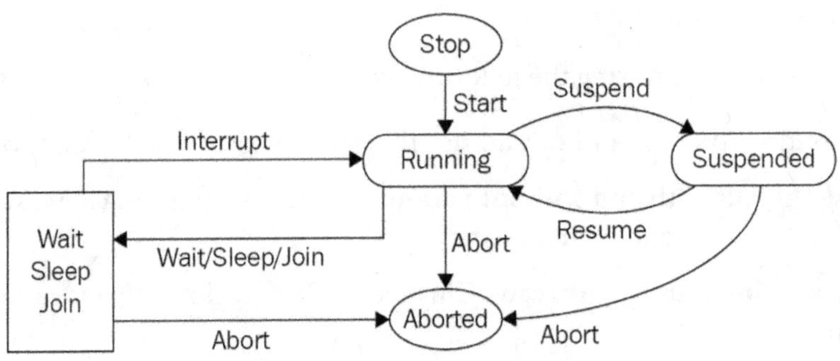

Figure 14.1 – The thread life cycle

When started, threads enter the **running** state. When running a thread, there is a possibility it will enter a **wait**, **sleep**, **join**, **stop**, or **suspended** state. A thread is suspended by calling the `Suspend` method, and calling the `Resume` method resumes a thread.

When the `Monitor.Wait(object obj)` method is called, the thread enters the `wait` state. A waiting thread will continue when the `Monitor.Pulse(object obj)` method is called, and you can make threads sleep by calling the `Thread.Sleep(int millisecondsTimeout)` method.

When you call the `Thread.Join()` method, it causes the thread to enter the `wait` state. The waiting thread will then continue once the dependent threads have completed running. If any dependent threads are canceled, the thread is aborted and enters the `stop` state. Once a thread has been completed or canceled, you cannot restart it.

> **Note**
>
> The `SYSLIB0006` compile-time warning will be raised by projects that target .NET 5 or higher if they call any of the `Thread.Abort` APIs. Microsoft recommends that you abort the `running` unit of work using `CancellationToken` instead. The `Thread.Abort` APIs are now obsolete.

In the next section, we will look at creating background and foreground threads with and without parameters.

Creating threads and using parameters

In this section, we look at the creation of threads. First, we will see how to create parameterless threads in the foreground and the background. Let's define both foreground and background threads as follows:

- **Foreground threads**: By default, threads run in the foreground. A process will continue to run if, at least, one foreground thread is running. Should the `Main` method be complete and the foreground thread is still running, the process will remain active until the foreground thread terminates.
- **Background threads**: Background threads are created in the same way as foreground threads. The main difference is that you must explicitly set the thread to run in the background.

The following code shows how to create and run a foreground thread:

```
var foregroundThread = new Thread(methodName);
foregroundThread.Start();
```

To create and run a background thread, you run the following code:

```
var backgroundThread = new Thread(methodName);
backgroundThread.IsBackground = true;
backgroundThread.Start();
```

Both versions of the code that generates foreground and background threads, that you have just seen, create threads without using parameters. The following code shows you how to create a thread using parameters:

```
static void ThreadCreationWithParameters()
{
    int result = 0;
    Thread thread = new Thread(() => { result = Add(1, 2); );
    thread.Start();
    thread.Join();
    Console.WriteLine($"The addition of 1 plus 2 is
        {result}." + $"");
}
static int Add(int a, int b)
{
    return a + b;
}
```

As you can see in the preceding code, the thread is used to sum two numbers and return the result. The thread calls the `Add` method and passes the two integers to be added. Both the method call and the result are placed within an anonymous function passed into the thread's constructor.

Creating multiple threads can be costly on performance. The performance of multiple-thread creation can be improved by using thread pools. Thread pools improve performance in multi-threaded applications by limiting the number of threads that should be created and managed.

When a new thread is created using a thread pool, it is kept there until it is needed. When required, the thread will run and complete its task. Once the task is completed, the thread will return to the thread pool for later reuse.

You can create a thread in a thread pool as follows:

```
ThreadPool
    .QueueUserWorkItem(
        new WaitCallback(ThreadPoolWorkerMethod)
    );
```

The thing to note when using a thread pool is that when first used, they have no history, but over time, they tune themselves to improve thread pool performance. For applications that use a large number of threads and put a heavy load on the CPU, it is possible that they will encounter a high startup cost. Threads have to be created and made available to the thread pool. This can cause the thread pool to have to wait until those threads are made available. A performance tweak you can make at startup is to set the minimum number of threads. The following code shows how to set the minimum number of threads:

```
const int WorkerThreads = 12;
const int CompletionPortThreads = 12;
ThreadPool.SetMinThreads(WorkerThreads,
   CompletionPortThreads);
```

The `WorkerThreads` value is the minimum number of worker threads created on demand by the `ThreadPool`. The `CompletionPortThreads` value is the number of asynchronous I/O threads created on demand by the `ThreadPool`.

In addition to setting the minimum number of threads, you can set the maximum number of threads as follows:

```
const int WorkerThreads = 12;
const int CompletionPortThreads = 12;
ThreadPool.SetMaxThreads(WorkerThreads, CompletionPortThreads);
```

In order for these settings to help with application performance, you need to set them correctly. Otherwise, you can end up creating too many threads and overscheduling tasks. This will reduce performance by increasing context switching, which will put more load on the CPU. The `ThreadPool` is intelligent enough to switch to an algorithm that will reduce the amount of work the CPU has to do once it gathers a history.

Before settings these values, it is a good idea to use performance monitoring to monitor the thread usage and context switching of your application. You can use performance counters tracing using the Contextual Visualizer, which is discussed in the following chapter. You can also use the `ThreadPool.GetMaxThreads` and `ThreadPool.GetMinThreads` methods to help you analyze the optimal values for setting the minimum and maximum numbers of worker threads and completion port threads.

You can also set a thread's priority. However, you have to be very careful about setting a thread priority as it can have a negative impact on other threads and other applications. Setting threads to a higher priority can starve lower priority threads, resulting in them rarely running.

Only when a fast response is required for an event, such as an exception, should you consider changing thread priority to a high value. When race conditions are encountered, you can legitimately lower a thread's priority. Threads that do not run for a while because of their lower priority will run at some point. This is because the dynamic priority of a thread is increased by Windows the longer it goes without running.

If you do change the priority of a thread, its priority will be reset on entry back into the pool. However, a thread may be used for several tasks. In this case, the thread will not return to the pool until these tasks are completed. If the priority is set incorrectly, then this can degrade both application performance and system-wide performance.

We now understand how to create and run threads. Let's turn our attention to pausing and interrupting threads.

Pausing and interrupting threads

In this section, we will look at pausing and interrupting threads. An example of why you would need to pause or interrupt a thread is if the code running is a debugger. If a thread is executing and it hits a breakpoint, it would need to be paused.

The most common way to pause/delay a thread is to call `Thread.Sleep(millisecondsDuration)`, but this may freeze the main thread and your users may think your program has stopped working, leading them to terminate it.

A better way to delay a thread is to let `Task.Delay(TimeSpan)` run in the background. This will allow the thread to work in the background and prevent the delayed thread from stopping the main thread from doing its work.

The following code shows how to delay a thread:

```
static void Main(string[] args)
{
    Console.WriteLine($"Current Time: {DateTime.Now}");
    var delay = Task.Delay(TimeSpan.FromSeconds(5));
    var duration = 0;
    while (!delay.IsCompleted)
    {
        duration++;
        Thread.Sleep(TimeSpan.FromSeconds(5));
        Console.WriteLine($"Slept for {seconds} seconds");
    }
    Console.WriteLine($"Delay End:{DateTime.Now} after
        {duration} seconds");
  }
}
```

We create the task with a time delay of five seconds. The loop keeps running until the time delay has been completed.

The `Interrupt` method is called to interrupt a thread that is in a blocked state of `wait`, `sleep`, or `join`. When the method is called, `ThreadInterruptedException` is raised. This exception is not raised when calling the `Interrupt` method on a thread not in a blocked state.

Destroying and canceling threads

Aborting threads is not a good idea as you don't always know the state of a thread. It can be made worse if the thread is part of a static constructor. Using `Thread.Abort` to abort a thread is one of the main reasons for application crashes. The `Thread.Abort` APIs are now obsolete. So, you are encouraged to use the cooperative cancellation pattern to periodically check for cancellations using `CancellationToken`.

Under normal circumstances, when a thread is aborted, it is destroyed. The cancellation of a thread also destroys the thread. Let's write some sample code that demonstrates the usage of `CancellationToken` to cancel a synchronous operation when it times out, as follows:

1. Start a new .NET 6 console application and call it CH14_Multithreading.

2. In the *Program.cs* file of the *CH14_Multithreading* project, add the following method:

```
static bool TryCallWithTimeout<TResult>(
      Func<CancellationToken, TResult> function,
      TimeSpan timeout,
      out TResult result
)
{
     var cancellationTokentSource =
         new CancellationTokenSource(timeout);
     try
     {
         result =
         function(cancellationTokentSource.Token);
         return true;
     }
     catch (TaskCanceledException)
     {
     }
     finally
     {
         cancellationTokentSource.Dispose();
     }
     result = default;
     return false;
}
```

This method receives a method to execute over a specified timeout period and returns a result. `SleepyMethod` is executed, but if it exceeds the timeout value, then `TaskCanceledException` is raised and then `CancellationTokenSource` is disposed of.

3. Add the `SleepyMethod` code as follows:

```
static int SleepyMethod(CancellationToken ct)
{
    for (var i = 0; i < 10; i++)
    {
        Thread.Sleep(TimeSpan.FromMilliseconds(500));
```

```
            if (ct.IsCancellationRequested) { throw new
                TaskCanceledException(); }
        }
        return 1234567890;
    }
```

The SleepMethod accepts CancellationToken as a parameter. It then loops ten times. During each iteration, it sleeps for half a second. Then, it checks to see whether cancellation has been requested. If cancellation has been requested, then TaskCanceledException is raised. Otherwise, the value of the method is returned.

4. Add the SynchronousThreadCancelation method as follows:

```
static void SyncrhonousThreadCancelation()
{
    TimeSpan timeoutTimeSpan = TimeSpan
        .FromMilliseconds(750);
    bool callResult = TryCallWithTimeout(
        SleepyMethod,
        timeoutTimeSpan,
        out int result
    );
    Console.WriteLine($"SleepyMethod() {
        (callResult ? "Executed" : "Cancelled" )
    }");
}
```

This method creates a timeout value of three-quarters of a second. It then calls the TryCallWithTimeout method, which returns a Boolean value. The parameters passed into the TryCallWithTimeout method are the following:

- SleepyMethod: The name of the method to be executed
- timoutTimeSpan: The duration the method is to run for before it times out
- result: Contains the result of CancellationToken

Once the call has been made, the name of the called method and its call result are sent to the console. In this code, we are not writing the result to the console window, but you can modify the code to do so.

5. At the top of the class, update the code as follows:

```
SyncrhonousThreadCancelation();
```

The preceding code calls our method and is an example of the cancellation of a synchronous operation.

6. Run the preceding code and the result should look something like the following:

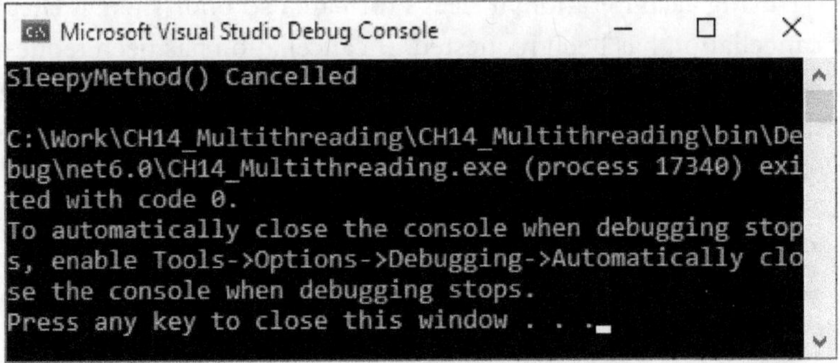

Figure 14.2 – Console output for our program showing that the thread was canceled

This concludes the topic of canceling and destroying threads. Let's now look at scheduling threads.

Scheduling threads

The `Thread.Start` method schedules a `Thread` to start. You can overload this method with different parameters. We will look at two examples in this section. The first example will call the `Thread.Start()` method without passing any parameters, and the second will call `Thread.Start(object)`.

We will now write the code as follows:

1. Add a class called `Job` as follows:

```
internal class Job
{
    public void Execute()
    {
        Console.WriteLine(
            "Execute() method execute.");
    }
    public void PrintMessage(object message)
```

```
        {
            Console.WriteLine($"Message: {message}");
        }
    }
```

This class provides two methods that will be used in our `Thread` scheduling examples. The `Execute` method is used with the parameterless `Thread.Start` method, and the `PrintMessage` function is used with the `Thread.Start` method that takes parameters.

2. In the `Program.cs` class, add the `SheduleThreadWithoutParameters` method as follows:

```
static void ScheduleThreadWithoutParameters()
{
    Job job = new();
    Thread thread =
        new Thread(new ThreadStart(job.Execute));
    thread.Start();
}
```

In the preceding code, we create a new instance of the `Job` class. Then, we create a new `Thread` passing a new `ThreadStart` instance into its constructor. Into the `ThreadStart` constructor, we pass `object.method` that we wish to execute, and then we start the thread.

3. Add the `ScheduleThreadWithParameters` method as follows:

```
static void ScheduleThreadWithParameters()
{
    Job job = new();
    var thread1 = new Thread(
        new ParameterizedThreadStart(
            job.PrintMessage
        )
    );
    var thread2 = new Thread(
        new ParameterizedThreadStart(
            job.PrintMessage
        )
    );
```

```
        thread1.Start("Hello, world!");
        thread2.Start("Goodbye, world!");
    }
```

In the preceding code, we created a new `Job` instance and two threads by calling the `ParameterizedThreadStart` class for each thread to execute a parameterized method on an object. We then start each of the threads.

4. Add a call to each of the methods at the top of the class and then run the preceding code. Your console should look like the following:

14.3 – Our parameterized thread output

Thread synchronization and locking

When using multiple threads in an application, you have to consider thread synchronization and locking. If you don't, you can end up with race conditions and deadlocks. There are several ways to synchronize threads. You can use interlocked methods and synchronization objects, such as `Monitor`, `Semaphore`, and `ManualResetEvent`.

> **Note**
> In *Chapter 8*, *Threading and Concurrency*, in the *Clean Code in C#* book, we provide a detailed discussion on threads covering using threads, thread safety, parallel threads using semaphores, thread synchronization and preventing deadlocks, and race conditions.

To synchronize your code, you can use a lock object as follows:

```
internal class LockMutexExample
{
public object _lockObject = new();
public void UsingLockObject()
{
```

```
lock( _ lockObject)
{
// Perform your unsafe code here.
}
}
}
```

When the locked code is entered, all of the other threads are barred from accessing the locked code. The only downside to this is that you can end up with a deadlock. This can be overcome by using a mutex as follows:

```
internal class LockMutextExample
{
    private static readonly Mutex _ mutex = new();
    public void UsingMutext()
    {
      try
      {
          _ mutex.WaitOne();
          // ... Do work here ...
      }
      finally
      {
          _ mutex.ReleaseMutex();
      }
    }
}
```

The preceding code declares a `Mutex` class-level variable. The code that needs protecting is then wrapped in a `try/catch` block. The current thread is blocked by the `WaitOne()` method until the wait handle receives a signal. `True` is then returned from the `WaitOne()` method upon a `Mutex` being signaled. The `Mutex` is then owned by the calling thread that can access protected resources. Once the protected resources are finished, the `Mutex` is released by calling `ReleaseMutext()`. Always call the `ReleaseMutext()` method in the final block to prevent resources from remaining locked if an exception is encountered.

Race conditions happen when the same resource is accessed by multiple threads that produce different outcomes based on their timings. A race condition can be avoided by using code such as the following:

```
Task
    .Run(() => Method1())
    .ContinueWith(task => Method2())
    .Wait();
```

The `Task` runs `Method1()` and then continues with `Method2()`. We then `Wait()` for the `Task` to complete its execution of `Method1()` and `Method2()` before continuing.

That concludes our look at multi-threaded programming. As you can see, there is not much to scheduling threads. Let's summarize what we have learned in this chapter.

Summary

In this chapter, we have come to an understanding of threads and the thread life cycle. We built some sample code that shows how to create threads with and without parameters. We also looked at running threads in the foreground and background.

Next, we looked at pausing and interrupting threads. Then, we moved on to destroying and canceling threads. You no longer use `Thread.Abort` in your code. `Thread.Abort` has been responsible for applications crashing at runtime. Instead, you use cancellation tokens. Canceling threads also destroys them.

We looked at scheduling threads with and without parameters. In the next chapter, we will be looking at parallel programming.

Finally, we looked at thread synchronization and locking using lock objects and mutexes and learned how to avoid deadlocks and race conditions.

It is now time to answer some questions to see how well you have retained the knowledge in this chapter. Once you have completed the questions, the *Further reading* section provides some external sources to further your knowledge on threads and multi-threaded programming.

Questions

1. What states can a thread be in?
2. Which part of the `Thread.Abort` API do you use to terminate a thread?
3. Which two locations can a thread be executed in?
4. What is the correct way to terminate a thread?
5. What method is used to schedule a thread?

Further reading

- Managing and implementing multi-threading: `https://subscription.packtpub.com/book/programming/9781789536577/6/ch06lvl1sec52/understanding-threads-and-the-threading-process`
- Pausing and interrupting threads: `https://docs.microsoft.com/en-us/dotnet/standard/threading/pausing-and-resuming-threads`
- How to terminate a thread in C#: `https://www.geeksforgeeks.org/how-to-terminate-a-thread-in-c-sharp/`
- How to destroy threads in C#: `https://www.tutorialspoint.com/How-to-destroy-threads-in-Chash`
- How to schedule threads in C#: `https://www.geeksforgeeks.org/how-to-schedule-a-thread-for-execution-in-c-sharp/#:~:text=%20How%20to%20schedule%20a%20thread%20for%20execution,1%20Start%20%28%29%202%20Start%2-0%28Object%29%20More%20`
- Understanding threads and the threading process: `https://subscription.packtpub.com/book/programming/9781789536577/6/ch06lvl1sec52/understanding-threads-and-the-threading-process`
- How to pause code execution in C#: `https://csharpsage.com/c-delay/`
- Pausing and interrupting threads: `https://docs.microsoft.com/en-us/dotnet/standard/threading/pausing-and-resuming-threads`

15
Parallel Programming

In this chapter, you will learn how to take advantage of the multiple CPU cores that are available in today's modern computers. You will learn how to process your code by distributing the work between processes concurrently, as well as how to use the **Task Parallel Library** (TPL) and **Parallel LINQ** (**PLINQ**) to run code in parallel. Throughout this book, you will learn how to use parallel data structures and use the Visual Studio debugger to diagnose tasks and parallel stacks. You will also learn about the Concurrency Visualizer.

In this chapter, we will cover the following topics:

- **Using the Task Parallel Library (TPL)**: In this section, we will compare parallel and non-parallel code and its effect on CPU core utilization using *perfmon*.
- **Using Parallel LINQ (PLINQ)**: In this section, we will look at PLINQ and how it can be used to execute LINQ statements with varying degrees of parallelism.
- **Programming parallel data structures**: In this section, we will review some of the thread-safe collections you can use for programming parallel data structures.
- **Benchmarking with BenchmarkDotNet**: In this section, we will look at benchmarking our parallel code and find that, in some instances, it can be faster than non-parallel code, and at other times, it can be slower.

- **Using lambda expressions with TPL and LINQ**: In this section, we will review a piece of code that uses lambda expressions to express the `Func` and `Action` delegates.

By the end of this chapter, you will be able to do the following:

- Use TPL and PLINQ for parallel programming tasks.
- Program parallel data structures.
- Diagnose issues with tasks and parallel data structures.
- Use lambda expressions in TPL and PLINQ queries.

Technical requirements

For this chapter, you will need the following:

- Visual Studio 2022
- This book's source code: `https://github.com/PacktPublishing/High-Performance-Programming-in-CSharp-and-.NET/tree/master/CH15`
- Concurrency Visualizer for Visual Studio 2022: `https://marketplace.visualstudio.com/items?itemName=Diagnostics.DiagnosticsConcurrencyVisualizer2022#overview`

Using the Task Parallel Library (TPL)

In this chapter, we will be working with TPL to enhance the performance of our programs by making use of the available processor power on a machine.

We learned how to write threads and execute them in *Chapter 14, Multi-Threaded Programming*. When multiple threads are running on a single processor, providing the illusion that they are running in parallel, they are running concurrently.

When threads run concurrently, the processor uses a scheduling algorithm and/or interrupts to determine the switching and prioritization between threads. Parallel programming, however, runs different threads on different processors so that threads execute in parallel to each other with a reduced need for switching and thread interrupts.

As its name suggests, TPL is used to run tasks in parallel. Tasks are run in parallel by running each task against a separate core of the computer's processor. So, for example, say your computer has four cores and you have four tasks. Each task would run on a separate core, and each task would be run parallel to the other three. This helps improve the overall performance of the code as you can have as many tasks executing in parallel as you have processor cores.

Also, if you have a big dataset that needs to process many records and store them in a variable, you can partition the task so that the records are split into different threads running on different processors. These are then synced backed together and stored in a variable.

> **Note**
>
> Code that cannot be parallelized will slow down parallel tasks, as will code that must be partitioned and scheduled by the task scheduler. It is always a good idea to profile your code to see if the methods you are employing will speed up or slow things down.

A good way to see the value in parallel programming is to compare a thread running on a single processor against the same code split between different processors. Let's write some code for this comparison:

1. Start a new console application and call it `CH15_ParallelProgramming`. Then, check the checkbox that says **do not use top-level statements**.
2. Add the following `using` statement:

```
using System.Threading.Tasks;
```

   This `using` statement gives us access to TPL.

3. Update the `Main` method in the `Program` class, as follows:

```
static void Main(string[] _)
{
  RunSingleProcessorExample();
}
```

   This method calls the `RunSingleProcessorExample` method.

4. Add the `RunSingleProcessorExample` method:

```
static void RunSingleProcessorExample()
{
  Thread thread = new(SingleProcessorExample);
  thread.Start();
}
```

This method creates a new thread and assigns it the `SingleProcessorExample` method, which it will invoke. The method is then invoked using the `Start` method.

5. Now, add `SingleProcessorMethod`:

```
static void SingleProcessorExample()
{
string output = "Index: ";
    for (int index = 0; index < 1000000; index++)
{
      Console.WriteLine($"{output}{index}");
}
    Console.ReadKey();
}
```

This method writes the value of the `for` loop index to the console window 1 million times and then pauses until it receives a user keypress.

6. Type `Performance Monitor` into your task bar's search area and open it. Then, remove the existing counter, and then add a counter to view the processor time for all the processors on your computer. If you need to, you can change the thickness of the lines.

7. Clear out the **Performance Monitor** screen and then run the console app. You should see something similar to the following:

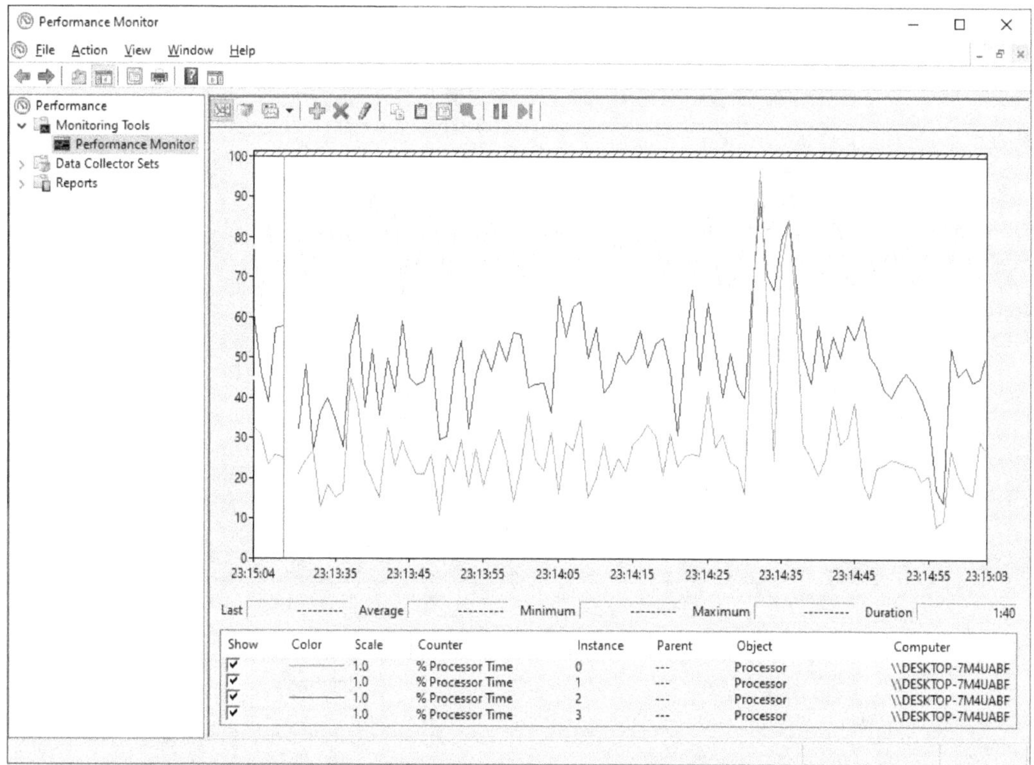

Figure 15.1 – Performance Monitor with our console application running

As you can see, processor instance 1 is the most utilized processor. What we need to do is modify the program to utilize all available processors.

8. Comment out the method call in the `Main` method and add the following code after the commented-out method:

```
Parallel.For(
    0, 1000000, x => MultipleProcessorExample(x)
);
```

This code uses a parallel `for` loop to process the `MultipleProcessorExample` method 1 million times.

9. Run the code again. You should see the following in Performance Monitor:

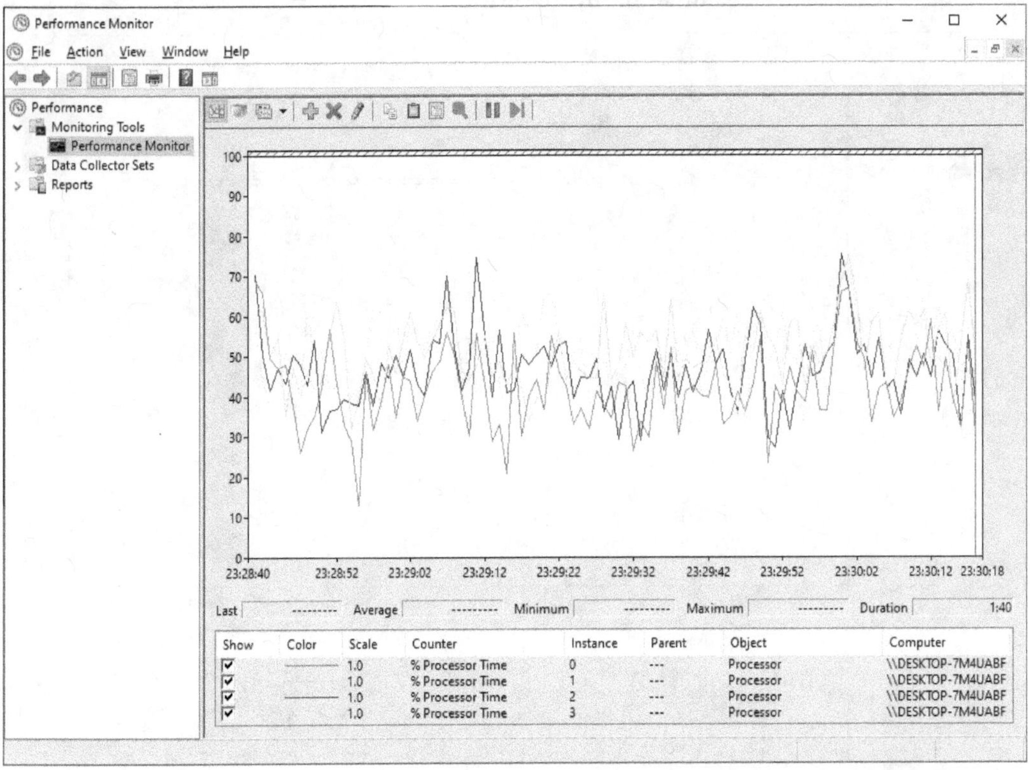

Figure 15.2 – Performance Monitor showing all our processors being used by our modified program

As you can see, with very minimal code, you can go from utilizing a single processor to utilizing all the processors using TPL. In previous chapters, you learned how to use *BenchmarkDotNET* to benchmark the performance of different variations of the same code. When deciding whether to turn your single processor code into multiple processor code, it's a good idea to benchmark. There is an overhead to using parallel code, so you need to ensure that parallel code will improve your program.

Now, let's learn how to use PLINQ.

Using Parallel LINQ (PLINQ)

In this section, you will learn how to convert your sequential LINQ queries into parallel LINQ using PLINQ. Take a look at the following code:

```
var productNames = GetProductNames();
var names = from name in productNames
```

```
                where name.Length > 8
                select name;
```

The preceding code calls the `GetProductNames` method and stores the results in the `productNames` variable. A LINQ statement is then performed on the `productNames` list to extract a list of all product names greater than eight characters in length. The result of this LINQ statement is then stored in the `names` variable.

The following code is identical to the preceding code, except we have modified it so that it operates in parallel across multiple processors:

```
var productNames = GetProductNames();
var names = from name in productNames.AsParallel()
            where name.Length > 8
            select name;
```

Here, we can see that the only change to the LINQ statement to get it to execute as parallel LINQ is to add the `AsParallel()` method call. The rest of the code stays the same.

If you want the data to be returned from the PLINQ statement, then suffix the `AsParallel()` call with the `AsOrdered()` call:

```
var productNames = GetProductNames();
var names = from name in productNames
              .AsParallel().AsOrdered()
            where name.Length > 8
            select name;
```

The preceding code will return a list of product names whose lengths are greater than `8` in alphabetical order.

PLINQ utilizes all the processors on the executing computer. However, you can limit the number of processors that are used by PLINQ using the `WithDegreeOfParallelism` call, passing in the number of processors you want to limit PLINQ being executed on:

```
var productNames = GetProductNames();
var names = from name in productNames
              .AsParallel()
              .WithDegreeOfParallelism(2)
            where name.Length > 8
            select name;
```

The preceding code has been limited to running on only two processors.

The following are some performance considerations when using PLINQ:

- Don't use PLINQ on single-core computers. This would result in slower performance than using standard LINQ.
- `AsOrdered()` will slow PLINQ down. Only use it if you need to. Benchmark alternative ordering techniques to see which is quickest, and then implement the quickest method.
- Employ production-sized datasets when developing and testing your PLINQ code. This will reveal performance issues sooner rather than later!
- Avoid using PLINQ on small collections since this could provide less performance. This is because PLINQ has been optimized for large datasets.

In the next section, we will consider some data structures that are suitable for parallel programming.

Programming parallel data structures

When we do parallel programming, we should always consider that we are using threads. Therefore, we should use data structures that are thread-safe.

For types that implement the `IProducerConsumerCollection<T>` interface, you should use the generic `BlockingCollection<T>` class, which provides bounding and blocking functionality. Use the `ConcurrentDictionary<TKey, TValue>` class for thread-safe dictionaries. For thread-safe FIFO queues, use the `ConcurrentQueue<T>` class. Use the `ConcurrentStack<T>` class for LIFO stacks. For a thread-safe implementation of a collection of elements, use the `ConcurrentBag<T>` class. Finally, for types to be used in a `BlockingCollection`, implement the `IProducerConsumerCollection<T>` class.

You can read more about thread-safe collections on the Microsoft Docs website: `https://docs.microsoft.com/en-us/dotnet/standard/collections/thread-safe/`.

Next, we'll look at benchmarking loops, LINQ, and PLINQ.

Benchmarking with BenchmarkDotNet

In this section, we will benchmark some methods to determine which method gives us the best performance. Keep in mind that there is some initial expense when running code in parallel. So, sometimes, parallel code may not be the best option for improving code performance. Let's get started:

1. Comment out the code in the `Main` method and add the following line:

   ```
   BenchmarkRunner.Run<Benchmarks>();
   ```

2. Add a class called `Benchmarks`.
3. Add the following `NuGet` packages:

   I. `BenchmarkDotNet`

   II. `LinqOptimizer.Csharp`

4. Add the `using` statements for each of the `NuGet` packages to the `Benchmarks` class.
5. Add the following code to set up our benchmarks:

   ```
   private short[] data;
   [GlobalSetup]
   public void GlobalSetup()
   {
         integers = new Int16[Int16.MaxValue];
         for (short x = 1; x <= integers.Length - 1; x++)
         {
         integers[x] = x;
         }

   }
   ```

   Here, we are declaring an array that's a short data type. The array is then initialized and filled with values. This array will be used by two of the following six methods.

6. Add the `StandardForLoopExample` method:

   ```
   [Benchmark]
   public void StandardForEachLoopExample()
   {
         foreach (int x in integers)
   ```

```
            Console.WriteLine($"Item {x}: {x}");
    }
```

The preceding code uses a standard `foreach` loop to loop through the values in the data array and then writes the value of the array at the given index to the console window.

7. Add the `ParallelForLoopExample` method:

```
[Benchmark]
public void ParallelForEachLoopExample()
{
        Parallel.ForEach(integers, x => {
            Console.WriteLine($"Item {x}: {x}");
        });
}
```

The preceding code does the same as the preceding code but executes the code using PLINQ.

8. Add the `UrlDownloader1` method:

```
public List<string> DownloadWebsites1()
        {
            List<string> websitesContent = new();
            HttpClient httpClient = new();

            string[]? websites = new[]
            {
            "https://docs.microsoft.com",
             "https://ownCloud.com",
             "https://www.oanda.com/uk-en/",
             "https://azure.microsoft.com/en-gb/"
            };

            foreach (string? website in websites)
            {
                Console.WriteLine($"Downloading of
                    {website} content has started.");
                string websiteContent =
```

```
                httpClient.GetStringAsync(website)
                .GetAwaiter().GetResult();
                websitesContent.Add(websiteContent);
                Console.WriteLine($"Downloading of
                    {website} content has finished.");
            }

            httpClient.Dispose();

            return websitesContent;
        }
```

The preceding code creates an array of URLs and downloads their content using a `foreach` loop.

9. Add the `UrlDownloader2` method:

```
[Benchmark]
        public List<string> DownloadWebsites2()
        {
            List<string> websitesContent = new();

            string[]? websites = new[]
                {
            "https://docs.microsoft.com",
            "https://ownCloud.com",
            "https://www.oanda.com/uk-en/",
            "https://azure.microsoft.com/en-gb/"
                };

            Task[]? downloadJobs = websites
                .Select(jobs => Task.Factory.StartNew(
                    state =>
                    {
                        using HttpClient? httpClient = new
                            HttpClient();
                        string? website = state == null ?
                            String.Empty : (string)state;
```

```
                    Console.WriteLine($"Downloading of
                      {website} content has started.");
                    string result =
                    httpClient.GetStringAsync(website)
                    .GetAwaiter().GetResult();
                    websitesContent.Add(result);
                    Console.WriteLine($"Downloading of
                    {website} content has finished.");
                }, jobs)
            )
            .ToArray();

        Task.WaitAll(downloadJobs);
        return websitesContent;
    }
```

The preceding code creates an array of URLs and downloads them as a set of tasks. The code waits for all the tasks to complete before the content is returned.

10. Add the `Urldownloader3` method:

```
[Benchmark]
    public List<string> DownloadWebsites3()
    {
        List<string> websitesContent = new();
        HttpClient httpClient = new();

        List<string> websites = new()
        {
        "https://docs.microsoft.com",
        "https://ownCloud.com",
         "https://www.oanda.com/uk-en/",
         "https://azure.microsoft.com/en-gb/"
        };

        websites.ForEach(website =>
        {
            Console.WriteLine($"Downloading of
```

```
                {website} content has started.");
            string result =
                httpClient.GetStringAsync(website)
                    .GetAwaiter().GetResult();
            websitesContent.Add(result);
            Console.WriteLine($"Downloading of
                {website} content has finished.");
        });

        httpClient.Dispose();

        return websitesContent;
    }
```

The preceding code uses a `Parallel.ForeEach` loop to download the contents of URLs stored in an array.

11. Make sure that your project is set to Release mode, and then run your program. The program will take some time to execute. However, once it has finished executing, you should see something similar to the following:

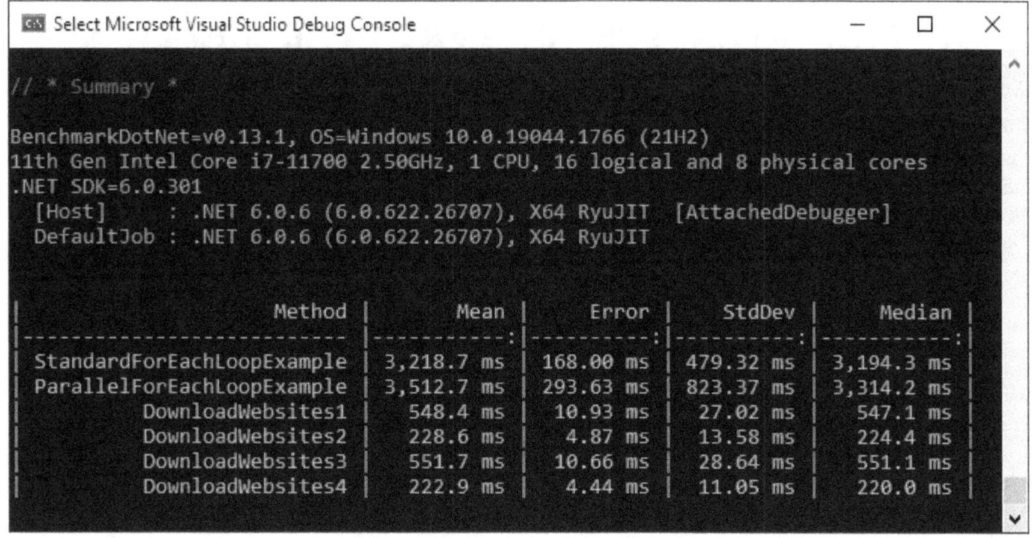

Figure 15.3 – BenchmarkDotNet results

Looking at the `ForEachLoop` examples, we can see that the standard `foreach` loop executed faster than our `Parallel.ForEach` loop. And so, in this example, using parallel code was slightly slower than using non-parallel code. But if the dataset was much larger and the data type was more complex, then the results could show that parallel code performs faster.

When looking at our `UrlDownloader` methods, `UrlDownloader4` uses the `Parallel.ForEach` loop, which is much faster than the two methods that use the `foreach` loop and `foreach` with lambda methods. However, the method that creates an array of tasks and waits for them all to complete is slightly faster than the `Parallel.ForEach` loop.

From these test results, we can see that we have different ways to perform the same actions, and each method's processing speed is different. In some cases, we have seen that parallel code is slower than non-parallel code, while in others, we have seen that parallel code is faster than non-parallel code.

When performance is an issue, you can use BenchmarkDotNet to test the efficiency of different approaches to the same task. Then, you can choose the most efficient option for the problem that you are trying to solve.

In the next section, we will learn how to use lambda expressions with TPL and LINQ.

Using lambda expressions with TPL and LINQ

There are several methods in TPL that take a `System.Func<TResult>` or `System.Action` delegate as an input parameter. These can be used to pass custom logic into a task, query, or parallel loop. Inline blocks can be used when creating delegates.

Use `Func` delegates to encapsulate methods that return a value and use `Action` delegates to encapsulate methods that do not return values. Let's review the following example:

```
static void FuncAction()
{
    int[] numbers = { 15, 10, 12, 17, 11, 13, 16,
        14, 18 };
    int additionResult = 0;

    try
    {
        Parallel.ForEach(
            numbers,
```

```
                () => 0,
                (number, currentState, addition) =>
                {
                    addition += number;
                    Console.WriteLine($"Thread:
                   {Thread.CurrentThread.
                    ManagedThreadId}, Number:
                    {number}, Addition: {addition}");
                    return addition;
                },
                (addition) => Interlocked.Add(ref
                     additionResult, addition)
            );
            Console.WriteLine($"Addition Result:
                {additionResult}");
        }
        catch (AggregateException e)
        {
            Console.WriteLine($"Aggregate Exception:
                FuncAction.\n{e.Message}");
        }
    }
```

The preceding code shows how to use the `Parallel.ForEach` method and a thread-local state. We expect the code to execute in parallel and sum up all the values stored in the `int` array. Each thread of the `Parallel.For` loop maintains a local addition variable. This addition variable is set to `0` when each thread is initialized. With each iteration, the addition is incremented with the number value. Once the thread has completed its task, the local sum for that thread is safely added to the global sum. The global sum is then printed out once the loop is complete.

The preceding code also demonstrates how to use lambda expressions to express both `Func` and `Action` delegates:

```
]Parallel.ForEach<TSource,TLocal>(IEnumerable<TSource>,
    Func<TLocal>, Func<TSource,ParallelLoopState,Tlocal
        ,TLocal>, Action<TLocal>).
```

In the next section, we will look at some parallel debugging tools.

Parallel debugging and profiling tools

In this section, we will look at three parallel application debugging and profiling tools. These are the **Parallel Stacks** window, the **Tasks** pane, and the Concurrency Visualizer. You will need to open the `CH15_ParallelProgrammingDebuggingAndProfilingSample` project for this. We will be using this project as we work through the next three sections.

The Parallel Stacks window

Run the program until it is paused by the debugger. Then, from the **Visual Studio** menu, select **Debug** | **Windows** | **Parallel Tasks**. This will display the **Parallel Tasks** window. You should see the following:

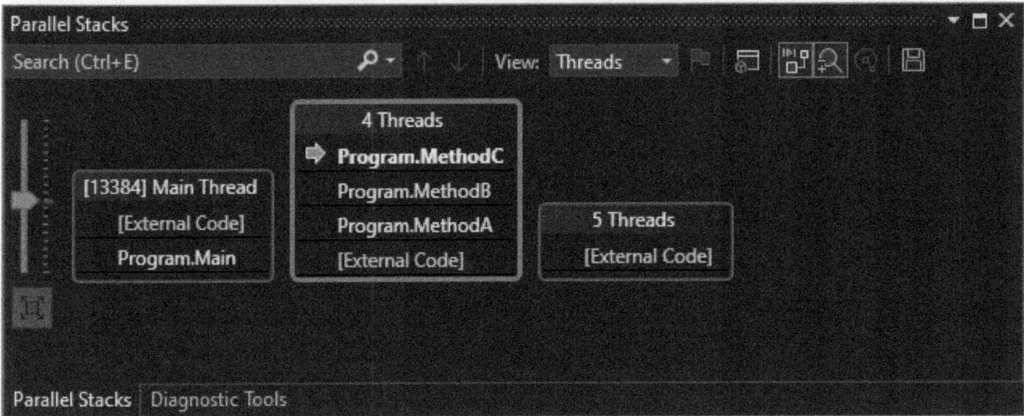

Figure 15.4 – The Parallel Stacks thread view

As you can see, our main thread is initiated via our `Program.Main` method. We can see that the debugger is paused in `Program.MethodC`. There are four threads – one each for methods A, B, and C, and a fourth in external code. There are also five threads running – these are external code threads.

If you hover over the methods, you will see the following popup:

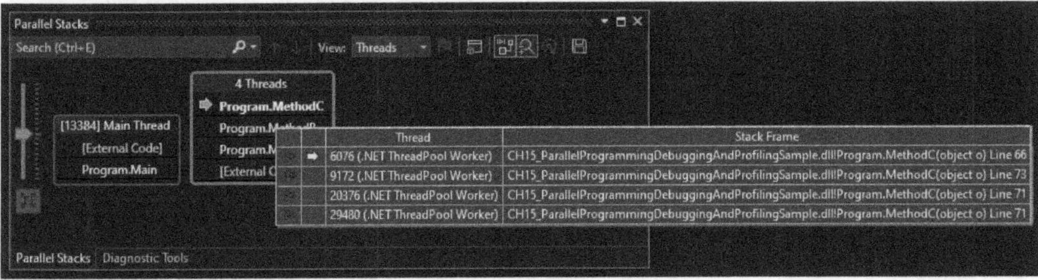

Figure 15.5 – The Parallel Stacks thread view with the Thread and Stack Frame view displayed

By hovering over each method group, you get to see a table of threads and their stack frames. These stack frames provide the method name and line number. The active stack frame of the current thread is identified by the yellow arrow. If you right-click while hovering over the stack frame, you can select what details to show, including parameter values, as shown here:

		Thread	Stack Frame
	➡	6076 (.NET ThreadPool Worker)	CH15_ParallelProgrammingDebuggingAndProfilingSample.dll!Program.MethodB(object obj = 1) Line 50
		9172 (.NET ThreadPool Worker)	CH15_ParallelProgrammingDebuggingAndProfilingSample.dll!Program.MethodB(object obj = 2) Line 50
		20376 (.NET ThreadPool Worker)	CH15_ParallelProgrammingDebuggingAndProfilingSample.dll!Program.MethodB(object obj = 3) Line 50
		29480 (.NET ThreadPool Worker)	CH15_ParallelProgrammingDebuggingAndProfilingSample.dll!Program.MethodB(object obj = 4) Line 50

Figure 15.6 – The Thread and Stack Frame view

Here, we can see the values of each of the parameters of our thread methods. Next, we will look at the **Tasks** window.

The Tasks window

To view the **Tasks** window, from the **Parallel Tasks** tab, select **Tasks** from the dropdown. You should see the following:

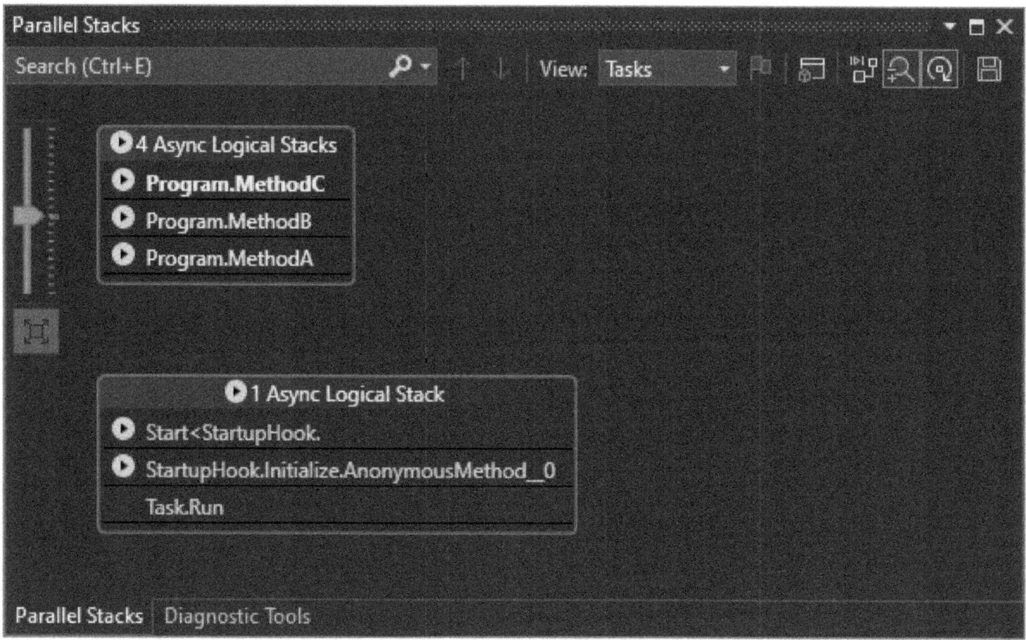

Figure 15.7 – The Tasks view

The preceding screenshot shows the async logical stacks. If you hover over each method, you will see the following window pop up, as you did with the threads view:

		Thread	Stack Frame
	➡	6076 (.NET ThreadPool Worker)	CH15_ParallelProgrammingDebuggingAndProfilingSample.dll!Program.MethodB(object obj = 1) Line 50
		9172 (.NET ThreadPool Worker)	CH15_ParallelProgrammingDebuggingAndProfilingSample.dll!Program.MethodB(object obj = 2) Line 50
		20376 (.NET ThreadPool Worker)	CH15_ParallelProgrammingDebuggingAndProfilingSample.dll!Program.MethodB(object obj = 3) Line 50
		29480 (.NET ThreadPool Worker)	CH15_ParallelProgrammingDebuggingAndProfilingSample.dll!Program.MethodB(object obj = 4) Line 50

Figure 15.8 – The Thread and Stack Frame view

From the **Visual Studio** menu, select **Debug | Windows | Tasks**. You should see the following pane:

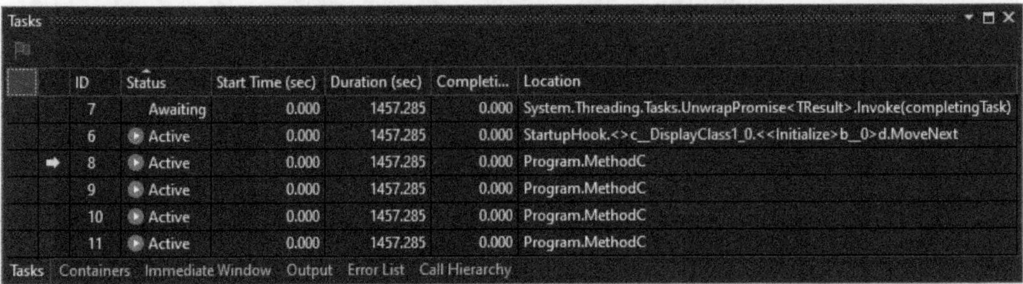

Tasks

	ID	Status	Start Time (sec)	Duration (sec)	Completi...	Location
	7	Awaiting	0.000	1457.285	0.000	System.Threading.Tasks.UnwrapPromise<TResult>.Invoke(completingTask)
	6	Active	0.000	1457.285	0.000	StartupHook.<>c__DisplayClass1_0.<<Initialize>b__0>d.MoveNext
➡	8	Active	0.000	1457.285	0.000	Program.MethodC
	9	Active	0.000	1457.285	0.000	Program.MethodC
	10	Active	0.000	1457.285	0.000	Program.MethodC
	11	Active	0.000	1457.285	0.000	Program.MethodC

Tasks Containers Immediate Window Output Error List Call Hierarchy

Figure 15.9 – The Tasks pane

This view shows you the various tasks and their states, along with other information. You can right-click on the columns to customize what columns you want to see. Clicking on a line should take you to the source location for you to view the code.

In the next section, we will look at the Concurrency Visualizer.

The Concurrency Visualizer

The Concurrency Visualizer is a command-line utility that allows you to collect traces from the command line. These can be viewed in the Concurrency Visualizer for Visual Studio 2022, which can be used on computers that don't have Visual Studio installed. Web projects are not supported by the Concurrency Visualizer; it relies on Windows event tracing.

By default, `CVCollectionCmd.exe` is installed in `C:\Program Files\Microsoft Visual Studio\2022\Preview\Common7\IDE\Extensions\rf2nfg00.o0t` and/or `C:\Program Files\Microsoft Visual Studio\2022\Community\Common7\IDE\Extensions\rf2nfg00.o0t`.

To begin collecting a trace, you can use a command such as the following:

```
C:\Program Files\Microsoft Visual
Studio\2022\Preview\Common7\IDE\Extensions\rf2nfg00.o0t\CVC
ollectionCmd.exe" /launch D:\dev\CH15_ParallelProgrammingDe
buggingAndProfilingSample\CH15_ParallelProgrammingDebugging
AndProfilingSample\bin\Debug\net6.0\CH15_ParallelProgrammin
gDebuggingAndProfilingSample.exe /outdir D:\Debugging
    \TraceData
```

This will start our application and log trace data to the location specified by the `/outdir` command-line argument. Several files will be generated by the tool. They will have `.etl` and `.cvtrace` file extensions.

From the **Visual Studio** menu, select **Analyze | Concurrency Visualizer | Open Trace** to view the generated trace file. You should see something similar to the following:

Figure 15.10 – The Contextual Visualizer Utilization tab

This screen shows the number of logical cores that are being utilized by the program you have traced. As you can see, my computer has 16 logical cores. Out of those 16, only 12 are being utilized. Clicking on the **Threads** tab gives you the following view:

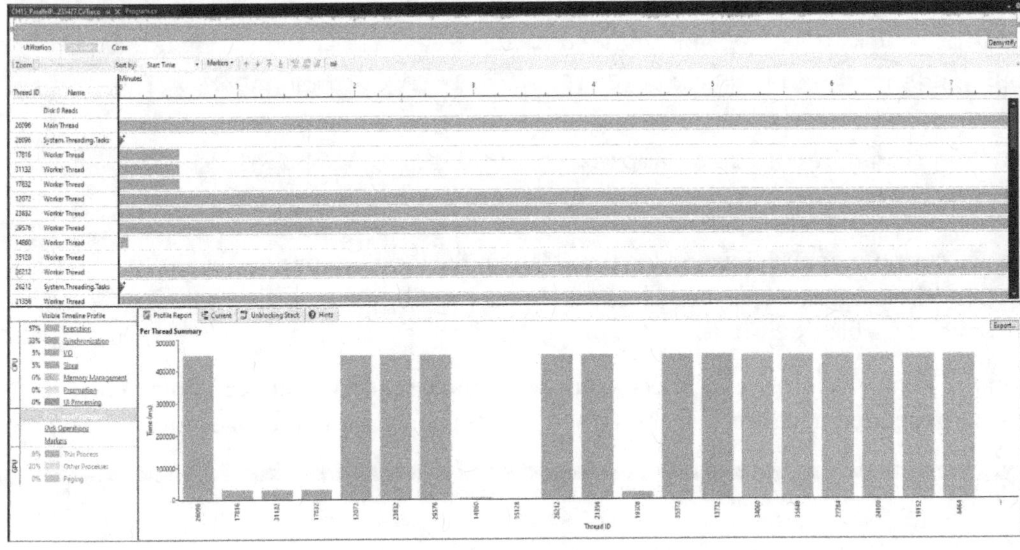

Figure 15.11 – The Contextual Visualizer Threads tab

This screen gives us a good, detailed breakdown of the threads that were used, their functionality, and the time they took to execute. Clicking on the **Cores** tab will display the following view:

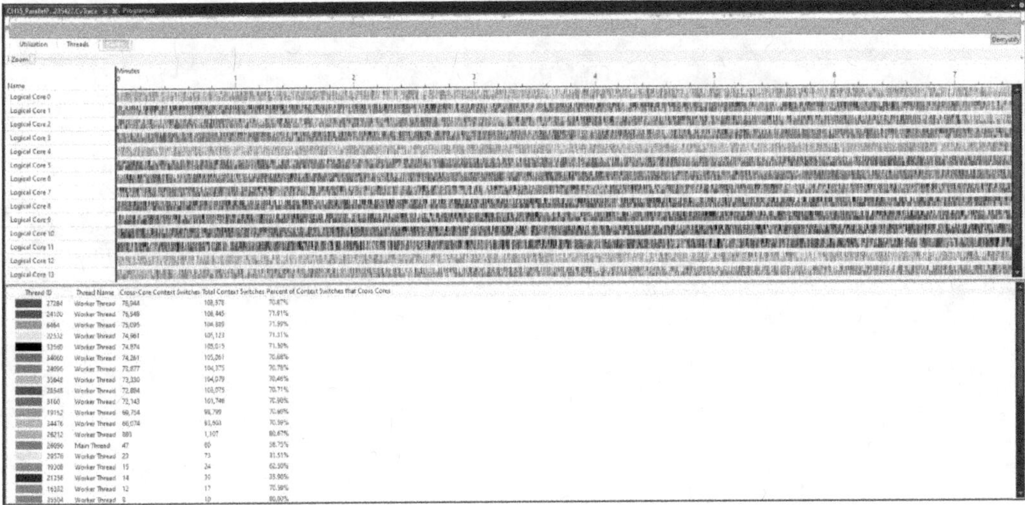

Figure 15.12 – The Contextual Visualizer Cores tab

This view shows the logical cores and their usage by the main thread and worker thread. You will see the thread ID, its name, the number of cross-core context switches, total context switches, and the percent of context switches.

> **Note**
>
> Microsoft provides a more detailed look into the Concurrency Visualizer. I have just provided you with a brief overview of the tool and how to use it. If you would like to learn more about how to use this tool, then you can view Microsoft's documentation at `https://docs.microsoft.com/en-us/visualstudio/profiling/concurrency-visualizer?view=vs-2022`.

With that, we've reached the end of this chapter. Now, let's summarize what we've learned.

Summary

In this chapter, we looked at how to use TPL and PLINQ to execute code in parallel. At this point, we understand that the main difference between TPL and PLINQ is that TPL does not efficiently utilize all the cores on a computer, whereas PLINQ does.

We also saw how we can view the computer's CPU utilization. Using PLINQ enables us to utilize all the cores of a CPU efficiently to improve code performance. However, when benchmarking parallel code, we saw that it is sometimes faster than non-parallel code, while other times, it is faster. Therefore, it pays to benchmark your code to see what method works best for you.

We also reviewed a piece of code that demonstrates the use of lambda expressions for expressing both `Func` and `Action` delegates.

Finally, we looked at debugging parallel applications with a code sample that employed the Parallel Tasks window, the Tasks pane, and the Concurrency Visualizer.

In the next chapter, we will look at asynchronous programming. But before we do, try and answer the questions to see how well you have retained what you have read. Then, check out the *Further reading* section to enhance your knowledge.

Questions

Answer the following questions to test your knowledge of this chapter:

1. What does TPL stand for?
2. What does PLINQ stand for?
3. What Windows program can you use to view CPU core usage?
4. Is parallel code always faster than non-parallel code?
5. How can you measure the code performance of parallel methods?

Further reading

To learn more about the topics that were covered in this chapter, take a look at the following resources:

- *Lambda Expressions in PLINQ and TPL*: `https://docs.microsoft.com/en-us/dotnet/standard/parallel-programming/lambda-expressions-in-plinq-and-tpl`
- *Task Parallel Library (TPL)*: `https://docs.microsoft.com/en-us/dotnet/standard/parallel-programming/task-parallel-library-tpl`
- *Introduction to PLINQ*: `https://docs.microsoft.com/en-us/dotnet/standard/parallel-programming/introduction-to-plinq`
- *Parallel Diagnostic Tools*: `https://docs.microsoft.com/en-us/dotnet/standard/parallel-programming/parallel-diagnostic-tools`
- *Debugging Async Code: Parallel Stacks for Tasks*: `https://devblogs.microsoft.com/visualstudio/debugging-async-code-parallel-stacks-for-tasks/`
- *Walkthrough: Debugging a Parallel Application in Visual Studio (C#, Visual Basic, C++)*: `https://docs.microsoft.com/en-us/visualstudio/debugger/walkthrough-debugging-a-parallel-application?view=vs-2022&tabs=csharp#main`

16
Asynchronous Programming

In this chapter, you will learn about the **Task-based Asynchronous Pattern** (**TAP**). You will learn how to program tasks asynchronously and how to access web resources using `async`, `await`, and `WhenAll`. You will also learn about different return types and extract the required results. Plus, you will learn how to correctly cancel asynchronous operations and perform asynchronous file reading and writing.

In this chapter, we will be covering the following topics:

- **Understanding the TAP model**: In this section, we provide a high-level overview of the TAP model.
- **Using async, await, and Task**: In this section, we will benchmark the performance of a method run synchronously (using `Task.Run`) and asynchronously.
- **Benchmarking GetAwaiter.GetResult(), .Result, and .Wait for both Task and ValueTask**: In this section, we benchmark the performance of an asynchronous operation using `GetAwaiter.GetResult()`, `.Result`, and `.Wait` for both `Task` and `ValueTask`.
- **Canceling asynchronous operations**: In this section, we write code that demonstrates asynchronous task cancellation.

- **Writing files asynchronously**: In this section, we write text to a file asynchronously.
- **Reading files asynchronously**: In this section, we read text from a file asynchronously.

After completing this chapter, you will be skilled in the following areas:

- Understanding the TAP model
- Processing web resources asynchronously
- Writing files asynchronously
- Reading files asynchronously

Technical requirements

You'll need Visual Studio to work on the code presented in this chapter.

All code from this chapter is placed on GitHub at `https://github.com/PacktPublishing/High-Performance-Programming-in-CSharp-and-.NET/tree/master/CH16`.

Understanding the TAP model

Before we begin, it is worth noting that there are three different models for dealing with asynchronous programming. These are as follows:

- The **Asynchronous Programming Model** (**APM**)
- The **Event-Based Asynchronous Pattern** (**EAP**) model
- The **Task Parallelism Library** (**TPL**)

APM uses `BeginMethod` to start the asynchronous process and `EndMethod` to complete the asynchronous process. EAP uses `MethodAsync` to start an asynchronous process, `CancelAsync` to handle the cancellation of an asynchronous operation, and a completed event handler to handle the completed asynchronous operation. Both these ways of performing asynchronous operations were replaced by TPL in C# 4.5.

TPL uses the `async` and `await` pattern. Asynchronous method names are suffixed with `async`. An asynchronous method usually returns an awaitable `Task` or `Task<Result>`. From .NET 4.5 onwards, you are advised to use TPL instead of using APM and EAP.

TAP's foundation types are the System.Thread.Tasks namespace, and the Task and Task<Tresult> classes via asynchronous operations. Microsoft advises that you should use TAP when starting new projects.

Naming, parameters, and return types

An asynchronous method using the TAP model prefixes the method signature with async Task for void methods, or async Task<Tresult>, async ValueTask, or async ValueTask<Tresult> for methods that return a value. The name of an asynchronous method that does not return a value should begin with a verb such as Begin or Process.

TAP method parameters should match and be in the same order as the parameters of synchronous counterpart methods. You should avoid entirely using out and ref parameters that are exempt from this rule. If you need to return data, use Tresult returned by Task<Tresult>. Use data structures to accommodate multiple return types. It is also worth considering adding cancellation tokens to TAP methods as parameters even if synchronous method counterparts don't have such tokens.

Combinator methods that work with multiple tasks where the intent is clear do not have to follow this naming pattern. WhenAll and WhenAny are examples of combinator methods.

Initiating asynchronous operations

You may wish to perform some synchronous tasks, such as validation and preparing the asynchronous operation for execution, at the start of an asynchronous method. If so, you are advised to keep these tasks to the minimum, and the time they take should be minimal. The reason is that such methods may be invoked from **User Interface** (**UI**) threads, and you don't want to cause your applications to hang or freeze momentarily.

Another reason for keeping synchronous operations to the minimum and for spending minimal time within asynchronous operations is that when you run concurrent asynchronous methods, long-running synchronous operations can and do decrease the benefits of concurrency.

Sometimes, it can take longer to prepare and launch an asynchronous operation than it can take to complete the same operation synchronously. In these situations, you can run the method synchronously and return a task.

Exceptions

Usage errors, such as passing `null` arguments, are the only errors that should be raised in asynchronous methods. You can prevent asynchronous methods from raising usage errors by modifying the calling code to ensure that erroneous arguments are not passed into the asynchronous methods. All other types of exceptions and errors should be assigned to the task being returned. Normally, one exception is returned by one task. But when there are multiple operations represented by a single task, multiple exceptions may be returned by a single task.

Optional cancellation

Cancellation of asynchronous method implementers and consumers is optional. An asynchronous method that can be canceled, exposes an overload method that accepts a `CancellationToken` that is named `cancellationToken` by convention.

Cancellation requests are monitored by the asynchronous operation. When a cancellation request is received, it may be honored. If cancellation results in unfinished work, a task in the `Canceled` state is returned with no available result and no exceptions.

The `Canceled` state is a completed task state, as are `RanToCompletion` and `Faulted`. When a task's state is either `Canceled`, `RanToCompletion`, or `Faulted`, the `IsCompleted` property returns `true`.

Continuations will continue to be scheduled and executed when a task is canceled unless the `NotOnCancelled` continuation option is specified. If this option is specified, then continuations will not be scheduled or executed when a task is canceled.

Asynchronous code waiting for canceled tasks via language features will continue to run but will receive an `OperationCanceledException` or one of its derivatives. And code that is blocked synchronously waiting on tasks through methods like `Wait` and `WaitAll` will continue to run with an exception.

TAP methods should return a `Canceled` task when a cancellation token has requested cancellation before the TAP method that accepts the token has been called. During the execution of an asynchronous operation, cancellation requests can be ignored. When returning a task, you will normally return the task with one of three states:

- `Canceled`: The operation has ended as a result of a cancellation request.
- `RanToCompletion`: A cancellation was requested but the operation was completed and produced a result.
- `Faulted`: A cancellation was requested that resulted in the generation of an exception.

If you are coding an asynchronous method and want to enable the operation to be canceled first and foremost, then there is no need to produce an overload method devoid of a `CancellationToken`. If you are coding an asynchronous method that cannot be canceled, then you do not have to provide an overload method that accepts a `CancellationToken`. These guidelines help the caller to know whether or not the target method can be canceled. When a method that accepts a `CancellationToken` is called by a consumer that has no desire to cancel the method call, `None` can be passed in for the `CancellationToken` argument, as this is functionally equivalent to the default `CancellationToken`.

Optional Progress Reporting

When asynchronous operations are running as part of a UI procedure, it can be beneficial to provide progress updates. This helps the end user to know that the program is still working.

The `IProgress<T>` interface is used to handle progress and is passed into an asynchronous method as a parameter that is conventionally called `progress`. Passing this interface into an asynchronous method can help prevent race conditions that can occur when event handlers are incorrectly registered once the operation has started, which can lead to missed updates. Another reason for passing in an interface is that consuming code can support various progress implementations. Only provide an `IProgress<T>` interface when progress notifications are supported by the TAP implementation.

An example that fits well with progress updates is the `FindFilesAsync` method, which returns a list of files meeting a particular search pattern. In this scenario, you could provide the percentage of work completed along with the current set of partial results. The information would be provided by some data type that is specific to your API. Such data types are conventionally suffixed with `ProgressInfo`.

TAP methods that provide a progress parameter should allow no progress reporting by allowing the progress parameter to be `null`. Progress should be reported to the `Progress<T>` object that implements the `IProgress<T>` interface synchronously. This enables the asynchronous method to quickly provide progress. Consumers can then determine how and where they want to handle the information provided by the progress update.

The `ProgressChanged` event is exposed by instances of the `Progress<T>` class. This event is raised every time a progress update is reported by the asynchronous operation. When a `Progress<T>` object is instantiated, the `ProgressChanged` event is raised on the captured `SynchronizationContext` object. A default context that targets the thread pool is used when there is no synchronization context available.

You can either register handlers for this event as you would any other event, and you can also provide the `Progress<T>` constructor with a single handler, for convenience. The single handler behaves the same as an event handler for the `ProgressChanged` event. During the execution of event handlers, delays to asynchronous operations are avoided by raising progress updates asynchronously.

Now that we have a high-level understanding of the task-based asynchronous pattern, in the next section, we will look at `async`, `await`, and `Task`.

async, await, and Task

In this section, we will be looking at the performance differences between running methods synchronously, using `Task.Run`, and asynchronously. An asynchronous method is identified by the `async` keyword.

The `await` keyword informs the runtime to wait at the specified line until the current task has been completed. It can only be used with a method that is prefixed with the `async` keyword.

The **Task Parallel Library** (**TPL**) can be found in the `System.Threading.Tasks` namespace. A task encapsulates threading in order to maximize the use of multiple cores on computer hardware.

Let's write a simple project to benchmark three different ways of calling a method. We will call the method synchronously using `Task.Run`, and asynchronously using `async/await`. We will be using `BenchmarkDotNet` to see how each method call type performs. We aim to show the performance advantage of using asynchronous calls over synchronous and `Task.Run` calls.

We perform the following steps to write our little program:

1. Start a new .NET 6.0 console application and call it `CH16_AsynchronousProgramming`.
2. Add the `BenchmarkDotNet` NuGet package.
3. Add a new class called `Benchmarks`, and in that class add the following method:

```
public static void LengthyTask()
{
    int y = 0;
    for (int x = 0; x < 10; x++)
        y++;
}
```

This method is our worker method. All it does is increment the `y` variable by one for ten iterations.

4. Add the `SynchronousMethod` to the class:

```
[Benchmark]
public void SychronousMethod()
{
      LengthyTask();
}
```

This method calls the `LengthyTask` method synchronously and is a benchmark.

5. Add the `TaskMethod` to the class:

```
[Benchmark]
public void TaskMethod()
{
     Task.Run(new Action(LengthyTask));
}
```

This method runs the `LengthyTask` method as a new `Action`, which is queued to run on the `ThreadPool`. A `Task` or `Task<Tresult>` handle is returned for that method.

6. Add the `AsynchronousTaskMethod` to the class:

```
[Benchmark]
public void AsynchronousTaskMethod()
{
     var data = async () => await Task.Run(new
         Action(LengthyTask));
}
```

This method runs the `LengthyTask` method as an action using `Task.Run` asynchronously, and await the completion of the method before it continues.

7. Our benchmark class is now complete. So, in the `Program.cs` file, replace the code with the following:

```
using BenchmarkDotNet.Running;
using CH16_AsynchronousProgramming;
Console.WriteLine("CH16 - Asynchronous Programming");
var summary = BenchmarkRunner.Run<Benchmarks>();
Console.ReadLine();
```

   This code will run our benchmarks and produce a report for us.

8. Make sure that the project is set to `Release` build.
9. Build the project.
10. Open a command window and execute the compiled executable file called `CH16_AsynchronousProgramming.exe` in the `bin\Release\net6.0` folder.
11. The benchmarks should start running, and once complete, you should see a report like the one shown in *Figure 16.1*:

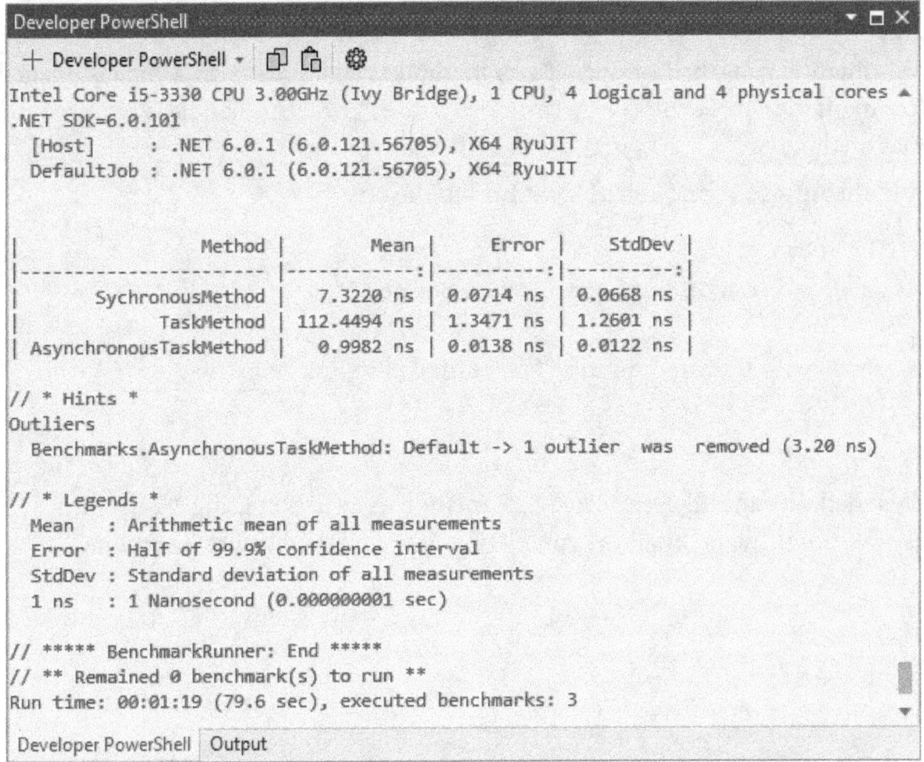

Figure 16.1 – The BenchmarkDotNet report for our CH16_AsynchronusProgramming Project

As you can see in *Figure 16.1*, running the `LengthyTask` method synchronously took `7.3220 ns` to complete. Using `Task.Run` took the longest time to run at `112.4494 ns`. And the fastest way to run the code was asynchronously, which only took `0.9982ns` to complete.

We can clearly see from those times that there is a clear performance benefit to running our code asynchronously, as it takes less overall time for our code to complete.

In the next section, we will compare the performance of `await` with `GetAwaiter.GetResult()`, `.Result`, and `.Wait`. We will cover both `Task` and `ValueTask`.

Benchmarking GetAwaiter.GetResult(), .Result, and .Wait for both Task and ValueTask

In this section, we will be writing some code to benchmark the `GetAwaiter.GetResult()`, `.Result`, and `.Wait` methods to see which method is best for obtaining the return value for both a `Task` and a `ValueTask`.

At `https://github.com/dotnet/BenchmarkDotNet/issues/236`, the `BenchmarkDotNet` maintainer called *adamsitnik* wrote in reply to *@i3arnon*:

"@i3arnon Thanks for the hint! I have measured `.Result` *vs* `.Wait` *vs* `GetAwaiter.GetResult()` *and it seems that for* `Tasks` *the* `GetAwaiter.GetResult()` *is also the fastest way to go. On the other hand, for* `ValueTask` *it was much more slower so I stayed with* `.Result` *for VT."*

And so, from the code that we will be writing, we should see that `.Result` should provide us with the best performance when working with a `ValueTask`. And `GetAwaiter.GetResult()` should give us the best performance when working with a `Task`.

We will now start writing our code. Please complete the following tasks in the `CH16_AsynchronousProgramming` project that we started in the previous section:

1. Open the `CH16_AsynchronousProgramming` project.
2. Open the `Benchmarks` class.
3. Add the following method that returns an `int`:

```
public static int LengthyTaskReturnsInt()
{
    int y = 0;
    for (int x = 0; x < 10; x++)
```

```
        y++;
      return y;
  }
```

In this code, we are incrementing the `y` variable and returning the result.

4. Add the `GetAwaiterGetResult` method:

```
[Benchmark]
public void GetAwaiterGetResult()
{
    int value = Task.Run(() =>
        LengthyTaskReturnsInt()).GetAwaiter()
          .GetResult();
}
```

This method benchmarks the time taken to return an `int` from a method using `GetAwaiter().GetResult()`.

5. Add the `Result` method:

```
[Benchmark]
public async Task Result()
{
    int value = await Task.Run(() =>
      LengthyTaskReturnsInt()).ConfigureAwait(false);
}
```

This method benchmarks the time taken to await the return of `int` from a method.

6. Add the `Wait` method:

```
[Benchmark]
public void Wait()
{
    Task.Run(() => LengthyTask()).Wait();
}
```

This method runs a lengthy task and waits for it to finish before continuing.

7. Add the `GetAwaiter` method:

```
[Benchmark]
public void GetAwaiter()
{
        Task.Run(() => LengthyTask()).GetAwaiter();
}
```

This method gets an awaiter used to await the task completion.

8. Build the project and run the executable via the command line. You should see a summary report like the one shown in *Figure 16.2*:

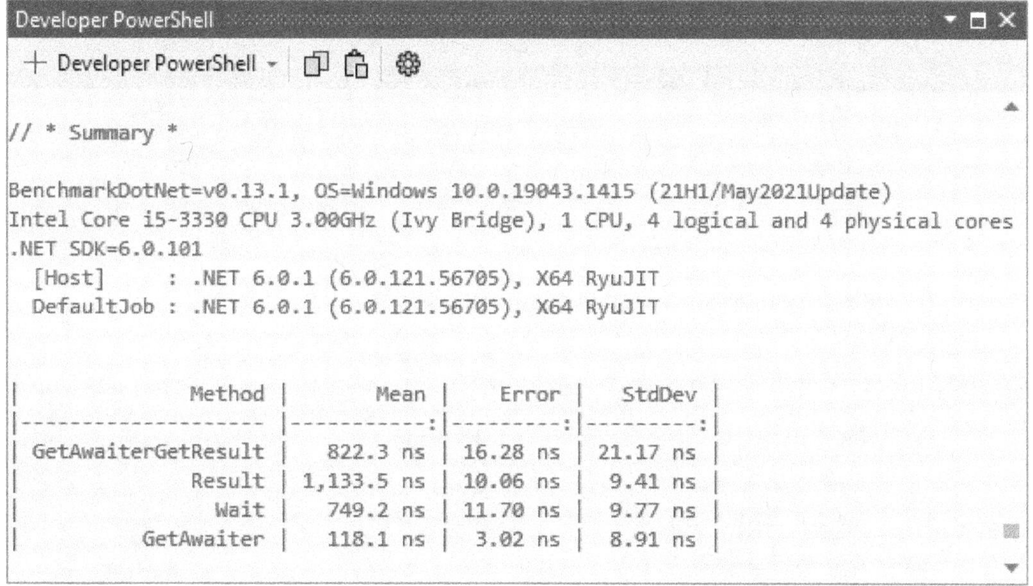

Figure 16.2 – The BenchmarkDotNet summary report for this section's methods

As we can see from these results, when returning a value from a `Task`, the `GetAwaiterGetResult` method operates much faster than the `Result` method. And when executing a long-running `Task`, the `GetAwaiter` method operates much more quickly than the `Wait` method.

In the next section, we will look at how we can speed up our code asynchronously when awaiting multiple tasks by using `WhenAll`.

Using async, await, and WhenAll

In this section, we will write some example code that demonstrates the use of `async`, `await`, and `WhenAll` and the effect on execution time.

If you have multiple tasks that are being executed in a method and you `await` each task, your code will work asynchronously, and the execution time will be expensive. You can circumvent this time expense with improved performance by using `WhenAll` to `await` all completed tasks before continuing. In the code we will be writing, you will see how `WhenAll` reduces the time taken to execute two asynchronous methods within a function when compared to awaiting each task in turn.

Let's work our way through the following tasks:

1. In the `Benchmarks` class still, add the following asynchronous method, which waits `300` milliseconds before returning an `int`:

```
private async Task<int> TaskOne()
{
        await Task.Delay(300);
        return 100;
}
```

   The `TaskOne` method is the first of our methods that will be run by our benchmarks.

2. Add the second of our asynchronous methods:

```
private async Task<string> TaskTwo()
{
        await Task.Delay(300);
        return "TaskTwo";
}
```

   The `TaskTwo` method waits for `300` milliseconds and then returns a `string`.

3. Firstly, we will benchmark running asynchronous tasks synchronously:

```
[Benchmark]
public async Task SynchronousAwait()
{
        int intValue = await TaskOne();
        string stringValue = await TaskTwo();
}
```

Here, we have two tasks and we `await` them both to complete before continuing.

4. Now, we'll add our method that will utilize `WhenAll`:

```
[Benchmark]
public async Task AsynchynchronousWhenAll()
{
    var taskOne = TaskOne();
    var taskTwo = TaskTwo();
    await Task.WhenAll(taskOne, taskTwo);
}
```

In this method, we create our two tasks, then we pass them into the `WhenAll` method as parameters. We do not continue until all tasks are complete.

5. Build and run your executable via the command line. You should see something like *Figure 16.3*:

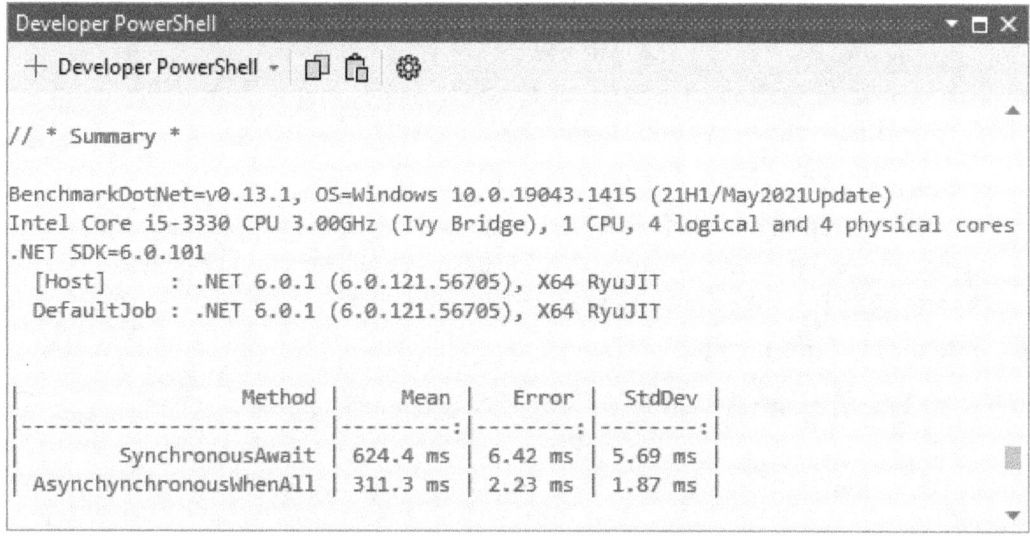

Figure 16.3 – The results of synchronous and asynchronous execution of multiple asynchronous calls

As you can see from the results of our benchmarking, using `WhenAll` executes multiple asynchronous tasks much faster than when you await them in turn. In the next section, we will look at canceling asynchronous tasks.

Canceling asynchronous operations

In this section, we will look at how we can cancel long-running asynchronous operations. Sometimes a task will take longer than it should do. A good example of this is fetching data from a website when it goes down. Asynchronous operations can take a long time before they are reset by the server due to something like `Error 404`, `Error 401`, or `Error 500` for example. And so, it pays to have the ability to cancel an asynchronous operation after a set period to prevent wasting an end user's time.

The code we will write will return the text from a website URL. We will assign a very short timeout. This timeout will cancel the task that is responsible for returning the website text. Follow these steps:

1. Open the `CH16_AsynchronousProgramming` project, and add a new class called `TaskCancellation`.
2. Add the `using System.Text;` statement.
3. Add the following two member variables:

```
private const string _website =
    "https://docs.microsoft.com";
private static readonly CancellationTokenSource
    _cancellationTokenSource = new();
```

   The `_website` variable holds the URL of the website whose page text we will be returning. And the `CancellationTokenSource` will be used to signal to a `CancellationToken` that it should be cancelled.

4. Add the following method:

```
private static readonly HttpClient HttpClient = new()
{
    MaxResponseContentBufferSize = 1000000
};
```

   Here, we declare a method that returns a `HttpClient` for our HTTP request. The `MaxResponseContentBufferSize` sets the number of bytes to buffer when reading the response content.

5. Now add the `ReturnWebsiteTextAsync` method:

```
private static async Task<string>
    ReturnWebsiteTextAsync()
{
```

```
        HttpResponseMessage response = await HttpClient
                .GetAsync(
                _website,
                _cancellationTokenSource.Token)
              .ConfigureAwait(false);
        byte[] contentAsByteArray = await response
            .Content
            .ReadAsByteArrayAsync(
                _cancellationTokenSource.Token)
              .ConfigureAwait(false);
        return Encoding.ASCII.GetString(
            contentAsByteArray
        );
    }
```

In this method, we declare `HttpResponseMessage`, which awaits an asynchronous task that returns the contents of a web page. The response is then read and converted into a byte array. This byte array is then transformed into an ASCII string and returned.

6. Now add the `Start` method:

```
    public static async Task Start()
    {
        Console.WriteLine("Task started.");
        try {
            _cancellationTokenSource.CancelAfter(3000);
              await ReturnWebsiteTextAsync()
                .ConfigureAwait(false);
        }
        catch (OperationCanceledException) {
          Console.WriteLine(
          "\nThe task has timed out and been cancelled.
            \n");
        }
        finally {
            _cancellationTokenSource.Dispose();
```

```
        }
        Console.WriteLine("Task completed.");
    }
```

In the `Start` method, we write a console message that states the task has started. We then set the cancellation time of `cancellationTokenSource` to 30 seconds, which is 3000 milliseconds. Then we `await` the call to the `ReturnWebsiteTextAsync`. If the process times out after the set timeout period, an `OperationCanceledException` is raised, which outputs a message to the console. Finally, `cancellationTokenSource` is disposed of and a console message is an output stating that the task is finished.

7. Comment out the benchmark running code in the `Program.cs` file, and add the following line:

```
TaskCancellation.Start().GetAwaiter();
```

8. Run the project and try it several times with different timeout periods to test the code completing successfully and returning text, and to test the operation timing out and raising an exception.

Running this code through a couple of times with timeouts of `3000` and `30000` will present an operation timeout exception and display the web page text, respectively. As you can see if you run the code yourself, it is very easy to write asynchronous tasks that are canceled after a set period.

In the next section, we will be writing code that shows how to write files asynchronously.

Writing files asynchronously

In this section, we will write text to a file asynchronously. Scenarios where asynchronous file writing can be useful include writing large volumes of text and data to files that will not be read immediately.

Use the following steps to write our code:

1. On your `C:\` drive, add a folder called `Temp` if one does not already exist.
2. Open the `CH16_AsynchronousProgramming` project.
3. Add a class called `FileReadWriteAsync`.

4. Add the following method:

```
public static async Task WriteTextAsync()
{
string filePath = @"C:\Temp\Greetings.txt";
string text = "Hello, World!";
byte[] encodedText =
    Encoding.Unicode.GetBytes(text);
using (FileStream fileStream = new FileStream(
                filePath,
                FileMode.Append,
                FileAccess.Write,
                FileShare.None,
                bufferSize: 4096,
                useAsync: true
            )
        )
{
        await fileStream.WriteAsync(
            encodedText, 0, encodedText.Length);
};
}
```

In the `WriteTextAsync` method, we declare a file path for a text file and a variable that contains the text to be written to the file. The text to be written gets converted into a byte array. A writable asynchronous file stream is then opened in append mode. Then we write the text to the file stream and close it.

In the next section, we continue in this class as we add our asynchronous read method that shows how to read a file asynchronously.

Reading files asynchronously

In this section, we will read text from a file asynchronously. We will be building upon the code from the previous section that writes the text to a file asynchronously.

The following steps will add our asynchronous read method and update the `Program.cs` file to run our asynchronous code:

1. In the `FileReadWriteAsync` class, add the following method:

```
public static async Task<string> ReadTextAsync()
{
    string filePath = @"C:\Temp\Greetings.txt";
    using (FileStream fileStream = new FileStream(
            filePath,
              FileMode.Open,
              FileAccess.Read,
              FileShare.Read,
              bufferSize: 4096,
              useAsync: true
        )
    )
    {
        StringBuilder sb = new StringBuilder();
        byte[] buffer = new byte[0x1000];
        int numRead;
        while (( numRead = await fileStream
          .ReadAsync(buffer, 0, buffer.Length)) != 0
        )
        {
            string text = Encoding.Unicode
                .GetString(buffer, 0, numRead);
            sb.Append(text);
        }
        return sb.ToString();
    }
}
```

Here, we define the path of the file that we need to read. Then we open a file stream in read mode with read access. Next, we define `StringBuilder` and byte array that will act as our buffer to store read data. We then read the stream until the read has been completed. During each iteration of the read, we read the text from the file, encode it into Unicode, and then append it to `StringBuilder`. Then, once the loop has finished and exits, we return the string from the method.

2. Open the `Program.cs` class.
3. Comment out the following lines:

```
//var summary = BenchmarkRunner.Run<Benchmarks>();
// TaskCancellation.Start().GetAwaiter();
```

We won't be needing these lines when we run our code.

4. Add the following lines of code:

```
FileReadWriteAsync.WriteTextAsync().GetAwaiter();
string data = FileReadWriteAsync.ReadTextAsync()
    .GetAwaiter().GetResult();
Console.WriteLine(data);
```

In this code, we call our methods that write text to a file asynchronously, read the text asynchronously into a variable, and then print the contents of the variable to the console.

5. Run the code, and you should see something like *Figure 16.3*:

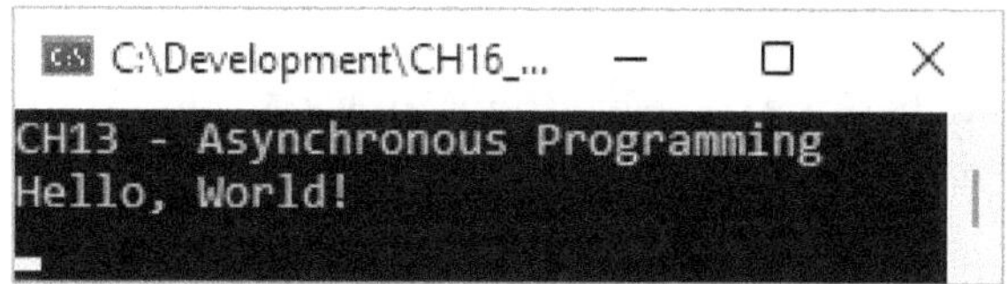

Figure 16.4 – The result of our asynchronous write and read code

As you can see from the screenshot, we have successfully written text asynchronously to a file, asynchronously read it from that file, and printed the contents to the console window.

In the next section, we will summarize what we have learned in this chapter.

Summary

In this chapter, we began with a high-level overview of the task-based asynchronous pattern. Things we covered were naming, parameters, return types, initializing asynchronous operations, exceptions, and optionally providing ways to report progress updates and cancel operations. We saw that we can have asynchronous operations that allow cancellation, and those that don't allow cancellation. Plus, we learned that when a cancellation has been requested, the cancellation will either go ahead or be ignored. Completed tasks can have a completed state of `Canceled`, `RanToCompletion`, or `Faulted`.

We then benchmarked three different ways of calling a method synchronously, using `Task.Run`, and asynchronously. Using `Task.Run` took the longest time, followed by running the method synchronously, and running the method asynchronously was by far the quickest way to run the method.

Then we benchmarked `GetAwaiter.GetResult()`, `Result`, and `Wait` for both `Task` and `TaskValue`. We saw that when returning a value from a `Task`, the `GetAwaiterGetResult` method operates much faster than the `Result` method. And when executing a long-running `Task`, the `GetAwaiter` method operates much more quickly than the `Wait` method.

Next, we looked at cancelling asynchronous operations. We coded an example that obtains the text from a website and outputs the text to the console. If the operation fails to complete within a set period of time, then it is cancelled.

In the final two sections, we wrote some code to demonstrate the writing and reading of text and data asynchronously.

To complete this chapter, there are some questions for you to answer to see how well you have retained what you have read and some further reading on asynchronous programming.

Thank you for purchasing this book. I hope you have enjoyed reading it, and that you have learned plenty of ways to improve your own code. Happy coding!

Questions

1. What does TAP stand for?
2. What parameter type identifies that an asynchronous operation can be canceled?
3. What parameter type is passed into an asynchronous task to provide progress updates?

4. Explain `async`, `await`, and `Task`.
5. How do you cancel an asynchronous operation?
6. How do you report on an asynchronous operation's progress?

Further reading

- Asynchronous programming; APM vs EAP: `https://stackoverflow.com/questions/11276314/asynchronous-programming-apm-vs-eap`
- Asynchronous programming: `https://docs.microsoft.com/en-us/dotnet/csharp/async`
- Introduction to async programming in C#: `https://auth0.com/blog/introduction-to-async-programming-in-csharp/`
- The performance characteristics of async methods in C#: `https://devblogs.microsoft.com/premier-developer/the-performance-characteristics-of-async-methods/`
- Exception Handling (Task Parallel Library): `https://docs.microsoft.com/en-us/dotnet/standard/parallel-programming/exception-handling-task-parallel-library`

Assessments

This section is for answers to questions from all chapters.

Chapter 1, Introducing C# 10.0 and .NET 6

1. Performance improvements in the garbage collector and JIT compiler, improved performance of text-based processing, faster regular expression processing, and performance of threading and asynchronous operations has been boosted. There have also been performance improvements to collections, LINQ, networking, and Blazor; plus, there are additional performance-based APIs and analyzers that are new to .NET 6.
2. You can now write top-level programs and use `init`-only properties and records. There are new pattern matching features and new expressions with targeted types. You can use covariant returns and perform native compilation.
3. `dotnet` and `ngen`.
4. Run the Microsoft Store app performance assessment. Follow Microsoft's advice based on the assessment to improve your app's performance, and address each of the highlighted issues found with your app.
5. Perform baseline measurements, begin optimizations by performing the refactoring with the largest overall impact, enable HTTP compression, reduce TCP/IP connection overheads, and use HTTP/2 over SSL.
6. Reading tasks to be completed by the reader at their discretion.
7. Coding tasks to be completed by the reader at their discretion.
8. Benchmarking tasks to be completed by the reader at their discretion.

Chapter 2, Implementing C# Interoperability

1. Platform invocation.
2. Explain what `P/Invoke` is.
3. It reminds the programmer that they are responsible for the safety of their code, since it is not managed by the .NET Framework.
4. There are three generations of objects: zero, one, and two. Normally, objects are added to generation zero and garbage is collected. But if they survive generation zero, they are promoted to generation one. Objects that survive generation one are promoted to generation two. If generations zero, one, and two are completely full and new objects are added, then you end up with `OutOfMemoryException`, and your application will crash.
5. The `fixed` keyword is used to ensure that objects referenced by pointers are not promoted by the garbage collector. Otherwise, the pointers would point to the wrong thing, causing bugs in the software.
6. BSTR.
7. IronPython, although other packages also exist.
8. Implement the disposable design pattern.
9. Set large fields to `null` when the object is being disposed of. This makes them unreachable, and they are released faster than if they were reclaimed non-deterministically. You will do this outside of the `conditional` block. See `https://docs.microsoft.com/en-us/dotnet/standard/garbage-collection/implementing-dispose`.

Chapter 3, Predefined Data Types and Memory Allocations

1. `bool`, `byte`, `char`, `DateTime`, `decimal`, `double`, `enum`, `float`, `int`, `long`, `sbyte`, `short`, `struct`, `value tuple`, `uint`, and `ulong`.
2. `object`, `string`, `delegate`, and `dynamic`.
3. Create an instance of the `static` type.
4. No. The same physical memory is used for both the stack and the heap.

5. Items are simply popped onto the stack when they are in use and immediately popped off the stack when they are no longer needed. Objects added to the heap need to be managed and object reference counters maintained. Items placed on the stack use both the stack and the heap, as items on the heap have pointer variables on the stack. So, there is more overhead to using the heap compared to the stack.
6. A string is placed on the heap. A variable is placed on the stack with the string's memory address. When another variable is assigned the same string, it is given the address of the string. So, multiple items on the stack will be pointing to the same string. However, if you append anything to the string, a new string is then created on the heap with a new memory address. The variable that is assigned the new string will have the memory address that points to the new string on the heap, so the original string is never updated.
7. Less than 80,000 bytes.
8. 80,000 bytes or higher.

Chapter 4, Memory Management

1. Three: generation 0, generation 1, and generation 2.
2. Objects less than 80,000 bytes are placed on the SOH.
3. Objects 80,000 bytes or more are placed on the LOH.
4. A strong reference is a reference that does not get garbage-collected.
5. A weak reference is a reference that does get garbage-collected.
6. Implement the `IDisposable` pattern.
7. Unsubscribe event listeners when they are no longer used. Dispose of event publishers or set them to null when they are no longer used.
8. `Marshal.ReleaseComObject(object)`.
9. Make sure that any allocated memory is deallocated. Use the `IDisposable` pattern to ensure that memory is cleaned up when the object is disposed of.

Chapter 5, Application Profiling and Tracing

1. Applications, assemblies, namespaces, types, methods, and fields.
2. Maintainability index, cyclomatic complexity, depth of inheritance, class coupling, lines of source code, and lines of executable code.

3. Dump location and time, the name of the process, processor architecture, exception information, OS and CLR version, and the names, versions, and physical paths of the loaded modules.
4. The name, path, optimized user code, symbol status, O (order), version, process, and AppDomain.
5. Microsoft Visual Studio 2022, and JetBrains dotTrace, dotMemory, and dotnet-counters.
6. We were able to list the .NET processes that can be monitored and counters that can be used to collect data. We obtained the .NET process identifiers and monitored them, and we collected, saved, and viewed data that we collected from the running .NET processes.

Chapter 6, The .NET Collections

1. `System.Collections`, `System.Collections.Generic`, `System.Collections.Concurrent`, and `System.Collections.Specialized`.
2. Big O notation is used to determine algorithmic efficiency.
3. Algorithmic efficiency determines how time scales with respect to input.
4. Benchmarking showed that using `IList<T>` was faster than using `List<T>`, and so using `IList<T>` is preferred over using `List<T>`.
5. You can use either. What you choose depends upon your performance requirements and what you are trying to achieve. There are trade-offs between using collections and arrays. Understanding these trade-offs will help you choose which option you should apply to your code.
6. Indexers enable objects in classes to be accessed in the same way as you access items in an array.
7. `IEnumerator<T>` is faster at iterating through in-memory collections than `IEnumerable<T>`.
8. In terms of memory and speed performance, querying the database and obtaining the enumerator is the fastest way to query a database and iterate through the results according to the benchmarks.
9. Use the `yield` keyword.

Chapter 7, LINQ Performance

1. Use the index rather than the `Last()` call for direct access to the last element in a collection. Avoid using the `let` keyword in your LINQ queries. Convert a list to an array to perform group by, and then return an enumerator.
2. The compiler generates more lines of code that take longer to run, and more memory is allocated at runtime than when the `let` keyword is not used.
3. Filter items starting with objects that have the least number of items, followed by the objects with an increasing number of items. Also, avoid using the `let` keyword.
4. Closures with parameters perform better than closures without parameters.

Chapter 8, File and Stream I/O

1. Absolute, relative, UNC, and DOS device.
2. In the registry editor, set `HKEY_LOCAL_MACHINE\SYSTEM\CurrentControlSet\Control\FileSystem\LongPathsEnabled` to `1`.
3. The most efficient way to calculate the size of a directory is to get `DirectoryInfo` for the directory, followed by the call to `GetFileSystemInfos()`. You then iterate through the result, adding the length of each `FileInfo` object to get the directory's size.
4. The most efficient method of moving files is to obtain `FileInfo` objects from the in-memory cache and then use the `FileInfo.MoveTo(string destination)` method to move the file.
5. When you encounter a non-recoverable exception before you exit the application.
6. `IOException`.
7. Local, Local Cache, Roaming, Temporary, and C:\ProgramData.
8. Users may only install the software for themselves when prompted. This will result in each logged-on person using the software having their own copy of the data, with the data located in the Microsoft VirtualStore under their logged-on account.
9. When multiple users log onto the same computer, and an application has been installed for just one user rather than all users, instead of the application data being stored under the centralized location of `C:\ProgramData`, it will be stored under Microsoft Virtual Store.
10. `C:\Users\%USERNAME%\AppData\Local\VirtualStore`.

Chapter 9, Enhancing the Performance of Networked Applications

1. Application layer, presentation layer, session layer, transport layer, network layer, data link layer, and physical layer.
2. HTTP, HTTPS, SSH, SSL, DHCP, DNS, FTP, TFTP, Telnet, SMTP, IMAP4, POP3, TCP, IP, UDP, Ethernet, and PPP.
3. TCP enables the transmitting and receiving of data that is guaranteed to be received. UDP only allows the transmission of data that is not guaranteed to be received.
4. Use the developer tools that are built into your browser.
5. gRPC is a cross-platform, cross-language, and cross-device framework for making remote procedure calls between applications. gRPC-Web is a proxy for browser-based RCP calls, as browser applications are unable to use gRPC directly.
6. Reduce the number of things the page is doing and the number of services the page calls. Reduce the size of images. Use file compression to reduce the size of files transmitted over a network. Cache network resources. Filter data on the server, divide it into pages, and return only the requested page of data.

Chapter 10, Setting Up Our Database Project

N/A.

Chapter 11, Benchmarking Relational Data Access Frameworks

1. Executing a stored procedure with Dapper.NET.
2. Executing a raw SQL statement with Dapper.NET.
3. Executing a stored procedure with ADO.NET.
4. Executing a stored procedure with ADO.NET.
5. Executing a stored procedure with ADO.NET.
6. Not necessarily. A hybrid approach may be better because you can maximize your data access performance for the data operations in question by using the most performant method from the frameworks you have selected to work with.

Chapter 12, Responsive User Interfaces

1. Configure the application for high-DPI awareness.
2. Configure the application to be long file path-aware.
3. Add a splash screen to the start of your application.
4. Run the long-running task as a background task.
5. Memory caching and distributed caching.
6. Use AJAX.
7. `WebSockets` and SignalR.
8. `SetSemanticFocus`, `Announce`, and `Font scaling`.
9. Add the `BlazorWebView` component to a page and point it to the root of your Blazor application.
10. `ProgressRing` and `ProgressBar`.

Chapter 13, Distributed Systems

1. Command query responsibility separation.
2. We may want to use one model for commands and another model for queries.
3. Event-driven programming.
4. We use events to trigger the execution of a serverless function, such as an Azure Durable Function.
5. A piece of software that is used to package an application and its dependencies that can be deployed to and executed in the cloud or on-premises.
6. To deploy third-party dependencies and legacy code.
7. Microservices in the form of functions that only run when they are required and that usually run in response to an event trigger.
8. Serverless functions can scale rapidly, and you only pay for the time the functions run. This can save money when compared to containers that need to be running most of the time.
9. Extensions to Azure Functions that enable the writing of stateful functions in a serverless environment. We can also use them to define workflows.
10. Activity, Orchestrator, Entity, and Client.

11. Aggregator (stateful entities), fan-out/fan-in, function chaining, human interaction, and monitoring (actors).
12. An infrastructure-as-code platform for managing microservices.
13. You can manage your microservices and their resources using C#, from creation to running, stopping, and deleting them.

Chapter 14, Multi-Threaded Programming

1. `Running`, `suspended`, `wait`, `sleep`, `join`, and `stop`.
2. You don't – this API is now obsolete.
3. Foreground and background.
4. Use `CancellationToken` to raise `TaskCanceledException` when a `CancellationTokenSource` operation times out.
5. `Thread.Start()` or `Thread.Start(object)`.

Chapter 15, Parallel Programming

1. Task Parallel Library.
2. Parallel LINQ Library.
3. Performance Monitor aka `perfmon`.
4. No.
5. Use `BenchmarkDotNet` to test the performance of various methods.

Chapter 16, Asynchronous Programming

1. Task-based asynchronous pattern.
2. `CancellationToken`.
3. `IProgress<T>`.
4. An asynchronous method is declared, with the `async` keyword preceding the method name. The `await` keyword precedes an asynchronous operation and prevents the continuation of any further code until the asynchronous operation is complete. `Task` is what an asynchronous method returns. For `void` methods, the return type is `Task`, and for methods that return a value, the return type is `Task<T>`.

5. Create a new `CancelationTokenSource` and then set the method of cancelation, such as `CancelAfter(3000)`.
6. Pass an `IProgress<T>` type into an asynchronous method as a parameter and add event handlers for the `ProgressChanged` event. Alternatively, you can pass a single handler into the `Progress<T>` constructor.

Index

Symbols

A

B

C

D

E

F

G

H

I

N

O

P

R

S

T

U

Packt.com

Subscribe to our online digital library for full access to over 7,000 books and videos, as well as industry leading tools to help you plan your personal development and advance your career. For more information, please visit our website.

Why subscribe?

- Spend less time learning and more time coding with practical eBooks and Videos from over 4,000 industry professionals
- Improve your learning with Skill Plans built especially for you
- Get a free eBook or video every month
- Fully searchable for easy access to vital information
- Copy and paste, print, and bookmark content

Did you know that Packt offers eBook versions of every book published, with PDF and ePub files available? You can upgrade to the eBook version at packt.com and as a print book customer, you are entitled to a discount on the eBook copy. Get in touch with us at customercare@packtpub.com for more details.

At www.packt.com, you can also read a collection of free technical articles, sign up for a range of free newsletters, and receive exclusive discounts and offers on Packt books and eBooks.

Other Books You May Enjoy

If you enjoyed this book, you may be interested in these other books by Packt:

C# 10 and .NET 6 – Modern Cross-Platform Development - Sixth Edition

Mark J. Price

ISBN: 9781801077361

- Build rich web experiences using Blazor, Razor Pages, the Model-View-Controller (MVC) pattern, and other features of ASP.NET Core
- Build your own types with object-oriented programming
- Write, test, and debug functions
- Query and manipulate data using LINQ
- Integrate and update databases in your apps using Entity Framework Core,
- Microsoft SQL Server, and SQLite
- Build and consume powerful services using the latest technologies, including gRPC and GraphQL
- Build cross-platform apps using XAML

Software Architecture with C# 10 and .NET 6 - Third Edition

Gabriel Baptista, Francesco Abbruzzese

ISBN: 9781803235257

- Use proven techniques to overcome real-world architectural challenges
- Apply architectural approaches such as layered architecture
- Leverage tools such as containers to manage microservices effectively
- Get up to speed with Azure features for delivering global solutions
- Program and maintain Azure Functions using C# 10
- Understand when it is best to use test-driven development (TDD)
- Implement microservices with ASP.NET Core in modern architectures
- Enrich your application with Artificial Intelligence
- Get the best of DevOps principles to enable CI/CD environments

Enterprise Application Development with C# 10 and .NET 6 - Second Edition

Ravindra Akella, Arun Kumar Tamirisa, Suneel Kumar Kunani, Bhupesh Guptha Muthiyalu

ISBN: 9781803232973

- Design enterprise apps by making the most of the latest features of .NET 6
- Discover different layers of an app, such as the data layer, API layer, and web layer
- Explore end-to-end architecture by implementing an enterprise web app using .NET and C# 10 and deploying it on Azure
- Focus on the core concepts of web application development and implement them in .NET 6
- Integrate the new .NET 6 health and performance check APIs into your app
- Explore MAUI and build an application targeting multiple platforms - Android, iOS, and Windows

Packt is searching for authors like you

If you're interested in becoming an author for Packt, please visit `authors.packtpub.com` and apply today. We have worked with thousands of developers and tech professionals, just like you, to help them share their insight with the global tech community. You can make a general application, apply for a specific hot topic that we are recruiting an author for, or submit your own idea.

Share Your Thoughts

Now you've finished *High-Performance Programming in C# and .NET*, we'd love to hear your thoughts! Scan the QR code below to go straight to the Amazon review page for this book and share your feedback or leave a review on the site that you purchased it from.

`https://packt.link/r/1-800-56471-6`

Your review is important to us and the tech community and will help us make sure we're delivering excellent quality content.

www.ingramcontent.com/pod-product-compliance
Lightning Source LLC
LaVergne TN
LVHW081505050326
832903LV00025B/1395

* 9 7 8 1 8 0 0 5 6 4 7 1 8 *